I0064811

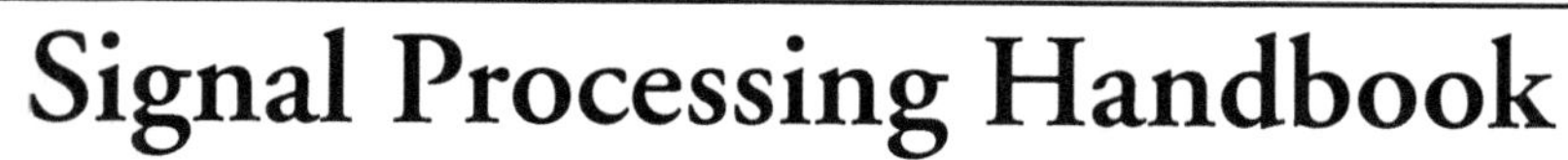

Signal Processing Handbook

Signal Processing Handbook

Editor: Amber Fowler

New York

Published by NY Research Press
118-35 Queens Blvd., Suite 400,
Forest Hills, NY 11375, USA
www.nyresearchpress.com

Signal Processing Handbook
Edited by Amber Fowler

International Standard Book Number: 978-1-63238-860-5 (Hardback)

Cataloging-in-Publication Data

Signal processing handbook / edited by Amber Fowler.
p. cm.
Includes bibliographical references and index.
ISBN 978-1-63238-860-5
1. Signal processing. 2. Signal theory (Telecommunication). 3. Signal processing--Equipment and supplies.
I. Fowler, Amber.
TK5102.9 .S54 2022
621.382 2--dc23

Contents

Preface

Signals can be broadly defined as functions that convey information regarding the attributes or behavior of some phenomenon, with respect to sound, images or biological measurements. Signal processing is a sub-field of electrical engineering, mathematics and information engineering that is concerned with the analysis, synthesis and modification of such signals. It can be of different types, such as analog signal processing, discrete-time signal processing, digital signal processing and continuous-time signal processing, among others. It has prominent applications in the fields of speech signal processing, video processing, audio signal processing, image processing, wireless communication, control systems, process control, array processing, etc. This book outlines the techniques and applications of signal processing in detail. It will also provide interesting topics for research which interested readers can take up. With state-of-the-art inputs by acclaimed experts of this field, this book targets students and professionals.

This book unites the global concepts and researches in an organized manner for a comprehensive understanding of the subject. It is a ripe text for all researchers, students, scientists or anyone else who is interested in acquiring a better knowledge of this dynamic field.

I extend my sincere thanks to the contributors for such eloquent research chapters. Finally, I thank my family for being a source of support and help.

Editor

1

An Advanced Partial Discharge Recognition Strategy of Power Cable

Xiaotian Bi, Ang Ren, Simeng Li, Mingming Han, and Qingquan Li

Shandong Provincial Key Laboratory of UHV Transmission Technology & Equipment, School of Electrical Engineering, Shandong University, Jinan 250061, China

Correspondence should be addressed to Qingquan Li; lqq@sdu.edu.cn

Academic Editor: John N. Sahalos

Detection and localization of partial discharge are very important in condition monitoring of power cables, so it is necessary to build an accurate recognizer to recognize the discharge types. In this paper, firstly, a power cable model based on FDTD simulation is built to get the typical discharge signals as training samples. Secondly, because the extraction of discharge signal features is crucial, fractal characteristics of the training samples are extracted and inputted into the recognizer. To make the results more accurate, multi-SVM recognizer made up of six Support Vector Machines (SVM) is proposed in this paper. The result of the multi-SVM recognizer is determined by the vote of the six SVM. Finally, the BP neural networks and ELM are compared with multi-SVM. The accuracy comparison shows that the multi-SVM recognizer has the best accuracy and stability, and it can recognize the discharge type efficiently.

1. Introduction

With the development of the power industry and urbanization in China, the power cable is used everywhere. As a result, the detection and localization of partial discharge are becoming more and more important in condition monitoring of power cables. Building an accurate model of cables helps analyzing the propagation characteristics of electromagnetic pulse and recognizing its type accurately when partial discharge happens in the cable. So, accurate model is necessary [1, 2].

In recognizing the types of partial discharge, the extraction of discharge signal features is the key [3]. Because of the large amounts of data from the measured graphics or waveform, they are too difficult to recognize directly. So they need to be transformed from original data to signal features. So far, researchers from home and abroad use statistical characteristic parameters, pulse characteristic parameters, moment features, or fractal characteristics to recognize discharge types. Because of their accuracy and less characteristic parameters, fractal characteristics are becoming more and more popular [4].

Back propagation (BP) network is a kind of widely used pattern recognizer [5]. It has some disadvantages such as network structure, local minimum, and over- or underlearning. Support Vector Machine (SVM) is a new kind of pattern recognition method which is proposed and developed in the recent decades. It can solve the small sample learning problems and solve the local minimum and over-study-learning or under-study-learning problems. Now the SVM has been widely used in the power system [6].

2. Support Vector Machine

SVM is a kind of data mining method based on statistical learning theory. It can handle the regression problems (time series analysis) and pattern recognition (classification problem and discriminant analysis) successfully [7].

The principle of SVM is to find an optimized classified hyperplane based on classification requirements. The hyperplane can maximize blank area on both its sides and, at the same time, guarantee the classification accuracy [8]. In theory, SVM can achieve optimal classification.

Assuming that the given samples are $\{(x_1, y_1), (x_2, y_2), \ldots, (x_l, y_l)\}$ and $x_i \in X \subseteq R^n$, $y_i \in X \subseteq R^n$. The number of samples is l. SVM defines a nonlinear feature space mapping function by inner product function $\varphi(\cdot)$.

The samples x are mapped into a high-dimensional space H. In the high-dimensional space, the linear regression function is constructed based on the principle of structural risk minimization [9]:

$$f(x) = \omega \cdot \phi(x) + b \tag{1}$$

where ω and b are the weight coefficient and the deviation, which can be obtained by minimizing the objective function as follows:

$$\min R(\omega) = \frac{1}{l}\sum_{i=1}^{l} \left|y_i - f(x_i)\right|_{\varepsilon} + C\|\omega\|^2, \tag{2}$$

where C is the generalization constant.

When using SVM, solving regression problem, introduce linear insensitive loss function ε. Cost function $|\cdot|_{\varepsilon}$ is Vapnik insensitive loss function [10]:

$$\left|y_i - f(x_i)\right|_{\varepsilon} = \begin{cases} \left|y_i - f(x_i)\right| - \varepsilon & \left|y_i - f(x_i)\right| < \varepsilon \\ 0 & \text{others.} \end{cases} \tag{3}$$

Considering the tolerated fitting deviation, the original problem can be transformed into structural risk minimization objective function problem by introducing two groups of nonnegative slack variables and using the principle of structural risk minimization. The optimization problem as (1) is transformed into a constrained minimization problem as follows:

$$\min \frac{1}{2}\|\omega\|^2 + C\frac{1}{l}\sum_{i=1}^{l}(\xi + \xi^*). \tag{4}$$

Constraints are

$$\begin{aligned} &y_i - \omega \cdot \phi(x_i) - b \le \varepsilon + \xi, \\ &\omega \cdot \phi(x_i) + b - y_i \le \varepsilon + \xi^*, \\ &\qquad \xi \ge 0,\ \xi^* \ge 0,\ i = 1, 2, \ldots, l-1, l. \end{aligned} \tag{5}$$

Then, build Lagrange function and transform inequality constraints into equality constraints as follows:

$$\begin{aligned} L = {} & \frac{1}{2}\|\omega\|^2 + C\sum_{i=1}^{l}(\xi_i + \xi_i^*) - \sum_{i=1}^{l}(\eta_i\xi_i + \eta_i^*\xi_i^*) \\ & - \sum_{i=1}^{l}\alpha_i(\varepsilon + \xi_i - y_i + \omega \cdot \phi(x_i) + b) \\ & - \sum_{i=1}^{l}\alpha_i^*(\varepsilon + \xi_i^* - y_i + \omega \cdot \phi(x_i) + b), \end{aligned} \tag{6}$$

where $\eta_i, \eta_i^*, \alpha_i, \alpha_i^*$ is Lagrange multipliers and it satisfies the nonnegative constraints as well:

$$\eta_i, \eta_i^*, \alpha_i, \alpha_i^* \ge 0. \tag{7}$$

The optimization problem as (6) can be solved in its dual form. According to the Karush-Kuhn-Tucker (KKT) conditions, the original problem can be transformed into the optimization objective function as follows [11]:

$$\begin{aligned} W(\alpha_i, \alpha_i^*) & \\ = {} & \sum_{i=1}^{l} y_i(\alpha_i - \alpha_i^*) - \varepsilon\sum_{i=1}^{l}(\alpha_i + \alpha_i^*) \\ & - \frac{1}{2}\sum_{i=1}^{l}\sum_{i=1}^{l}(\alpha_i - \alpha_i^*)(\alpha_j - \alpha_j^*)(\phi(x_i) \cdot \phi(x_j)). \end{aligned} \tag{8}$$

Constraints are

$$\sum_{i=1}^{l}(\alpha_i - \alpha_i^*) = 0, \quad \alpha_i, \alpha_i^* \in [0, C]. \tag{9}$$

Maximize (8) as

$$f(x) = \sum_{i=1}^{l}(\alpha_i - \alpha_i^*)(\phi(x_i) \cdot \phi(x)) + b, \tag{10}$$

where α_i, α_i^* is the introduced nonnegative Lagrange multiplier, $K(x_i, x_j)$ is the kernel function based on Mercer condition, and the inner product kernel function is as follows [12]:

$$K(x_i, x) = \phi(x_i) \cdot \phi(x). \tag{11}$$

The introduction of kernel function takes the place of dot product in the high-dimensional space, avoiding the problem of nonlinear mapping function which reduces the computation and complexity significantly. Then the result is as follows:

$$f(x) = \sum_{i=1}^{l}(\alpha_i - \alpha_i^*) \cdot K(x_i, x) + b. \tag{12}$$

3. Fractal Method

Most objects in nature are very complex and irregular. When the object has some similarities between the local and global, it can be viewed as fractal. The fractal dimension, as quantitative characterization and basic parameter of the fractal, is an important principle of the fractal theory. According to different definitions and calculation methods, box dimension and information dimension are often used in the fractal calculation [13].

To the Point Set $\Omega \subset R_n$, if it can be covered by $N(r)$ n-dimensional hypercube whose side length is r, then the box dimension of the Point Set is

$$D_B = \lim_{r \to 0}\frac{\ln N(r)}{\ln(1/r)}. \tag{13}$$

Because the box dimension is not able to reflect the unevenness of geometric objects, a box with one or several points may have the same weights. The information

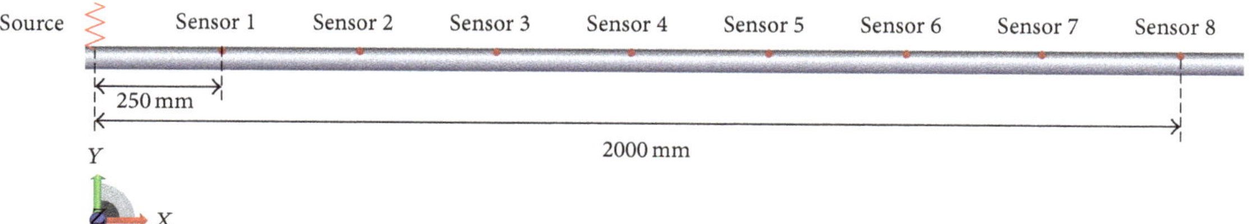

FIGURE 1: XLPE cable simulation model and the distribution of the sensors.

dimension has some advantages in this situation [14]. If the possibility of the Set Point falling into the kth hypercubes is $P_k(r)$, define the number of information as entropy, which can accurate the system status to level r, then

$$I(r) = -\sum_{k=1}^{N(r)} P_k(r) \ln P_k(r). \tag{14}$$

So, the information dimension of Point Set Ω is

$$D_I = \lim_{r \to 0} \frac{I(r)}{\ln(1/r)}. \tag{15}$$

4. Simulations

4.1. Finite Difference Time Domain (FDTD). Finite Difference Time Domain (FDTD) is a numerical calculation method in solving the time domain electromagnetic problem. It finishes finite difference discretization in time and space domain based on Maxwell equations and then builds the central finite difference equations whose accuracy is second-order. FDTD can simulate any kind of electromagnetic structure according to the electromagnetic parameters and medium parameters of the model [15]. In isotropic media, two curl equations of Maxwell equations are

$$\begin{aligned} \nabla \times \vec{E} &= -\mu \frac{\partial \vec{H}}{\partial t} - \sigma^* \vec{H}, \\ \nabla \times \vec{H} &= \varepsilon \frac{\partial \vec{E}}{\partial t} + \sigma \vec{E}, \end{aligned} \tag{16}$$

where ε is the permittivity, F·m^{-1}; μ is the magnetic permeability, H·m^{-1}; σ is the conductivity, S·m^{-1}; and σ^* is the magnetoconductivity, Ω·m^{-1}.

In suitable boundary conditions and initial conditions, FDTD can give the time domain characteristic of electromagnetic wave by solving the differential Maxwell equations, which makes it easier for us to analyze the discharge problems in the XLPE power cables [16].

4.2. Model of XLPE Cables. XLPE cables have the typical coaxial structure. To find out the characteristics of this structure, FDTD simulation program is applied. Figure 2 shows the model of the real 30 kV XLPE cable. The partial discharges source located in the cavity 1 mm from the left side of the port is a discharge voltage pulse along the radial distribution of the cavity. In the coaxial cavity, we placed sensors every 250 mm (red points in Figure 1).

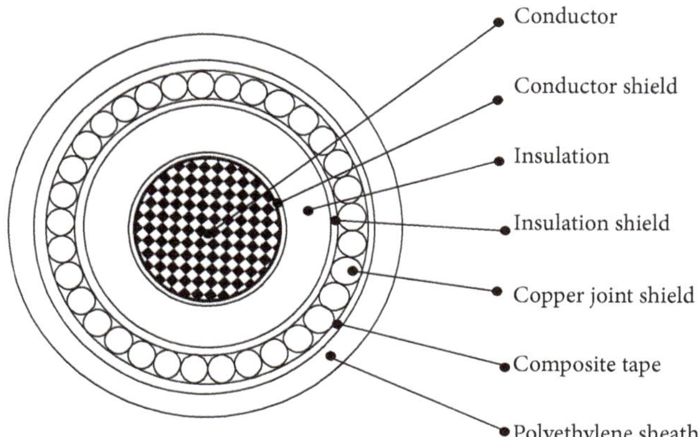

FIGURE 2: Structure of the model.

The structure and the materials electrical characteristics of the model are shown as in Figure 2 and Table 1.

4.3. Partial Discharge Source. Gaussian pulse is used to simulate the partial discharge voltage pulse:

$$u(t) = U_0 e^{-(t-t_0)^2/2\sigma^2}, \tag{17}$$

where U_0 is the amplitude of the pulse voltage; t_0 represents the time-delay constant; and σ is the decay time constant. Assuming that the amplitude of the voltage pulse is 1000 V, $\sigma = 0.17$ ns, and $t_0 = 5$ ns, Figure 3 shows the pulse voltage waveform, where the rise time t_r of pulse is 2 ns, the fall time t_f is 2 ns, and the pulse width t_w is 5 ns.

4.4. Simulation Results. In order to study propagation characteristics of the electromagnetic signal in the coaxial waveguide, signals are taken from sensor 1 to sensor 8. Field signal which only contains TEM wave measured by sensor 1 is shown in Figure 4.

Results show that the propagation of the signal from sensor 1 to sensor 8 has caused its amplitude to weaken from 9799.4 mV/m to 7485.6 mV/m, reduced by 2.34 dB, and the signal waveform is changed significantly. However, the amplitude and waveform of the signal components below 400 MHz which only contain TEM wave are not changed. It indicates that when propagating in the coaxial waveguide, the attenuation of TEM wave is small and its waveform will remain basically unchanged, while the amplitude of electromagnetic wave decays a lot and the waveform changes because of the dispersion effect of higher-order mode waves.

TABLE 1: Materials electrical characteristics of the model.

Structure parts	Outer diameter/mm	Material	Relative permittivity	Conductivity
Conductor	20.4	Copper	—	5.8×10^7
Conductor shield	21.8	Semiconductor	30.0	2
Insulation	30.1	Silicone rubber	3.2	0
Insulation shield	31.5	Conductive silicon rubber	90.0	0.5
Copper joint shield	34.9	Copper	—	5.8×10^7
Composite tape	35.5	PVC	3.0	1×10^{-13}
Polyethylene sheath	40.0	XLPE	2.3	0

TABLE 2: Real data of discharge signals recorded.

	Information dimension	Box dimension	Discharge type
Discharge 1	(0.51, 1.52)	(0.35, 1.86)	Creeping discharge
Discharge 2	(1.42, 1.15)	(1.67, 0.85)	External corona discharge
Discharge 3	(1.63, 1.63)	(1.44, 1.35)	Floating electrode discharge
Discharge 4	(1.69, 1.06)	(1.81, 0.77)	External corona discharge
Discharge 5	(0.46, 1.78)	(0.45, 1.55)	Creeping discharge
Discharge 6	(0.55, 1.65)	(0.37, 1.61)	Creeping discharge
Discharge 7	(0.61, 1.14)	(0.77, 1.27)	Internal air discharge
Discharge 8	(0.75, 1.02)	(0.65, 1.18)	Internal air discharge
Discharge 9	(1.77, 1.67)	(1.27, 1.54)	Floating electrode discharge
Discharge 10	(0.47, 1.68)	(0.28, 1.61)	Creeping discharge

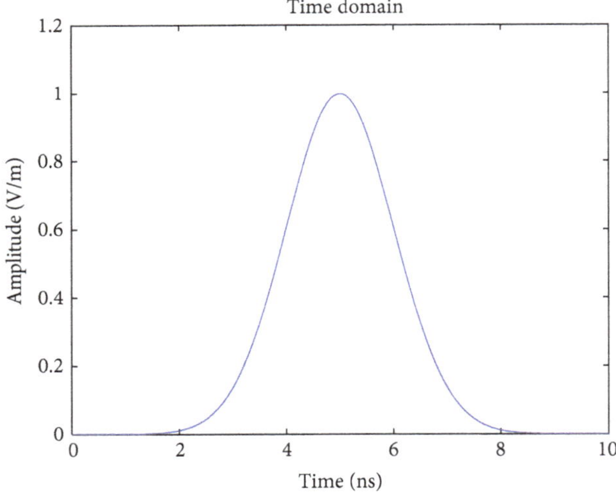

FIGURE 3: Waveform of partial discharge voltage pulse.

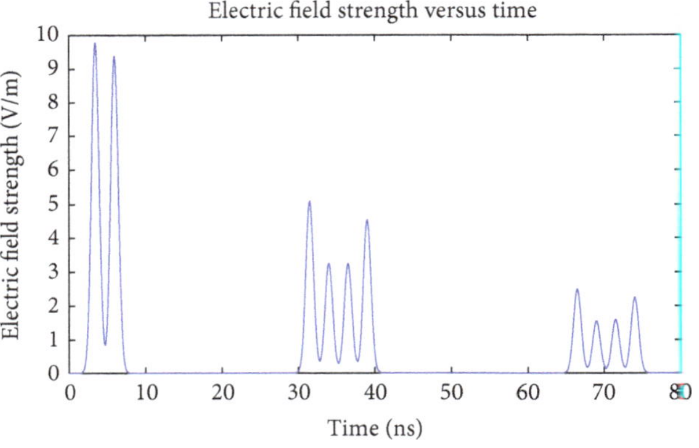

FIGURE 4: Waveform of sensor 1.

4.5. Real Data of Discharge Signals Recorded in 30 kV Power Cable. The simulation data may have some deviation. In order to make the result more accurate, a list of discharge signals from a real power cable line are monitored and recorded. The data recording system is shown as Figure 5.

The band of the measuring impedance is 45~250 kHz, and the setting of the monitor is 500 kS/s with 1 MB memory capacity and threshold trigger. So the monitor can store up to 100 cycles of discharge waveform. Their information dimension and box dimension are calculated and recorded, as shown in Table 2.

5. Training and Recognition Based on Multi-SVM

In order to make the multi-SVM recognizer more accurate, the simulation data are used as training samples to train the SVM and the measured data are used as testing samples to verify its accuracy.

When training SVM recognizer, SVM model and its Lagrange multipliers can be determined by training the samples and solving the quadratic programming equations. The flowchart of multi-SVM recognizer is shown as Figure 6.

5.1. Set Typical Discharge in Simulation to Get Training Samples for Multi-SVM. Studies show that there are approximately four types of PD in XLPE cables: creeping

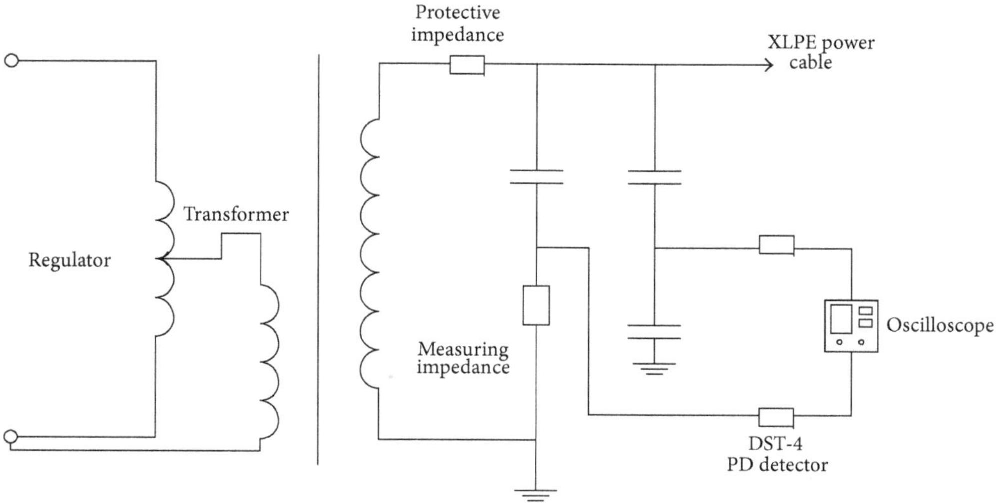

FIGURE 5: Real data of discharge signal recording system.

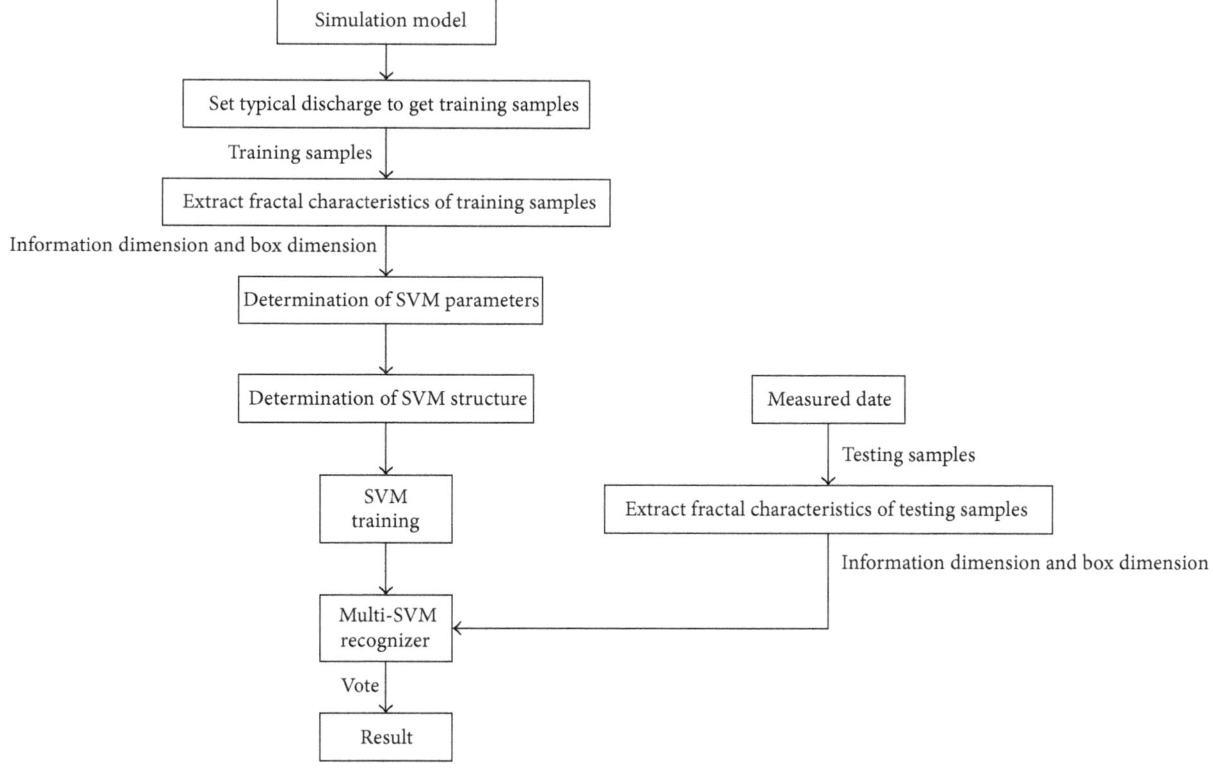

FIGURE 6: Flowchart of SVM recognizer.

discharge, floating electrode discharge, internal air discharge, and external corona discharge. According to their different characteristics, four PD faults are set in the simulation model.

(1) Creeping Discharge Model. To simulate the creeping discharge, a long thin wire is fixed on the joint surface of the main insulation and silicone rubber. Its length is 30 mm and its diameter is 0.5 mm.

(2) Floating Electrode Discharge Model. To simulate the floating electrode discharge model, a small round copper is fixed between the XLPE insulation of cable accessories and the main insulation. It has a diameter of 12 mm and a thickness of 0.5 mm.

(3) Internal Air Discharge Model. To simulate the internal air discharge model, a needle is fixed between the metal

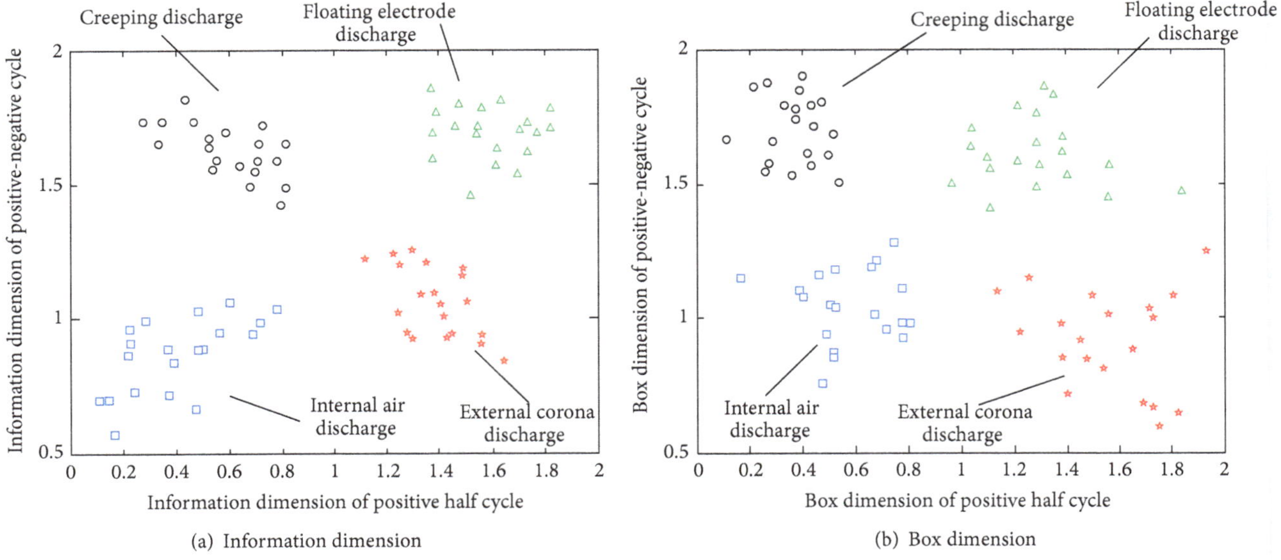

(a) Information dimension (b) Box dimension

FIGURE 7: Fractal characteristics of partial discharge.

shield and the outer semiconductor layer. It has a diameter of 0.2 mm and a length of 2 mm.

(4) External Corona Discharge Model. To simulate the external corona discharge model, a needle is fixed on the high voltage conductor. It has a diameter of 0.1 mm and a length of 10 mm.

5.2. Extract Fractal Characteristics of Training Samples. Gather 80 groups of data (20 groups of each discharge type) from cable partial discharge simulations and process them according to the minimum box counting method. Four fractal characteristics are found from grayscale images: box dimension of the positive half cycle D_B^+, box dimension of the negative half cycle D_B^-, information dimension of the positive half cycle D_I^+, and information dimension of the negative half cycle D_I^-. The results are shown as in Figure 7.

Figure 7 shows box dimension and information dimension of different types of discharge gathered in different areas. So fractal characteristics such as box dimension and information dimension have a strong ability to distinguish between different types of discharges [17, 18].

5.3. Determination of Multi-SVM Parameters. As we know, it is difficult for SVM to solve partial discharge problems. So in this paper the algorithm proposed is multi-SVM which is made up of six SVM. The principle to build multi-SVM recognizer proposed in this paper is to combine six one-to-one SVM together.

Use RBF as multi-SVM method's kernel function:

$$K(x, y) = \exp\left(-\frac{\|x - y\|_2^2}{\sigma^2}\right). \quad (18)$$

5.4. Determination of Multi-SVM Structure. Comparatively speaking, there are two different kinds of branching algorithms: one-to-one algorithms and one-to-many algorithms [19]. For one-to-one algorithm, each SVM recognizer only has two kinds of status [20]. So, one-to-one algorithm is easy to be trained and its decision boundary is relatively simple. Its classification accuracy is much better than the one-to-many algorithm as well. Therefore, one-to-one algorithm is used in this paper.

Six one-to-one SVM are proposed in this paper to recognize four types of discharges. They are defined as SVM1, SVM2, SVM3, SVM4, SVM5, and SVM6. The multi-SVM recognizer proposed in this paper is made up of these six SVM. The discharge type is determined by the vote shown in Table 3. The vote results of four discharge types are shown as in Table 4.

5.5. SVM Training and Recognizer Structure. Use the 80 groups of data (20 groups of each discharge type) from cable partial discharge simulations to train the SVM recognizer. After training, the number of support vectors of each SVM is shown as in Table 5.

Take SVM2 as an example. The structure of SVM2 is shown as in Figure 8. $x_1, x_2, \ldots, x_{18}$ is the support vectors of SVM2 and $\alpha_1 - \alpha_1^*, \alpha_2 - \alpha_2^*, \ldots, \alpha_{18} - \alpha_{18}^*$ is the weights shown as Table 6.

5.6. Extract Fractal Characteristics of Testing Samples. Use the 10 groups of real data recorded in 30 kV power cable as testing samples to verify the effectiveness of the multi-SVM recognizer proposed in this paper. After normalizing the data, the fractal characteristics of testing samples are shown as in Figure 9.

5.7. Vote Results. After inputting the fractal characteristics of the testing samples into the multi-SVM, the vote results are shown as in Table 7.

TABLE 3: Determine the discharge type by vote.

Recognizer	Creeping discharge	Floating electrode discharge	Internal air discharge	External corona discharge
SVM1	0	0	1	−1
SVM2	1	0	−1	0
SVM3	1	0	0	−1
SVM4	0	1	−1	0
SVM5	0	1	0	−1
SVM6	1	−1	0	0

TABLE 4: Vote results of four discharge types.

Discharge type	Vote results
Creeping discharge	SVM1 + SVM2 + SVM3 + SVM4 + SVM5 + SVM6
Floating electrode discharge	SVM1 + SVM2 + SVM3 + SVM4 + SVM5 − SVM6
Internal air discharge	SVM1 − SVM2 + SVM3 − SVM4 + SVM5 + SVM6
External corona discharge	−SVM1 + SVM2 − SVM3 + SVM4 − SVM5 + SVM6

TABLE 5: The number of support vectors.

SVM1	SVM2	SVM3	SVM4	SVM5	SVM6
40	18	23	19	35	24

Compared with Table 2, all the recognition results from multi-SVM are correct. So the multi-SVM recognizer proposed in this paper is effective in recognizing the discharge type of 30 kV XLPE power cables.

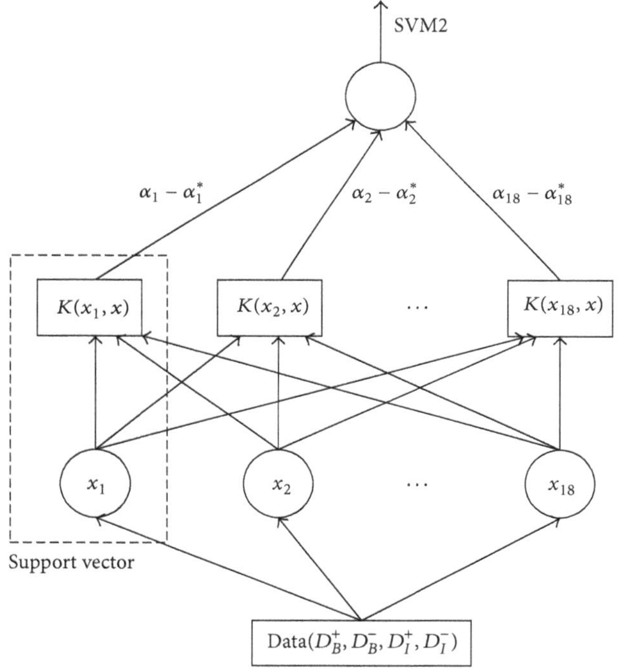

FIGURE 8: The structure of SVM2.

5.8. Accuracy Comparison. There are many kinds of intelligent algorithm such as BP neural networks and ELM. In order to prove the accuracy of multi-SVM, systematical and comprehensive comparisons are made in this paper. BP neural networks and ELM are applied into the recognizer instead of multi-SVM to recognize the discharge type.

5.8.1. BP Neural Networks Recognizer. BP neural networks are widely used in many aspects. In this paper, calculate the fractal calculation parameters ($D_B^+, D_B^-, D_I^+, D_I^-$) of discharge waveform from simulation as the BP neural network input vector X; the discharge type recognized is the output Y of the BP neural network. The hidden layer is a single neuron structure, which uses supervised learning method. The number of hidden layer neurons adjusts in the learning process and finally get the best BP neural network recognizer model.

The number of input layer neurons is k, the number of hidden layer neurons is g, and the number of output layer neurons is m. Use standard hyperbolic tangent function as the transfer function of the hidden layer. The details of BP neural network are shown in Table 8. The structure of BP neural network is shown in Figure 10.

In the setting, the target error is 10^{-4}; after 400 times of training, the BP neural network finishes. The dropping of deviation by training is shown as in Figure 11. Finally, the number of input layer neurons is $k = 6$, the number of hidden layer neurons is $g = 9$, and the number of output layer neurons is $m = 1$.

Use 100 groups of simulation data to test the accuracy of BP neural network; the results are shown as in Table 9.

5.8.2. ELM Recognizer. In the BP neural networks, the parameters of hidden layer are determined by large numbers of iterations which will take a lot of time and the results may be unsatisfying as well. In order to improve the performance of the network, ELM is proposed by Huang G.B. Etc [21–23]. Extreme Learning Machine (ELM) is a fast training algorithm for networks with single hidden layer. Its parameters of

Table 6: Weights of support vectors of SVM2.

$\alpha_1 - \alpha_1^*$	$\alpha_2 - \alpha_2^*$	$\alpha_3 - \alpha_3^*$	$\alpha_4 - \alpha_4^*$	$\alpha_5 - \alpha_5^*$	$\alpha_6 - \alpha_6^*$	$\alpha_7 - \alpha_7^*$	$\alpha_8 - \alpha_8^*$	$\alpha_9 - \alpha_9^*$
1.32	0.52	0.48	0.74	0.14	0.96	1.25	0.36	0.24
$\alpha_{10} - \alpha_{10}^*$	$\alpha_{11} - \alpha_{11}^*$	$\alpha_{12} - \alpha_{12}^*$	$\alpha_{13} - \alpha_{13}^*$	$\alpha_{14} - \alpha_{14}^*$	$\alpha_{15} - \alpha_{15}^*$	$\alpha_{16} - \alpha_{16}^*$	$\alpha_{17} - \alpha_{17}^*$	$\alpha_{18} - \alpha_{18}^*$
0.26	0.44	0.61	1.73	0.81	0.41	0.52	1.16	0.85

Table 7: Vote results of testing samples.

	Creeping discharge	Floating electrode discharge	Internal air discharge	External corona discharge	Discharge type result
Discharge 1	**3**	0	0	0	Creeping discharge
Discharge 2	0	0	0	**3**	External corona discharge
Discharge 3	0	**3**	0	0	Floating electrode discharge
Discharge 4	0	0	0	**3**	External corona discharge
Discharge 5	**3**	0	0	0	Creeping discharge
Discharge 6	**3**	0	0	0	Creeping discharge
Discharge 7	0	0	**3**	0	Internal air discharge
Discharge 8	0	0	**3**	0	Internal air discharge
Discharge 9	0	**3**	0	0	Floating electrode discharge
Discharge 10	**3**	0	0	0	Creeping discharge

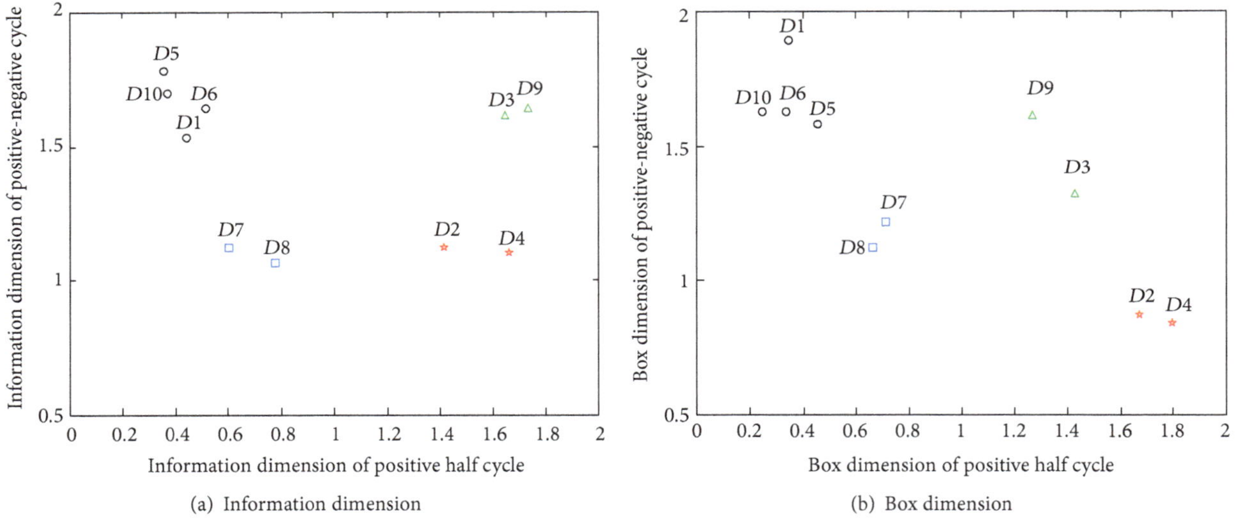

(a) Information dimension

(b) Box dimension

Figure 9: Fractal characteristics of testing samples.

hidden layer are determined randomly without any iteration, which significantly reduce the adjusting time for network parameters. The speed of training increases greatly and the results are better.

ELM Algorithm

(a) Input: training samples made up of fractal calculation parameters (D_B^+, D_B^-, D_I^+, D_I^-) of discharge signals; transfer function of the hidden layer; number of input layer neurons L.

(b) Parameters of hidden layer generated randomly (a_i, b_i) $i = 1, \ldots, L$.

(c) Calculating hidden layer output matrix H (H must be full rank).

(d) Output: optimized weight of network β: $\beta = H * T$.

Use 100 groups of simulation data to test the accuracy of ELM; the results are shown as in Table 10.

5.8.3. Multi-SVM Recognizer. In order to test the accuracy of the multi-SVM recognizer proposed in this paper, use 100 groups of simulation data to test the accuracy of multi-SVM; the results are shown as in Table 11.

TABLE 8: Detailed expression of BP neural networks.

	Input layer neurons	Hidden layer neurons	Output layer neurons
Input	$\{X_1, X_2, \ldots, X_n\}$	$\{P_1, P_2, \ldots, P_k\}$	$\{R_1, R_2, \ldots, R_g\}$
Output	$\{P_1, P_2, \ldots, P_k\}$	$\{R_1, R_2, \ldots, R_g\}$	$\{Y_1, Y_2, \ldots, Y_m\}$
Neurons model	$X_1, X_2, \ldots, X_n$; $k_1, k_2, \ldots, k_n$; Input layer neurons; Σ; $\rightarrow P_i$	$P_1, P_2, \ldots, P_n$; $e_1, e_2, \ldots, e_n$; b; Hidden layer neurons; Σ; Temp; $f(\cdot)$; $\rightarrow R_i$	$R_1, R_2, \ldots, R_n$; $w_1, w_2, \ldots, w_n$; Output layer neurons; Σ; $\rightarrow Y_i$
Neurons expression	$P_i = \sum_{i=1}^{n} (X_i \cdot k_i)$	$R_i = \dfrac{1 - e^{\sum_{i=1}^{k}(P_i \cdot e_i) - b}}{1 + e^{\sum_{i=1}^{k}(P_i \cdot e_i) - b}}$	$Y_i = \sum_{i=1}^{g} (R_i \cdot w_i)$

TABLE 9: Results of BP neural networks.

	Discharge type			
	Creeping discharge	Floating electrode discharge	Internal air discharge	External corona discharge
Test sample number	100	100	100	100
Correct results number	78	76	80	75
Accuracy	78%	76%	80%	75%

TABLE 10: Results of ELM.

	Discharge type			
	Creeping discharge	Floating electrode discharge	Internal air discharge	External corona discharge
Test sample number	100	100	100	100
Correct results number	90	88	91	86
Accuracy	90%	88%	91%	86%

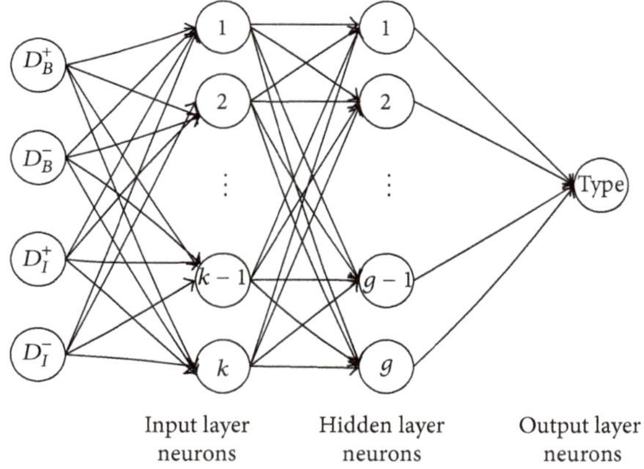

FIGURE 10: Structure of BP neural networks.

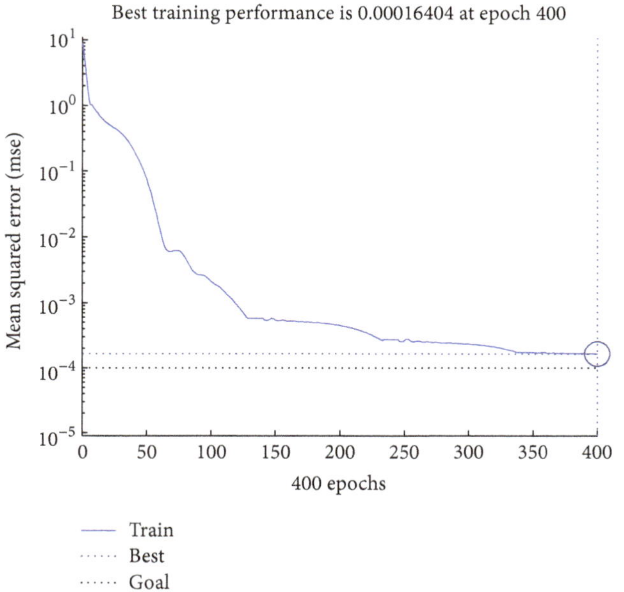

FIGURE 11: Dropping of deviation by training.

5.8.4. Comparison. The accuracy comparison between BP neural networks, SLM, and SVM is shown as in Table 12. Compared with BP neural networks, ELM and multi-SVM both perform excellently (accuracy of each type over 85%). As a matter of fact, from the view of accuracy, multi-SVM recognizer is even better (accuracy of each type over 90%).

The effectiveness of multi-SVM recognizer is satisfied. The accuracy of multi-SVM recognizer of all types is over 90%, especially the recognition of internal air discharge (96%). The result shows that the fractal characteristics of partial discharge signals have a strong ability to describe pattern and the recognizer is effective.

Table 11: Results of multi-SVM.

	Discharge type			
	Creeping discharge	Floating electrode discharge	Internal air discharge	External corona discharge
Test sample number	100	100	100	100
Correct results number	91	92	96	91
Accuracy	91%	92%	96%	91%

Table 12: Accuracy comparison.

Algorithm	Discharge type			
	Creeping discharge	Floating electrode discharge	Internal air discharge	External corona discharge
BP neural networks	78%	76%	80%	75%
ELM	90%	88%	91%	86%
Multi-SVM	91%	92%	96%	91%

6. Conclusion

Power cable is a very important part of the modern power system. Monitoring of power cables is directly related to the safety and stability of power system. So, it is necessary to build an accurate recognizer to recognize the discharge types correctly. The power cable model based on FDTD simulation is built to get the typical discharge signals as training samples, and fractal characteristics of the training samples are extracted and inputted into the recognizer. To make the results more accurate, multi-SVM recognizer is proposed in this paper. Finally, the BP neural networks recognizer and ELM recognizer are compared with multi-SVM recognizer proposed in this paper. The accuracy comparison shows that multi-SVM recognizer performs best (accuracy of each type over 90%), especially the recognition of internal air discharge (96%). The result shows that the fractal characteristics of partial discharge signals have a strong ability to describe pattern and the multi-SVM recognizer is effective.

Conflict of Interests

The authors declare that there is no conflict of interests regarding the republication of this paper.

References

[1] D. Wang, *The study of transmission characteristics of the particle discharge signals based on three phase cross-bonded cables [Ph.D. thesis]*, North China Electric Power University, Beijing, China, 2008.

[2] K. Takeuchi, N. Amemiya, T. Nakamura, O. Maruyama, and T. Ohkuma, "Model for electromagnetic field analysis of superconducting power transmission cable comprising spiraled coated conductors," *Superconductor Science and Technology*, vol. 24, no. 8, Article ID 085014, 2011.

[3] R. Ambikairajah, B. T. Phung, J. Ravishankar, T. R. Blackburn, and Z. Liu, "Detection of partial discharge signals in high voltage XLPE cables using time domain features," in *Proceedings of the 30th Electrical Insulation Conference (EIC '11)*, vol. 47, pp. 364–367, June 2011.

[4] A. Beroual and V.-H. Dang, "Fractal analysis of lightning impulse surface discharges propagating over pressboard immersed in mineral and vegetable oils," *IEEE Transactions on Dielectrics and Electrical Insulation*, vol. 20, no. 4, pp. 1402–1408, 2013.

[5] X. Chen, Y. Cheng, Z. Zhu, B. Yue, and X. Xie, "Insulating fault diagnosis of XLPE power cables using multi-parameter based on artificial neural networks," in *Advances in Neural Network—ISNN 2005*, vol. 3498 of *Lecture Notes in Computer Science*, pp. 609–615, Springer, Berlin, Germany, 2005.

[6] M. Wang and X. W. Li, "Power cable fault recognition using the improved PSO-SVM algorithm," *Applied Mechanics & Materials*, vol. 427–429, pp. 830–833, 2013.

[7] L. Nanni, A. Lumini, D. Gupta, and A. Garg, "Identifying bacterial virulent proteins by fusing a set of classifiers based on variants of Chou's Pseudo amino acid composition and on evolutionary information," *IEEE/ACM Transactions on Computational Biology and Bioinformatics*, vol. 9, no. 2, pp. 467–475, 2012.

[8] B.-J. M. Webb-Robertson, W. R. Cannon, C. S. Oehmen et al., "A support vector machine model for the prediction of proteotypic peptides for accurate mass and time proteomics," *Bioinformatics*, vol. 24, no. 13, pp. 1503–1509, 2008.

[9] M. Johannes, J. C. Brase, H. Fröhlich et al., "Integration of pathway knowledge into a reweighted recursive feature elimination approach for risk stratification of cancer patients," *Bioinformatics*, vol. 26, no. 17, pp. 2136–2144, 2010.

[10] H.-J. Dai, P.-T. Lai, and R. T.-H. Tsai, "Multistage gene normalization and SVM-based ranking for protein interactor extraction in full-text articles," *IEEE/ACM Transactions on Computational Biology and Bioinformatics*, vol. 7, no. 3, pp. 412–420, 2010.

[11] X. Xu, D. Niu, P. Wang, Y. Lu, and H. Xia, "The weighted support vector machine based on hybrid swarm intelligence optimization for icing prediction of transmission line," *Mathematical Problems in Engineering*, vol. 2015, Article ID 798325, 9 pages, 2015.

[12] J. A. Hunter, P. L. Lewin, L. Hao, C. Walton, and M. Michel, "Autonomous classification of PD sources within three-phase 11 kV PILC cables," *IEEE Transactions on Dielectrics and Electrical Insulation*, vol. 20, no. 6, pp. 2117–2124, 2013.

[13] W. Gang and T. Ju, "Study of minimum box-Counting method for image fractal dimension estimation," in *Proceedings of the China International Conference on Electricity Distribution (CICED '08)*, pp. 1–5, December 2008.

[14] H.-C. Chen, "Fractal features-based partial discharge pattern recognition using extension method," in *Proceedings of the 4th International Conference on Intelligent Systems Design and Engineering Applications*, pp. 274–277, IEEE, January 2012.

[15] L. Qi and X. Cui, "Electromagnetic transient analysis of cable system based on FDTD method," *Acta Electronica Sinica*, vol. 34, pp. 525–529, 2006.

[16] R. Papazyan, P. Pettersson, and D. Pommerenke, "Wave propagation on power cables with special regard to metallic screen design," *IEEE Transactions on Dielectrics and Electrical Insulation*, vol. 14, no. 2, pp. 409–416, 2007.

[17] M. Wang, P. Wang, J.-S. Lin, X. Li, and X. Qin, "Nonlinear inertia classification model and application," *Mathematical Problems in Engineering*, vol. 2014, Article ID 987686, 9 pages, 2014.

[18] R. V. Maheswari, P. Subburaj, B. Vigneshwaran, and L. Kalaivani, "Non linear support vector machine based partial discharge patterns recognition using fractal features," *Journal of Intelligent & Fuzzy Systems*, vol. 27, no. 5, pp. 2649–2664, 2014.

[19] B. L. Liu, "Study on the fault diagnosis of turbine based on support vector machine," *Applied Mechanics and Materials*, vol. 55-57, pp. 1803–1806, 2011.

[20] J. Cao, S. Sun, and X. Duan, "A multi-classification algorithm based on support vectors," in *Proceedings of the IEEE 3rd International Conference on Information Science and Technology (ICIST '13)*, pp. 305–307, IEEE, March 2013.

[21] C. Q. Men and W. J. Wang, "A randomized ELM speedup algorithm," *Neurocomputing*, vol. 159, pp. 78–83, 2015.

[22] R. Ahila, V. Sadasivam, and K. Manimala, "An integrated PSO for parameter determination and feature selection of ELM and its application in classification of power system disturbances," *Applied Soft Computing*, vol. 32, pp. 23–37, 2015.

[23] H. Chen, J. Peng, Y. Zhou, L. Li, and Z. Pan, "Extreme learning machine for ranking: generalization analysis and applications," *Neural Networks*, vol. 53, pp. 119–126, 2014.

2

Cross-Layer QoS Scheme based on Priority Differentiation in Multihop Wireless Networks

Mingjiu Wang[1,2] and Shu Fan[1,2]

[1] *Collaborative Innovation Center for Network Security Enforcement and Public Security Informatization, Criminal Investigation Police University of China, Shenyang 110854, China*
[2] *Audio-Visual and Image Technology Department, Criminal Investigation Police University of China, Shenyang 110854, China*

Correspondence should be addressed to Shu Fan; fanshufs@sina.com

Academic Editor: Rajesh Khanna

To support different QoS requirements of diverse types of services, a cross-layer QoS scheme providing different QoS guarantees is designed. This scheme sets values of service priorities according to services' data arrival rates and required end-to-end delay to endow different services with diverse scheduling priorities. To support QoS requirements better and maintain fairness, this scheme introduces delay and throughput weight coefficients. The methods of calculating the coefficients are also proposed. Through decomposing the optimization problem that uses weighted network utility as its optimization objective using Lyapunov optimization technique, this scheme can simultaneously support different QoS requirements of various services. The throughput utility optimality of the scheme is also proved. To reduce the computational complexity of the scheme, a distributed media access control scheme is proposed. A power control algorithm for this cross-layer scheme is also designed, and this algorithm transforms the power control into the solution of a multivariate equation. The simulation results evaluated with Matlab show that, compared with the existing works, the algorithm presented in this paper can simultaneously satisfy the delay demands of different services with maintaining high throughput.

1. Introduction

With various multimedia applications that have diverse QoS requirements appearing in multihop wireless networks [1], how to satisfy different QoS demands of diverse services has become a hotspot research issue. Considering the sharing of wireless channels among nodes and the diversity of QoS requirements of different services, the way of assigning transmission priorities to applications according to their QoS demands is an effective method to utilize wireless resources and provide QoS guarantees. Some algorithms following this line of thinking have been proposed. EDCA of IEEE 802.11e [2] standard defines four access categories corresponding to voice, video, best effort, and background to endow different applications with diverse priorities in media access control. Reference [3] adopts the 802.11e Access Control (AC) queue structure. Control packets used to route are prioritized according to the type of traffic associated with them to ensure that high priority packets are not penalized by the control packets. In [4], delay time of transmitting data is chosen as QoS metric. Packets in queues of flows with higher QoS level are delivered with higher priority. To support efficient video transmission, the scheduling algorithm in [5] assigns priority depending on the types of video frame. This video-based scheduling algorithm is combined with 802.11e protocol. Designed for video transport, the policy of [6] calculates the values of the counters depending on the delay estimation and the importance of packets. Under this policy, the packets with the lowest value of the counters gain the transmission opportunity. However, as there is no corresponding routing and flow control scheme combined with the above priority-based scheduling algorithms, congestion of high priority data packets may occur. Service-differentiation routing algorithms [7, 8] that select routes with different approaches depending on the type of traffic to ensure that packets with higher priority will be transmitted on higher quality links are also proposed. However, these algorithms may cause unbalanced distribution of packets in the network.

Different from the above layered QoS schemes, the Backpressure policy [9] is designed by applying Lyapunov optimization technique joint routes and schedules. The policy can also be combined with flow control [10] to ensure that the admitted rate injected into the network layer lies within the network capacity region, as well as combined with MAC [11], TCP [12], and application layers [13]. Due to its throughput-optimal characteristic for different network structures, the backpressure cross-layer control scheme has been a promising scheme to provide QoS guarantees. There are still few researches of Backpressure policies designed to support different QoS requirements of different types of traffic [14, 15]. In [14], services are divided into different classes according to their QoS demands. QoS requirements are supported through solving the optimization problem with the objective of maximizing weighted utility of different classes and constraints of QoS demands of each class. However, under the condition of high traffic loads, the fairness of the policy will decline. Reference [15] proposes a Backpressure cross-layer algorithm which ensures that the delays of flows are proportional to the priorities of services with keeping optimal throughput utility. However, the work of [15] did not consider the situation that services have different arrival data rates.

In this paper, we consider both arrival data rate and QoS demands when setting priorities. The effect of priorities of services on services' QoS performance is also studied. We propose a cross-layer QoS scheme which can provide QoS guarantees for different types of services simultaneously. The key contributions of this paper can be summarized as follows.

(i) The paper proposes a Lyapunov optimization technique-based cross-layer scheme which can satisfy different QoS requirements of various applications with priority differentiation. The method of how to calculate services' priorities is also designed.

(ii) The paper introduces throughput weight coefficient and delay weight coefficient that are updated according to QoS performance to meet QoS demands better and maintain fairness.

(iii) To reduce the computational complexity, a distributed media access control scheme is proposed. A power control algorithm to keep the data transmission rates of all wireless links being equal is also designed. This power control algorithm treats the power control as the solution of a multivariate equation.

(iv) The performance in terms of utility optimality is demonstrated with rigorous theoretical analyses. The policy is shown that it can achieve a time average throughput utility which can be arbitrarily close to the optimal value.

The structure of the rest of the paper is as follows. Section 2 introduces the system model and problem formulation. In Section 3, the algorithm is designed using Lyapunov optimization. The performance analyses of the proposed algorithm are present in Section 4. Simulation results are given in Section 5. Conclusions are provided in Section 6.

2. Model and Problem Formulation

2.1. Network Model. Consider a multihop wireless network consisting of several nodes. Let the network be modeled by a directed connectivity graph $G(N, L)$, where N is the set of nodes and $(i, j) \in L$ represents a unidirectional wireless link between node i and node j. M denotes the set of unicast sessions m between source-destination pairs in the network. K denotes the set of services k in each session. N_s is the set of source nodes $s_m^{(k)}$ of service k in session m. N_d is the set of destination nodes $d_m^{(k)}$ of service k in session m. Packets generated in the source nodes traverse multiple wireless hops before arriving at the destination nodes. The system is assumed to run in a time-slotted fashion. There are two channels including common control channel and data channel which use different communication frequencies in the network. Each node can broadcast control packets consisting of channel access negotiation information, lengths of queues, and weight values of nodes on the common control channel. Each node can gain control information by monitoring the control channel. The data channel is used for data communication. In this model scheduling will be subjected to the following constraints [16]:

$$\sum_{j:(n,j)\in L} \alpha_{nj}(t) + \sum_{i:(i,n)\in L} \alpha_{in}(t) \le 1. \quad (1)$$

$\alpha_{nj}(t) \in \{0, 1\}$ is used to indicate whether link (n, j) is used to transmit packets in time slot t. $\alpha_{nj}(t) = 1$ if $P_{nj}(t) > 0$, and $\alpha_{nj}(t) = 0$ if $P_{nj}(t) < 0$. $P_{nj}(t)$ denotes the transmit power from node n to node j in time slot t. Constraint (1) means that each node can either transmit or receive data on data channel at the same time. The SINR (Signal to Interference plus Noise Ratio) of link (x, y) at node y in time slot t is calculated as follows:

$$\text{SINR}_{xy}(t) = \frac{G_{xy}P_{xy}(t)}{\sum_{z\ne x, z\ne y} G_{zy}P_{zh}(t) + n_y}. \quad (2)$$

Node x is the sending node, and node y is the destination node of packets from node x. Node z denotes the neighbor nodes of node x. When node z sends packets, node h is the destination node of packets from node z. G_{xy} denotes the transmit loss from node x to node y. n_y is the receiver noise at node y. The achievable capacity of link (x, y) in time slot t is calculated as follows:

$$C_{xy}(t) = B \cdot \log_2\left(1 + \text{SINR}_{xy}(t)\right). \quad (3)$$

B represents the bandwidth of the data channel. There are two necessary constraints for the successful data transmission on link (x, y) to be satisfied. The first constraint can be expressed as

$$\text{SINR}_{xy}(t) = \frac{G_{xy}P_{xy}(t)}{\sum_{z\ne x, z\ne y} G_{zy}P_{zh}(t) + n_y} \ge \text{SINR}^{\text{th}}. \quad (4)$$

This constraint states that the SINR of link (x, y) at node y must be above the predefined SINR threshold SINR^{th}.

However, if the new link (x, y) is built, the transmission power from node x to y may result in additional interference at the receiving node q of existing link (p, q), and the SINR of link (p, q) at node q will decrease. To make sure that the new transmission will not impair the existing transmissions and the SINR of each existing link keeps being above the predefined SINR threshold SINR^{th}, the second constraint is expressed as

$$\begin{aligned}\text{SINR}_{pq}(t) &= \frac{G_{pq}P_{pq}(t)}{\sum_{z\neq p,z\neq q} G_{zq}P_{zh}(t) + n_q + \Delta I_{xq}(t)} \\ &= \frac{G_{pq}P_{pq}(t)}{\sum_{z\neq p,z\neq q} G_{zq}P_{zh}(t) + n_q + G_{xq}P_{xy}(t)} \\ &\geq \text{SINR}^{\text{th}}.\end{aligned} \tag{5}$$

$\Delta I_{xq}(t)$ denotes the additional interference at node q caused by the data transmission from node x to y in time slot t. According to constraints (4) and (5), the maximum and minimum transmit power of node x to node y in time slot t can be written as

$$\begin{aligned}P_{xy}^{\min}(t) &= \frac{\left(\sum_{z\neq x,z\neq y} G_{zy}P_{zh}(t) + n_y\right)\cdot \text{SINR}^{\text{th}}}{G_{xy}}, \\ P_{xy}^{\max}(t) & \\ &= \frac{G_{pq}P_{pq}(t)/\text{SINR}^{\text{th}} - \sum_{z\neq p,z\neq q} G_{zq}P_{zh}(t) - n_q}{G_{xq}}.\end{aligned} \tag{6}$$

2.2. Virtual Queue at the Transport Layer. $A_m^{(k)}(t) \in [0, A_{\max}^{(m)(k)}]$ denotes the arrival rate of service k in session m injected into the transport layer from the application layer at source node. $A_{\max}^{(m)(k)}$ is the maximum arrival rate of session m. $r_m^{(k)}(t) \in [0, A_m^{(k)}(t)]$ is the admitted rate of session m injected into the network layer. $\eta_m^{(k)}(t) \in [0, A_{\max}^{(m)(k)}]$ is an auxiliary variable called the virtual input rate. There is a virtual queue for every service k in session m at the service's source node. The virtual queue at the transport layer of source node $s_m^{(k)}$ is denoted by $Y_m^{(k)}$ that is updated as follows:

$$Y_m^{(k)}(t+1) = \max\left[Y_m^{(k)}(t) - r_m^{(k)}(t), 0\right] + \eta_m^{(k)}(t). \tag{7}$$

If each virtual queue $Y_m^{(k)}$ is guaranteed to be stable, according to the necessity and sufficiency for queue stability [17], it is apparent that $\overline{\eta_m^{(k)}} \leq \overline{r_m^{(k)}}$, where the time average value of time-varying variable $x(t)$ is denoted by $\overline{x} = \lim_{t\to\infty}(1/t)\sum_{\tau=0}^{t-1} E(x(\tau))$. Therefore, the lower bound of $\overline{r_m^{(k)}}$ can be derived from $\overline{\eta_m^{(k)}}$ which is calculable.

2.3. Data Queue at the Network Layer. The data backlog queue for service k in session m at the network layer of node n is denoted by $Q_n^{(m)(k)}(t)$. In each slot t, the queue is updated as

$$\begin{aligned}Q_n^{(m)(k)}(t+1) = \max&\left[Q_n^{(m)(k)}(t) - \sum_{i\in O(n)} \mu_{ni}^{(m)(k)}(t), 0\right] \\ &+ \sum_{j\in I(n)} \mu_{jn}^{(m)(k)}(t) + 1_{\{n=s_m^{(k)}\}} r_m^{(k)}(t),\end{aligned} \tag{8}$$

where $O(n)$ represents the set of nodes with $(n, i) \in L$. $I(n)$ represents the set of nodes with $(j, n) \in L$. $\mu_{ij}^{(m)(k)}(t)$ is the amount of data of service k in session m to be forwarded from nodes i to j in time slot t. $1_{\{n=s_m^{(k)}\}}$ is an indicator function that denotes 1 if $n = s_m^{(k)}$ and denotes 0 otherwise. In addition, $\sum_{m\in M}\sum_{k\in K} \mu_{ij}^{(m)(k)}(t)$ must not be greater than the transmission capacity of link (i, j) in time slot t.

2.4. Design of Priorities of Services. β_k represents the priority of service k, which is used to denote the importance degree of service k in scheduling. β_k is calculated using the method as follows:

$$\beta_k = \frac{A_k/D_k^{\text{th}}}{A_{\text{basic}}/D_{\text{basic}}^{\text{th}}}. \tag{9}$$

A_k is the average data arrival rate of service k. D_k^{th} represents the maximum allowable end-to-end delay bound of service k. A_{basic} denotes the basic average data arrival rate. $D_{\text{basic}}^{\text{th}}$ is the basic allowable end-to-end delay. A_{basic} and $D_{\text{basic}}^{\text{th}}$ can be calculated as

$$\begin{aligned}A_{\text{basic}} &= \min_{k\in K} A_k, \\ D_{\text{basic}}^{\text{th}} &= \max_{k\in K} D_k^{\text{th}}.\end{aligned} \tag{10}$$

2.5. Design of Throughput and Delay Weight Coefficients. σ_k^D represents delay weight coefficient of service k. In every T_0 interval, the destination nodes of the same service calculate the average end-to-end delay of their corresponding service and the delay weight coefficient. As an example of service k, in destination nodes of service k, σ_k^D is calculated as

$$\sigma_k^D = \begin{cases}\exp\left(D_k - D_k^{\text{th}}\right) & \text{if } D_k > D_k^{\text{th}} \\ 1 & \text{if } D_k \leq D_k^{\text{th}}.\end{cases} \tag{11}$$

Here, D_k is the average end-to-end delay of service k in interval T_0. Similarly, the throughput weight coefficient of service k, σ_k^T is calculated as follows:

$$\sigma_k^T = \begin{cases}\exp\left(\text{Th}_k - \text{Th}_k^{\text{th}}\right) & \text{if } \text{Th}_k < \text{Th}_k^{\text{th}} \\ 1 & \text{if } \text{Th}_k \geq \text{Th}_k^{\text{th}}.\end{cases} \tag{12}$$

Th_k represents the average throughput of service k in interval T_0. Th_k^{th} denotes the required average throughput of service k.

Through introducing throughput and delay weight coefficients into the optimization objective, the QoS performances of services are considered in the optimization. According to the calculation methods above, we can find that if the QoS performances of a service including average end-to-end delay and average throughput in the interval do not reach the threshold values, the delay and throughput weight coefficients will increase sharply. Meanwhile, the transmitted probability of the packets of this service will increase, which helps to support QoS requirements better.

2.6. Throughput Utility Optimization Problem. Similar to the design of utility function in [18], let the utility function of service k in session m, $U_m^{(k)}(\cdot)$, be a concave, differentiable, and nondecreasing utility function with $U_m^{(k)}(0) = 0$. The throughput utility maximization problem $P1$ can be defined as follows:

$$\begin{aligned} \text{maximize} \quad & \sum_{m\in M}\sum_{k\in K}\sigma_k^D\sigma_k^T U_m^{(k)}\left(\overline{\eta_m^{(k)}}\right) \\ \text{subject to} \quad & \overline{r}\in\Lambda, \\ & 0\le\overline{r}\le\lambda, \\ & \overline{\eta}\le\overline{r}, \\ & \sum_{j:(n,j)\in L}\alpha_{nj}(t)+\sum_{i:(i,n)\in L}\alpha_{in}(t)\le 1, \\ & P_{xy}^{\min}(t)\le P_{xy}(t)\le P_{xy}^{\max}(t), \\ & \forall x,y\in N. \end{aligned} \tag{13}$$

Similar to the definition in Section 2.2, $\overline{x}$ is the time average value of time-varying variable $x(t)$, and $\overline{x}$ is calculated according to $\overline{x} = \lim_{t\to\infty}(1/t)\sum_{\tau=0}^{t-1}E(x(\tau))$. Here, $\overline{r} = (\overline{r_m^{(k)}})$, $\overline{\eta} = (\overline{\eta_m^{(k)}})$, $\lambda_m^{(k)} = E\{A_m^{(k)}(t)\}$, and $\lambda = (\lambda_m^{(k)})$. Λ is the capacity region of the network. The constraint $\overline{r}\in\Lambda$ is used to guarantee the stability of the network.

3. Dynamic Algorithm via Lyapunov Optimization

Lyapunov optimization technique is applied to solve $P1$. $Q_n^{(m)(k)}(t)$ $(\forall n\ne d_m,\ m\in M,\ k\in K)$ and $Y_m^{(k)}(t)$ $(m\in M,\ k\in K)$ are used in the dynamic algorithm. Let $\Theta(t) = [Q(t), Y(t)]$ be the network state vector in time slot t. Define the Lyapunov function as

$$\begin{aligned} L(\Theta(t)) = \frac{1}{2}\Bigg[& \sum_{m\in M}\sum_{k\in K}\left(Y_m^{(k)}(t)\right)^2 \\ & + \sum_{n\ne d_m^{(k)}}\sum_{m\in M}\sum_{k\in K}\beta_k\left(Q_n^{(m)(k)}(t)\right)^2\Bigg]. \end{aligned} \tag{14}$$

The conditional Lyapunov drift in time slot t is

$$\Delta(\Theta(t)) = E\{L(\Theta(t+1)) - L(\Theta(t)) \mid \Theta(t)\}. \tag{15}$$

To maximize a lower bound for $\sum_{m\in M}\sum_{k\in K}\sigma_k^D\sigma_k^T U_m^{(k)}(\overline{\eta_m^{(k)}})$, the drift-plus -penalty function can be defined as

$$\begin{aligned} & \Delta_V(\Theta(t)) \\ & = \Delta\Theta(t) - V \\ & \cdot E\left\{\sum_{m\in M}\sum_{k\in K}\sigma_k^D\sigma_k^T U_m^{(k)}\left(\eta_m^{(k)}(t)\right) \mid \Theta(t)\right\}, \end{aligned} \tag{16}$$

where V is the weight of utility defined by user. The following inequality can be derived:

$$E\{\Delta_V(\Theta(t))\} \le B - \Psi_1(t) - \Psi_2(t) - \Psi_3(t); \tag{17}$$

here, $\Psi_1(t)$, $\Psi_2(t)$, and $\Psi_3(t)$ can be evaluated as follows:

$$\begin{aligned} \Psi_1(t) &= \sum_{m\in M}\sum_{k\in K}\left[V\cdot\sigma_k^D\sigma_k^T U_m^{(k)}\left(\eta_m^{(k)}(t)\right) - Y_m^{(k)}(t)\cdot\eta_m^{(k)}(t)\right], \\ \Psi_2(t) &= \sum_{m\in M}\sum_{k\in K}r_m^{(k)}(t)\left[Y_m^{(k)}(t) - \beta_k\cdot Q_n^{(m)(k)}(t)\cdot 1_{\{n=s_m^{(k)}\}}\right], \\ \Psi_3&(t) \\ &= \sum_{n\ne d_m^{(k)}}\sum_{m\in M}\sum_{k\in K}\sum_{i\in O(n)}\mu_{ni}^{(m)(k)}(t)\cdot\beta_k\cdot\left[Q_n^{(m)(k)}(t) - Q_i^{(m)(k)}(t)\right]. \end{aligned} \tag{18}$$

B is a constant and satisfies

$$B \ge \sum_{n\ne d_m^{(k)}}\sum_{m\in M}\sum_{k\in K}\beta_k\left[\left(\sum_{i\in O(n)}\mu_{ni}^{(m)(k)}(t)\right)^2 + \left(\sum_{j\in I(n)}\mu_{jn}^{(m)(k)}(t) + r_m^{(k)}(t)\right)^2\right] + \sum_{m\in M}\sum_{k\in K}\left[\left(r_m^{(k)}(t)\right)^2 + \left(\eta_m^{(k)}(t)\right)^2\right]. \tag{19}$$

Assume that in this paper the transmit capacity of each link is a constant value $C_{\max}$ by using a power control algorithm. According to $r_m^{(k)}(t)\in[0, A_m^{(k)}(t)]$, $\eta_m^{(k)}(t)\in[0, A_{\max}^{(m)(k)}]$, and $C_{\max}$, constant B must exist.

The algorithm CADSP (Cross-Layer Algorithm with Differentiated Service Prioritization) scheme is based on the drift-plus-penalty framework [17]. The main design principle of the algorithm is to minimize the right-hand side of (17). This scheme consists of three parts which are joint flow control, routing, and scheduling scheme, medium access control scheme, and power control algorithm.

3.1. Joint Flow Control, Routing, and Scheduling Scheme. This scheme includes four parts as follows.

Source Rate Control. For sessions $m \in M$ and $k \in K$ at source node $s_m^{(k)}$, the admitted rate $r_m^{(k)}(t)$ is chosen to solve

$$\begin{aligned} &\text{maximize} \quad r_m^{(k)}(t)\left[Y_m^{(k)}(t) - \beta_k \cdot Q_{s_m^{(k)}}^{(m)(k)}(t)\right] \\ &\text{subject to} \quad 0 \le r_m^{(k)}(t) \le A_m^{(k)}(t). \end{aligned} \tag{20}$$

Problem (20) is a linear optimization problem, and if $Y_m^{(k)}(t) > \beta_k \cdot Q_{s_m^{(k)}}^{(m)(k)}(t)$, $r_m^{(k)}(t)$ is set to be $A_m^{(k)}(t)$; otherwise it is set to be zero.

Virtual Input Rate Control. For sessions $m \in M$ and $k \in K$ at source node $s_m^{(k)}$, the virtual input rate $\eta_m^{(k)}(t)$ is chosen to solve

$$\begin{aligned} &\text{maximize} \quad V \cdot \sigma_k^D \sigma_k^T U_m^{(k)}\left(\eta_m^{(k)}(t)\right) - Y_m^{(k)}(t) \cdot \eta_m^{(k)}(t) \\ &\text{subject to} \quad 0 \le \eta_m^{(k)}(t) \le A_{\max}^{(m)(k)}. \end{aligned} \tag{21}$$

If $U_m^{(k)}(\cdot)$ is strictly concave and twice differentiable, (21) is a concave maximization problem with linear constraint. $\eta_m^{(k)}(\cdot)$ can be chosen by

$$\eta_m^{(k)}(t) = \max\left[\min\left[U_m^{\prime-1(k)}\left(\frac{Y_m^{(k)}(t)}{(V\sigma_k^D\sigma_k^T)}\right), A_{\max}^{(m)(k)}\right], 0\right], \tag{22}$$

where $U_m^{\prime-1(k)}(\cdot)$ is the inverse function of $U_m^{\prime(k)}(\cdot)$ that is the first-order derivative of $U_m^{(k)}(\cdot)$. Since the utility function $U_m^{(k)}(\cdot)$ is strictly concave and twice differentiable, $U_m^{\prime(k)}(\cdot)$ must be a monotonic function, and therefore, $U_m^{\prime-1(k)}(\cdot)$ must exist. If $U_m^{(k)}(\cdot)$ is a linear function, let us suppose $U_m^{(k)}(x) = bx$. $\eta_m^{(k)}(\cdot)$ can be calculated as

$$\eta_m^{(k)}(t) = \begin{cases} A_{\max}^{(m)(k)} & \text{if } Vb\sigma_k^D\sigma_k^T > Y_m^{(k)}(t) \\ 0 & \text{if } Vb\sigma_k^D\sigma_k^T \le Y_m^{(k)}(t). \end{cases} \tag{23}$$

Joint Routing and Scheduling. At the node $n \ne d_m^{(k)}$, routing and scheduling decisions for each service k in session m can be made by solving the following:

$$\begin{aligned} &\text{maximize} \quad \sum_{n\ne d_m^{(k)}} \sum_{m\in M} \sum_{k\in K} \sum_{i\in O(n)} \mu_{ni}^{(m)(k)}(t) \cdot \beta_k \cdot \left[Q_n^{(m)(k)}(t) - Q_i^{(m)(k)}(t)\right] \\ &\text{subject to} \quad 0 \le \mu_{ab}^{(m)(k)}(t) \le C_{ab}(t), \quad \forall a, b \in N, \\ &\qquad \sum_{j:(n,j)\in L} \alpha_{nj}(t) + \sum_{i:(i,n)\in L} \alpha_{in}(t) \le 1, \\ &\qquad P_{xy}^{\min}(t) \le P_{xy}(t) \le P_{xy}^{\max}(t), \quad \forall x, y \in N. \end{aligned} \tag{24}$$

$C_{ab}(t)$ denotes the capacity of link (a, b) in time slot t, and $C_{ab}(t)$ is calculated according to (3). The first constraint of (24) indicates that the amount of data to be forwarded from one node to another node in a time slot should not be greater than the capacity of the link between these two nodes in time slot t. The second constraint of (24) is built according to constraint (1) given in Section 2.1. The third constraint of (24) is built according to constraint (6) given in Section 2.1.

First, the best service k^* and the best session m^* whose data should be transmitted on link (n, i) can be chosen as

$$(m^*, k^*) = \underset{m\in M, k\in K}{\arg\max}\left[\beta_k \cdot \left(Q_n^{(m)(k)}(t) - Q_i^{(m)(k)}(t)\right)\right]. \tag{25}$$

The weight value of link (n, i) is calculated using the following method as

$$w_{ni} = \beta_{k^*} \cdot \left(Q_n^{(m^*)(k^*)}(t) - Q_i^{(m^*)(k^*)}(t)\right). \tag{26}$$

So the joint routing and scheduling problem can be reduced to the following problem:

$$\begin{aligned} &\text{maximize} \quad \sum_{n\ne d_{m^*}^{(k^*)}} \sum_{i\in O(n)} \mu_{ni}^{(m^*)(k^*)}(t) \cdot w_{ni} \\ &\text{subject to} \quad 0 \le \mu_{ab}^{(m)(k)}(t) \le C_{ab}(t), \quad \forall a, b \in N, \\ &\qquad \sum_{j:(n,j)\in L} \alpha_{nj}(t) + \sum_{i:(i,n)\in L} \alpha_{in}(t) \le 1, \\ &\qquad P_{xy}^{\min}(t) \le P_{xy}(t) \le P_{xy}^{\max}(t), \\ &\qquad \forall x, y \in N. \end{aligned} \tag{27}$$

Transmission rates $\mu_{ni}^{(m^*)(k^*)}(t)$ are chosen based on (27) which is a hard problem for solving as it requires global knowledge and centralized algorithm. We define P_c as the set of transmit powers on each link and define I_c as the set of links which can be used for data transmission simultaneously when using P_c as the set of transmit powers. I is defined

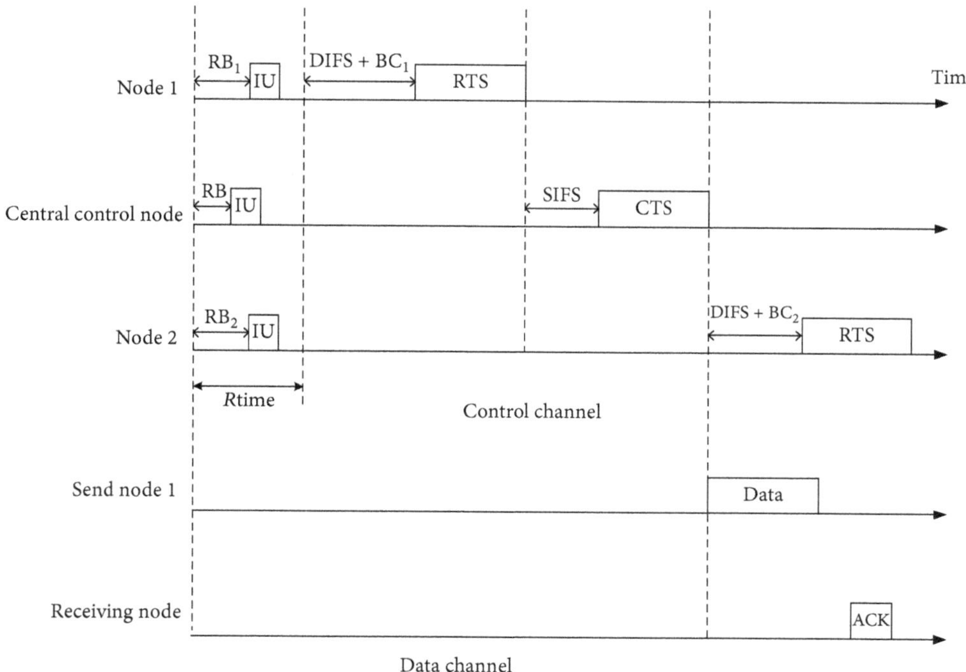

FIGURE 1: Medium access control logic.

as the set of (P_c, I_c). In each slot, $(P_c^*, I_c^*) \in I$ that maximizes $\sum_{n \neq d_{m^*}^{(k^*)}} \sum_{i \in O(n)} \mu_{ni}^{(m^*)(k^*)}(t) \cdot w_{ni}$ is chosen as the set of scheduled links and the set of transmit powers.

Update of Queues. $Y(t)$ and $Q(t)$ are updated using (7) and (8) in each time slot.

3.2. Distributed Medium Access Control Scheme. Solving (27) is a NP-hard problem whose computation complexity is $O(N^2)$ where N denote the number of nodes in the network. Obviously, the computation complexity will increase shapely with the increase of N. To reduce the computation complexity, a distributed medium access control scheme for routing and link scheduling is proposed in this section. The design principle of this distributed scheme is that nodes with higher weight values will get higher probabilities of accessing the medium and transmitting data. When the runtime is long enough, the distributed media access control scheme plays the same role to the GMS (Greedy Maximal Scheduling) algorithm which is a central algorithm and whose capacity region can reach 1/2 capacity region of MWM (Maximal Weighted Matching) [19] which is the basic of the central cross-layer routing and scheduling scheme proposed in Section 3.1.

The medium access control scheme is implemented in a time-slotted fashion on the common control channel. The way that nodes contend to access the control channel is similar to IEEE 802.11 two-way RTS and CTS handshake. The medium access control logic is illustrated in Figure 1. The details of the scheme are as follows. (i) There is a central control node which implements the power control algorithm and records state information of existing links, including transmit power, positions of nodes, and noises at the receiving nodes. (ii) At the beginning of each slot, each node trying to send data chooses a random waiting time RB $\in [0, R\text{time}]$. The value of Rtime is calculated in the central control node. It relates with number of nodes in the network. (iii) Each node sends IU packet that includes information about weight value, the next hop node chosen, current position, and noise on the control channel after waiting for RB. For the send node n, the receiving node of node n is $i^* = \arg\max_{i \in N} w_{ni}$, and the weight value of node n is $w_n = w_{ni^*}$. Each node also monitors the IU packets from other nodes to gain the weight values of other nodes. (iv) Every backlogged node i calculates its contention window CW_i and backoff counter BC_i [20] as follows:

$$\text{CW}_i = -\varphi \cdot \frac{w_i}{\sum_{n \in N} w_n} + \gamma, \quad \varphi > 0, \ \gamma > 0. \tag{28}$$

BC_i is randomly chosen from the range $[0, 2^{\text{CW}_i - 1}]$. (v) After Rtime from the beginning of the slot, each backlogged node i continues monitoring the control channel. If the node i senses an idle control channel for a period of DIFS + BC_i, it can send RTS packet which includes $Q_i^{(m^*)(k^*)}(t)$. RTS packet also includes the information about the receiving node in plan. (vi) After receiving RTS packet from node i, the central control node checks if the receiving node in plan of node i is in transmission. The control node also implements the power control algorithm to decide if the new link is allowed to be established. If the new link is allowed to be established, the control node responds with a CTS packet that includes the new transmit powers and transmission time lengths of all send nodes after a period of SIFS; otherwise, the control node responds with a NCTS that includes decision that the new link is not allowed to be established. (vii) The send nodes update the transmit powers after receiving CTS packet. The receiving node in plan of node i prepares for data reception and responds with a ACK packet after the successful data reception. Without considering the weight value of node i, idle nodes update their contention windows and backoff counters after receiving CTS packet. (viii) Node i and other idle nodes begin to monitor the control channel for further negotiation after receiving NCTS packet. (ix) The maximum times that each node is allowed to send RTS packets in a time slot is three.

3.3. Power Control Algorithm. The power control algorithm is implemented in the central control node of the network. The design objective of the algorithm is to ensure that the SINR at every receiving node is SINR^{th}. Assume that there have been n links in the network. The links from send nodes to receiving nodes are represented by $(x_1, y_1), (x_2, y_2), \ldots, (x_n, y_n)$. When node x_{n+1} tries to transmit data packets to node y_{n+1}, it send RTS packet on the control channel. After receiving RTS packet from x_{n+1} by monitoring the common control channel, the central control node begins the computation to check the transmit powers $(P_{x_1y_1}, P_{x_2y_2}, \ldots, P_{x_{n+1}y_{n+1}})$ of all send nodes $(x_1, x_2, \ldots, x_{n+1})$ which can guarantee that the following equalities exist:

$$\frac{P_{x_iy_i} \cdot G_{x_iy_i}}{n_{y_i} + \sum_{j=1}^{n+1} P_{x_jy_j} \cdot G_{x_jy_i} \cdot 1_{\{j\neq i\}}} = \text{SINR}^{\text{th}}, \tag{29}$$

where $i \in [1, n+1]$. If we can get $(P_{x_1y_1}, P_{x_2y_2}, \ldots, P_{x_{n+1}y_{n+1}})$, the new link (x_{n+1}, y_{n+1}) can be established. Equation (29) can be transformed into multivariate equations as follows:

$$GX = S, \tag{30}$$

where

$$G = \begin{pmatrix} G_{x_1y_1} & -G_{x_2y_1} \cdot \text{SINR}^{\text{th}} & -G_{x_3y_1} \cdot \text{SINR}^{\text{th}} & \cdots & -G_{x_{n+1}y_1} \cdot \text{SINR}^{\text{th}} \\ -G_{x_1y_2} \cdot \text{SINR}^{\text{th}} & G_{x_2y_2} & -G_{x_3y_2} \cdot \text{SINR}^{\text{th}} & \cdots & -G_{x_{n+1}y_2} \cdot \text{SINR}^{\text{th}} \\ -G_{x_1y_3} \cdot \text{SINR}^{\text{th}} & -G_{x_2y_3} \cdot \text{SINR}^{\text{th}} & G_{x_3y_3} & \cdots & -G_{x_{n+1}y_3} \cdot \text{SINR}^{\text{th}} \\ \cdots & \cdots & \cdots \quad \cdots & & \cdots \\ -G_{x_1y_{n+1}} \cdot \text{SINR}^{\text{th}} & -G_{x_2y_{n+1}} \cdot \text{SINR}^{\text{th}} & \cdots \quad \cdots & & G_{x_{n+1}y_{n+1}} \end{pmatrix},$$
$$X = \left(P_{x_1y_1}, P_{x_2y_2}, \ldots, P_{x_{n+1}y_{n+1}}\right)^T,$$
$$S = \text{SINR}^{\text{th}} \cdot \left(n_{y_1}, n_{y_2}, \ldots, n_{y_{n+1}}\right)^T. \tag{31}$$

If $0 < P_{x_iy_i} \le P_{x_i,\max}$, the link (x_{n+1}, y_{n+1}) is allowed to be established. Here $i \in [1, n+1]$, and $P_{x_i,\max}$ represents the maximum transmit power that node x_i can support. On the common control channel, the central control node will broadcast $(P_{x_1y_1}, P_{x_2y_2}, \ldots, P_{x_{n+1}y_{n+1}})$ which are new transmit powers of the send nodes.

4. Performance Analysis

Theorem 1 (algorithm performance). *Define* $\varphi(r) = \sum_{m\in M}\sum_{k\in K}\sigma_k^D\sigma_k^T \cdot U_m^{(k)}(r_m^{(k)})$ *and the optimization problem P2 as*

$$\begin{aligned} \text{maximize} \quad & \varphi(r) \\ \text{subject to} \quad & \overline{r} \in \Lambda, \\ & 0 \le \overline{r} \le \lambda, \\ & \sum_{j:(n,j)\in L} \alpha_{nj}(t) + \sum_{i:(i,n)\in L} \alpha_{in}(t) \le 1, \\ & P_{xy}^{min}(t) \le P_{xy}(t) \le P_{xy}^{max}(t), \\ & \forall x, y \in N. \end{aligned} \tag{32}$$

Define φ^* *as the optimal value of* $\varphi(r)$ *and* $r_m^{*(k)}$ *as the solution of P2. Under the implementation of central CADSP scheme proposed in Section 3.1, the following inequality holds:*

$$\sum_{m\in M}\sum_{k\in K}\sigma_k^D\sigma_k^T \cdot U_m^{(k)}\left(\overline{r_m^{(k)}}\right) \ge \varphi^* - \frac{B}{V}. \tag{33}$$

Proof. According to Lemma 4 in [18], similar to Theorem 3 in [16], the following inequality holds when $\tau = 0, 1, \ldots, T-1$:

$$\sum_{\tau=0}^{T-1} E\{\Delta\Theta(\tau)\} - V \cdot \sum_{\tau=0}^{T-1}\left(\sum_{m\in M}\sum_{k\in K}\sigma_k^D\sigma_k^T U_m^{(k)}\left(\eta_m^{(k)}(\tau)\right)\right) \le B \cdot T - V \cdot T \sum_{m\in M}\sum_{k\in K}\sigma_k^D\sigma_k^T U_m^{(k)}\left(\eta_m^{*(k)}\right)$$

$$-\sum_{\tau=0}^{T-1}\sum_{m\in M}\sum_{k\in K}Y_m^{(k)}(\tau)\left(r_m^{*(k)}-\eta_m^{*(k)}\right)$$

$$-\sum_{\tau=0}^{T-1}\sum_{n\neq d_m^{(k)}}\sum_{m\in M}\sum_{k\in K}Q_n^{(m)(k)}(\tau)\left(\sum_{i\in O(n)}\mu_{ni}^{*(m)(k)}(\tau)-\sum_{j\in I(n)}\mu_{jn}^{*(m)(k)}(\tau)-1_{\{n=s_m^{(k)}\}}r_m^{*(k)}\right). \tag{34}$$

According to the equalities which can be got, $\sum_{i\in O(n)}\mu_{ni}^{*(m)(k)}(\tau)-\sum_{j\in I(n)}\mu_{jn}^{*(m)(k)}(\tau) = 1_{\{n=s_m^{(k)}\}}r_m^{*(k)}$ and $r_m^{*(k)} = \eta_m^{*(k)}$, (34) can be transformed into following inequality:

$$\sum_{m\in M}\sum_{k\in K}\sigma_k^D\sigma_k^T U_m^{(k)}\left(\overline{\eta_m^{(k)}}\right)\geq\varphi^*-\frac{B}{V}. \tag{35}$$

As CADSP scheme can guarantee that $\overline{\eta}\leq\overline{r}$ and function $U_m^{(k)}(\cdot)$ is nondecreasing, the following inequality can be got:

$$\sum_{m\in M}\sum_{k\in K}\sigma_k^D\sigma_k^T\cdot U_m^{(k)}\left(\overline{r_m^{(k)}}\right)\geq\varphi^*-\frac{B}{V}. \tag{36}$$

Inequality (36) means the overall throughput utility achieved by the algorithm in this paper is within a constant gap from the optimum value. □

5. Simulation

5.1. Simulation Setup. The network for simulations is considered as a network with 20 nodes randomly distributed in a square of 900 m^2. There are two unicast sessions with randomly chosen sources and destinations. Each session includes three services. Data are injected at the source nodes following Poisson arrivals. The simulation time lasts 10000 time slots. All the initial queue sizes are set to be 0. The throughput utility function is $U(x) = \log(x+1)$. Table 1 summarizes the simulation parameters.

In this paper, the performance of CADSP is compared with Backpressure scheme [21] and PDA-PMF scheme [6]. Backpressure scheme is a classical joint routing and scheduling algorithm that can provide throughput utility optimality. PDA-PMF scheme is a services-differentiated scheduling policy. In this simulation, PDA-PMF scheme is combined with AODV routing algorithm.

5.2. Simulation of Services with Different Delay Requirements. In this section, the average data arrival rates of all the services are the same. The maximum allowable end-to-end delay bounds of services are set as $D_1^{\text{th}} = 0.8$ s, $D_2^{\text{th}} = 2$ s, and $D_3^{\text{th}} = 4$ s. The required average throughputs of services are set as $\text{Th}_1^{\text{th}} = \text{Th}_2^{\text{th}} = \text{Th}_3^{\text{th}} = 0.5\times10^5$ bits/s.

We compare against the three solutions by varying the average data arrival rate and plot the average throughput of service 1 in Figure 2, which shows that CADSP outperforms Backpressure and PDA-PMF. When the average data arrival rate is lower than 5×10^5 bits/s, the three schemes obtain similar throughput performance. However, with higher average data arrival rate, CADSP and PDA-PMF perform much better than Backpressure, since service 1 under CADSP and PDA-PMF is assigned the highest priority in the three services, and it can get more transmission opportunities than service 1 under Backpressure which has the same priority as the other two services. As Backpressure-based algorithm has throughput optimality, CADSP performs better than PDA-PMF in throughput performance.

Figure 3 shows the average end-to-end delay performance of service 1 for the three solutions. When the average data arrival rate is lower than 5.5×10^5 bits/s, PDA-PMF performs best. But when the average data arrival rate is above 6.5×10^5 bits/s, the average end-to-end delay of service 1 under PDA-PMF is higher than the maximum allowable end-to-end delay bound of service 1. The average end-to-end delay of CADSP is always below the maximum allowable end-to-end delay bound of service 1, since CADSP using delay weight coefficient can provide better delay guarantee. When the average data arrival rate is in the range from 0.5×10^5 bits/s to 4×10^5 bits/s, the average end-to-end delay of CADSP and Backpressure decreases. The reason is that the end-to-end delay will be high if the traffic load is too low for the formation of "queue length pressure" from source nodes to destination nodes.

The performances of the three solutions in terms of average throughput of service 2 are compared in Figure 4, which shows that CADSP outperforms Backpressure and PDA-PMF. When the average data arrival rate is lower than 4×10^5 bits/s, the three schemes obtain similar throughput performance. However, with higher average data arrival rate, CADSP performs better than Backpressure and PDA-PMF. From Figure 2 we can see that when traffic load is high, CADSP can still maintain good performance in terms of average throughput for service 2.

Figure 5 shows the average end-to-end delay performance of service 2 for the three solutions. Since the priority of service 2 in CADSP is not as high as service 1, the average end-to-end delay performance deteriorates. However, the average end-to-end delay of service 2 is still maintained being lower than the maximum allowable end-to-end delay bound of service 2 by using delay weight coefficient. We can also see that, in the condition of low traffic load, the performance of average end-to-end delay of PDA-PMF is the best.

In Figure 6, the average throughput of service 3 of the three solutions is compared. From the figure we can see

Table 1: Simulation parameters.

Parameter	Value
Bandwidth	1 MHz
Slot time	20 ms
$SINR^{th}$	20 dB
Packet length	1000 bits
DIFS	50 μs
SIFS	20 μs
*R*time	50 μs
Maximum times allowed to send RTS	3
φ	10
γ	10

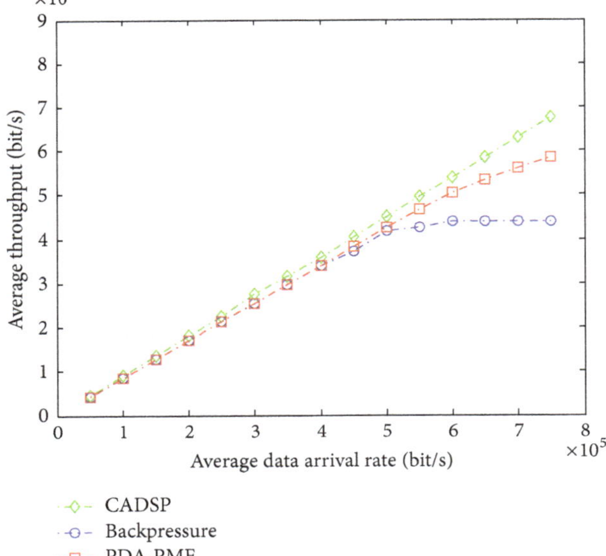

Figure 2: Comparison of average throughput of service 1 in multihop network.

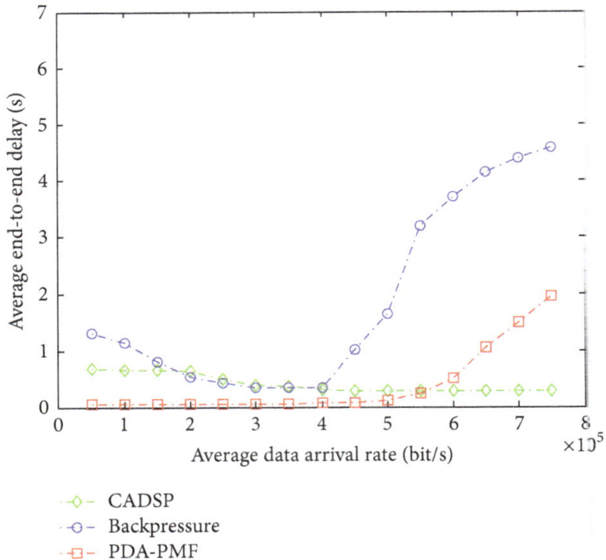

Figure 3: Comparison of average end-to-end delay of service 1 in multihop network.

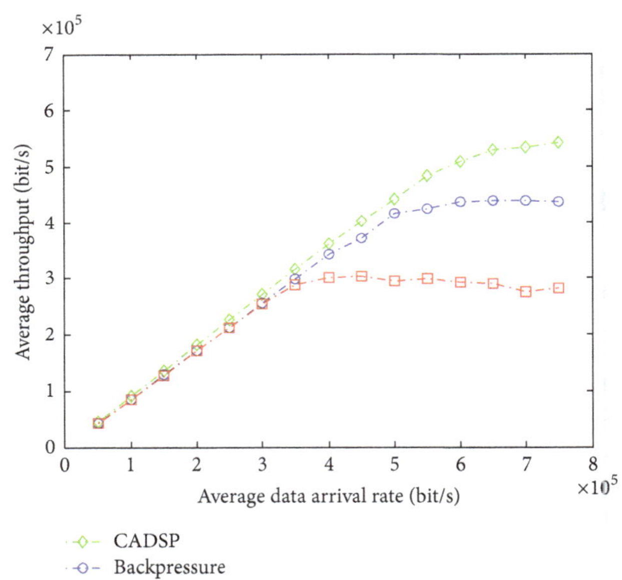

Figure 4: Comparison of average throughput of service 2 in multihop network.

that Backpressure with the same priority for each service outperforms CADSP and PDA-PMF. Since service 3 in CADSP and PDA-PMF is scheduled with the lowest priority, their average throughput cannot increase with the increase of average data arrival rate when traffic load is high. However, the average throughput of service 3 of CADSP is always higher than the required average throughput of service 3 through using throughput weight coefficient.

The average end-to-end delay performance of service 3 for the three solutions can be seen from Figure 7. Though CADSP performs worse than Backpressure in most conditions, its average end-to-end delay is maintained being lower than the maximum allowable end-to-end delay bound of service 3.

From the simulation results above, we can see that CADSP can support QoS requirements of all services.

We plot the throughput of the three solutions in Figure 8, which shows that Backpressure outperforms CADSP and PDA-PMF. The reason is that Backpressure can provide the throughput optimality, while the throughput optimality of CADSP is destroyed by introducing throughput and delay weight coefficients into the optimization objective.

5.3. Simulation of Services with Different Average Data Arrival Rates. The average data arrival rate of service 1 is four times that of service 3. The average data arrival rate of service 2 is two times that of service 3. The maximum allowable end-to-end delay bounds of services are set as $D_1^{th} = D_2^{th} = D_3^{th} = 2$ s. The required average throughputs of services are set as $Th_1^{th} = Th_2^{th} = Th_3^{th} = 0.5 \times 10^5$ bits/s.

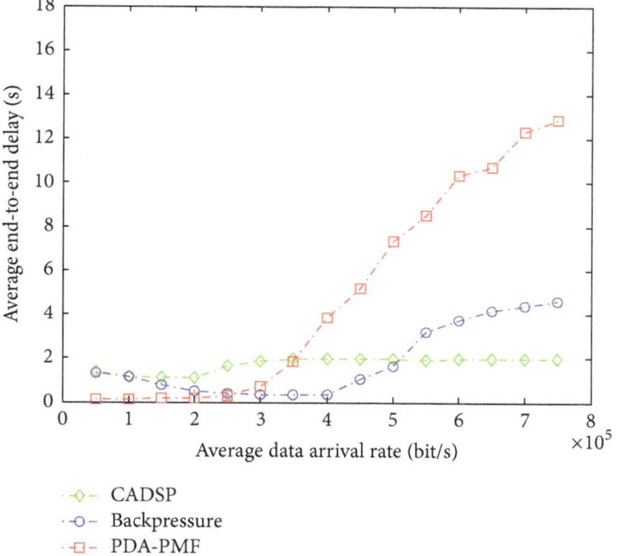

FIGURE 5: Comparison of average end-to-end delay of service 2 in multihop network.

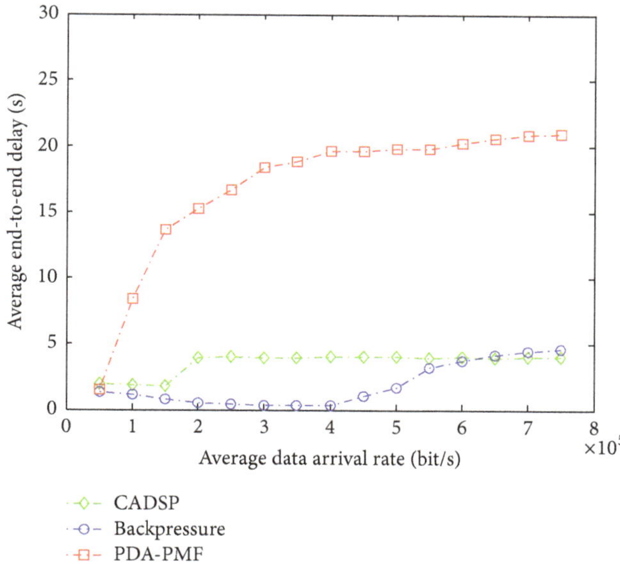

FIGURE 7: Comparison of average end-to-end delay of service 3 in multihop network.

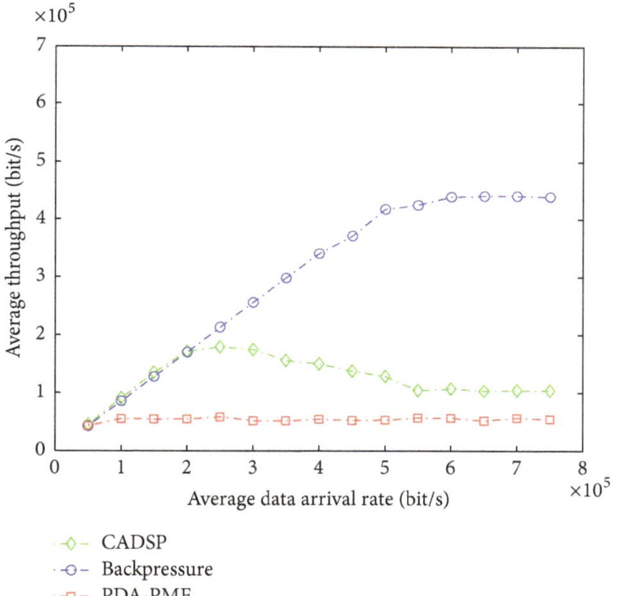

FIGURE 6: Comparison of average throughput of service 3 in multihop network.

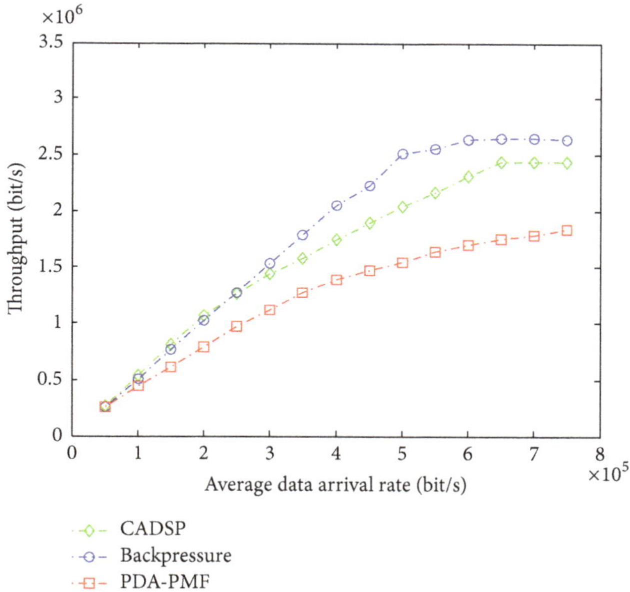

FIGURE 8: Comparison of throughput in multihop network.

In Figure 9 the average throughputs of the three services using CADSP are compared. From the figure we can see that the ratio among the average throughputs of the three services is close to the ratio among the average arrival data rates of the three services.

The average end-to-end delay performances of the three services using CADSP are compared in Figure 10. The performances in terms of average end-to-end delay of service 1 and service 2 are lower than the maximum allowable end-to-end delay bound. When the average data arrival rate is lower than 2×10^5 bits/s, the average end-to-end delay of service 3 is higher than the maximum allowable end-to-end delay bound of service 3. The reason is that the average data arrival rate of service 3 is too low to form the "queue length pressure" to push packets of service 3 from source node to destination node, and this increases the average end-to-end delay of service 3.

6. Conclusions

This paper proposed a cross-layer QoS scheme which can provide different QoS guarantees for diverse types of services.

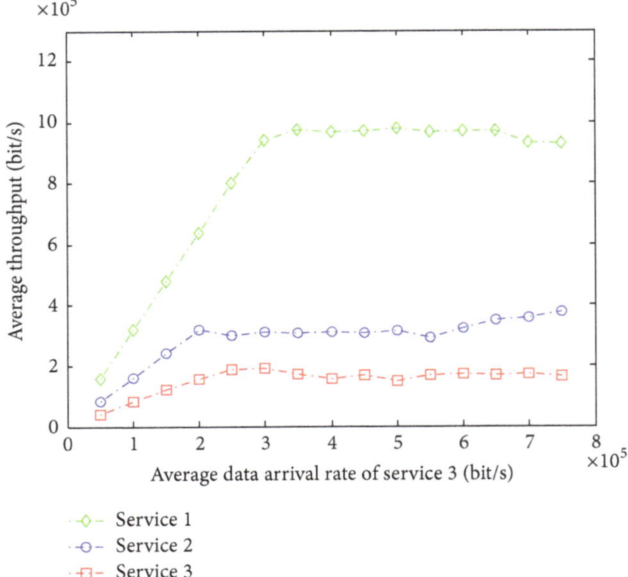

FIGURE 9: Comparison of average throughput in multihop network.

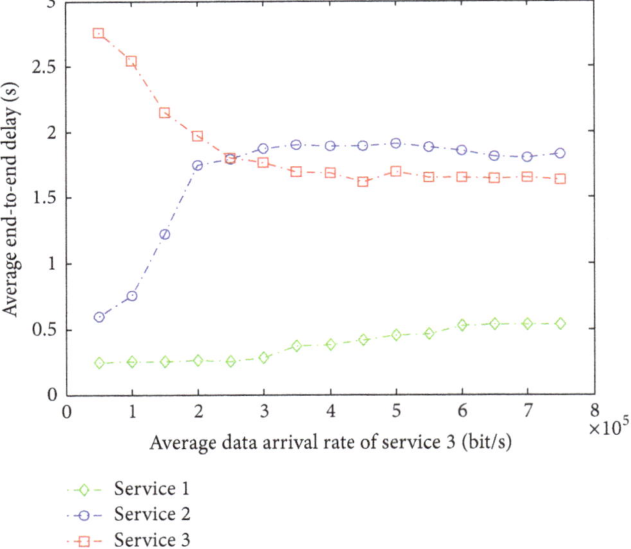

FIGURE 10: Comparison of average end-to-end delay in multihop network.

Through setting priorities of services depending on services' data arrival rates and end-to-end delay demands, services with higher QoS demands can gain better QoS performance. The delay and throughput weight coefficients in the objective of the optimization problem help to maintain fairness of the policy and make the scheme support QoS requirements better. The throughput utility optimality of the scheme is kept. A distributed medium access control scheme and a power control algorithm are designed to reduce the computational complexity of the scheme. Compared with the existing works, the policy presented in this paper can simultaneously support the delay requirements of different services and maintain higher throughput.

Conflicts of Interest

The authors declare that there are no conflicts of interest regarding the publication of the manuscript.

References

[1] S. Ehsan and B. Hamdaoui, "A survey on energy-efficient routing techniques with QoS assurances for wireless multimedia sensor networks," *IEEE Communications Surveys & Tutorials*, vol. 14, no. 2, pp. 265–278, 2012.

[2] IEEE std 802.11-2007 (2007) IEEE standard for information technology-telecommunications and information exchange between systems-local and metropolitan area networks-specific requirements-part 11: Wireless LAN Medium Access Control (MAC) and Physical Layer (PHY) specifications (Revision of IEEE Std 802.11-1999), article C1-1184.

[3] Y. Qin, L. Li, X. Zhong, Y. Yang, and C. L. Gwee, "A Cross-Layer QoS Design with Energy and Traffic Balance Aware for Different Types of Traffic in MANETs," *Wireless Personal Communications*, vol. 85, no. 3, pp. 1429–1449, 2015.

[4] C.-Y. Liu, B. Fu, and H.-J. Huang, "Delay minimization and priority scheduling in wireless mesh networks," *Wireless Networks*, vol. 20, no. 7, pp. 1955–1965, 2014.

[5] G. Adam, C. Bouras, A. Gkamas, V. Kapoulas, G. Kioumourtzis, and N. Tavoularis, "Cross-layer mechanism for efficient video transmission over mobile ad hoc networks," in *Proceedings of the 2011 3rd International Workshop on Cross Layer Design, IWCLD 2011*, pp. 1–5, December 2011.

[6] H. Wang and G. Liu, "Priority and delay aware packet management framework for real-time video transport over 802.11e WLANs," *Multimedia Tools and Applications*, vol. 69, no. 3, pp. 621–641, 2014.

[7] A. R. Lari and B. Akbari, "Network-adaptive multipath video delivery over wireless multimedia sensor networks based on packet and path priority scheduling," in *Proceedings of the 5th International Conference on Broadband Wireless Computing, Communication and Applications (BWCCA '10)*, pp. 351–356, Fukuoka, Japan, November 2010.

[8] D. Djenouri and I. Balasingham, "Traffic-differentiation-based modular QoS localized routing for wireless sensor networks," *IEEE Transactions on Mobile Computing*, vol. 10, no. 6, pp. 797–809, 2011.

[9] L. Tassiulas and A. Ephremides, "Stability properties of constrained queueing systems and scheduling policies for maximum throughput in multihop radio networks," *Institute of Electrical and Electronics Engineers Transactions on Automatic Control*, vol. 37, no. 12, pp. 1936–1948, 1992.

[10] M. J. Neely, E. Modiano, and C.-P. Li, "Fairness and optimal stochastic control for heterogeneous networks," *IEEE/ACM Transactions on Networking*, vol. 16, no. 2, pp. 396–409, 2008.

[11] L. Jiang and J. Walrand, "A distributed CSMA algorithm for throughput and utility maximization in wireless networks," *IEEE/ACM Transactions on Networking*, vol. 18, no. 3, pp. 960–972, 2010.

[12] H. Seferoglu and E. Modiano, "TCP-aware backpressure routing and scheduling," in *Proceedings of the IEEE Information Theory and Applications Workshop (ITA '14)*, pp. 1–9, San Diego, Calif, USA, February 2014.

[13] E. Anifantis, E. Stai, V. Karyotis, and S. Papavassiliou, "Exploiting social features for improving cognitive radio infrastructures and social services via combined MRF and back pressure cross-layer resource allocation," *Computational Social Networks*, vol. 1, article 4, 2014.

[14] G. A. Shah, V. C. Gungor, and O. B. Akan, "A cross-layer QoS-aware communication framework in cognitive radio sensor networks for smart grid applications," *IEEE Transactions on Industrial Informatics*, vol. 9, no. 3, pp. 1477–1485, 2013.

[15] A. Zhou, M. Liu, Z. Li, and E. Dutkiewicz, "Cross-layer design for proportional delay differentiation and network utility maximization in multi-hop wireless networks," *IEEE Transactions on Wireless Communications*, vol. 11, no. 4, pp. 1446–1455, 2012.

[16] S. Fan and H. Zhao, "Cross-layer control with worst case delay guarantees in multihop wireless networks," *Journal of Electrical and Computer Engineering*, vol. 2016, Article ID 5762851, 10 pages, 2016.

[17] M. J. Neely, *Stochastic Network Optimization with Application to Communication and Queueing Systems*, Morgan & Claypool Publishers, 2010.

[18] M. J. Neely, "Opportunistic scheduling with worst case delay guarantees in single and multi-hop networks," in *Proceedings of the IEEE INFOCOM*, pp. 1728–1736, Shanghai, China, April 2011.

[19] X. Lin and N. B. Shroff, "The impact of imperfect scheduling on cross-layer congestion control in wireless networks," *IEEE/ACM Transactions on Networking*, vol. 14, no. 2, pp. 302–315, 2006.

[20] L. Ding, T. Melodia, S. N. Batalama, J. D. Matyjas, and M. J. Medley, "Cross-layer routing and dynamic spectrum allocation in cognitive radio ad hoc networks," *IEEE Transactions on Vehicular Technology*, vol. 59, no. 4, pp. 1969–1979, 2010.

[21] L. Georgiadis, M. J. Neely, and L. Tassiulas, "Resource allocation and cross-layer control in wireless networks," *Foundations and Trends in Networking*, vol. 1, no. 1, pp. 1–144, 2006.

3

Design and Analysis of a Low Cost Wave Generator based on Direct Digital Synthesis

Jian Qi,[1] Qun Sun,[1] Xiaoliang Wu,[2] Chong Wang,[1] and Linlin Chen[1]

[1]*School of Mechanical & Automotive Engineering, Liaocheng University, Liaocheng 252059, China*
[2]*Department of Medical Equipment, Liaocheng People's Hospital, Liaocheng 252000, China*

Correspondence should be addressed to Qun Sun; sunxiaoqun97@163.com

Academic Editor: Raj Senani

Signal generators are widely used in experimental courses of universities. However, most of the commercial tests signal generators are expensive and bulky. In addition, a majority of them are in a fixed working mode with many little-used signals. In order to improve this situation, a small sized and highly accurate economic signal generator based on DDS technology has been developed, which is capable of providing wave signals commonly used in experiments. Firstly, it is introduced the basic principles of DDS and is determined the overall scheme of the signal generator. Then, it proposes a design of the hardware, which include power supply module, display module, keyboard module, waveform generating module based on DDS chip, and the minimum system module based on C8051F010. The signal generator was designed to output sine and square waveforms, and the other achieved performances included the frequency range 0.1 Hz–12.5 MHz, the frequency resolution 0.05 Hz–0.1 Hz, the output amplitude 1.0–4.5 V, the frequency accuracy $K_{f\min} = 94.12\%$ and $K_{f\max} = 99.99\%$, and the signal distortion $R_{\mathrm{THDmin}} = 0.638\%$ and $R_{\mathrm{THDmax}} = 11.67\%$.

1. Introduction

Signal generators are widely utilized in experimental courses [1–8]. Furthermore, square wave and sine wave signals generated by signal generators are extensively used in a wide range of applications, usually as a standard signal in electronic circuit testing, parameter measurement, or demonstration in experimental courses. However, due to the high cost, fixed working mode and poor extensibility combine the programmable functions for generating arbitrary waveforms and other functions cannot be fully played out in the teaching experiments of common signal generator; a cheap and small signal generator which can meet common signal output functions and be suitable for experimental courses was needed [9–12]. Based on DDS technology, an economic signal generator with small size and high precision was developed. Square wave signal with tunable frequency, pulse width, and duration and high precision sine signal with adjustable amplitude and frequency can be produced to satisfy the requirements in teaching experiments. Some reports focus on developing signal generator by DDS technology in theory. There are still few reports about manufacturing of the signal generator by DDS technology.

2. The Overall Design Scheme of the Signal Generator

2.1. Basic Principle of DDS Technology. The basic principle of DDS is using the phase concept to carry out frequency synthesis [2], which allows signals changing with phase to be obtained according to the variation of given signals under different amplitude. Phase accumulator is formed by cascaded N-bit adder with N-bit accumulator register. For each arriving clock pulse f_{MCLK}, the adder sums up the control word (M) with the accumulated phase data produced by the phase accumulator register, and the result is sent back to the input port of the accumulator register, so that the adder continuously sums up the frequency control word under the effect of the subsequent clock pulse. The phase adder continuously performs linear adding of the frequency control word in an accumulative fashion under the effect of

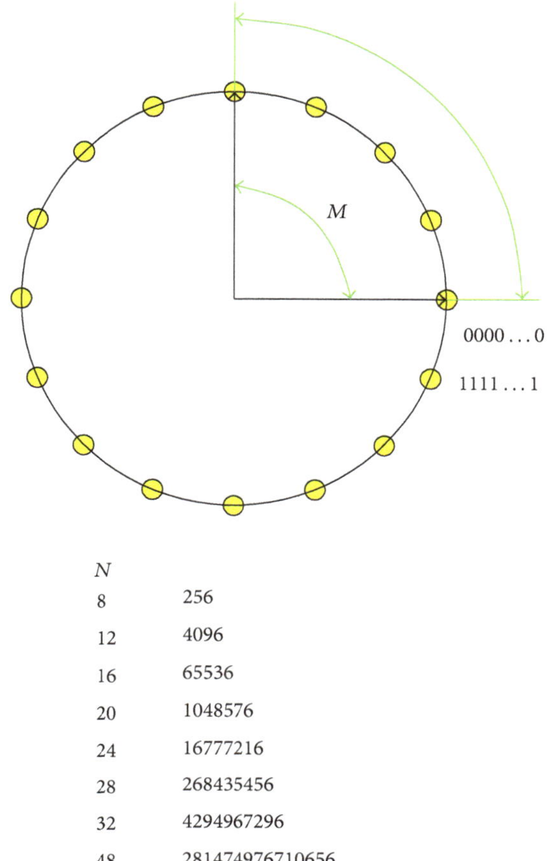

FIGURE 1: Diagrammatic sketch of the operating principle of phase accumulator.

the clock input. As shown in Figure 1, a sine wave is regarded as oscillation of a phase cycle in a rotating vector manner, and each given point on the phase wheel corresponds to an angular division point on a sine wave in a $0\sim2\pi$ cycle. Every time a vector turns around a phase circle at a constant speed, a complete cycle of sine wave is produced. It can be seen that the phase accumulator adds the control word once following each clock input, the output data of the phase accumulator is the phase of the synthesized signal, and the output frequency of the phase accumulator is the signal frequency of the DDS output. The frequency control word determines the number of division points that can jump on the phase cycle in each time clock cycle. More division points lead to faster overflow of the phase accumulator and shorter time to complete equal sine wave cycles; thus changing the value of the control word (M) can alter the output frequency [4]. The output data of the phase accumulator serves as the phase sampling address of a waveform memory (ROM). A phase/amplitude conversion is completed after retrieving the binary waveform sample values stored in the waveform memory using a lookup table.

2.2. The Overall Program Design of Signal Generator. The functional structure of the signal generator block diagram, as shown in Figure 2, mainly consists of a power supply system,

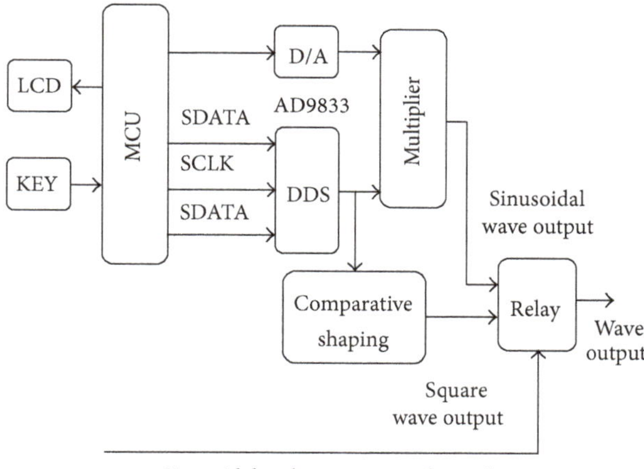

FIGURE 2: Block diagram of the signal generator based on DDS.

SCM system, DDS waveform generator module, amplitude adjustment module, square wave generator module, relay output module, and so on. The power supply system provides 2.5 V, 3.3 V, ±5 V, and ±15 V power supply voltage. The +5 V input of the power supply system goes through an integrated three-terminal voltage regulator to obtain 2.5 V and 3.3 V voltage and then uses a DC-DC power supply module for conversion to +5 V and +15 V. The SCM system is employed to control the operation of the human-machine interface, read keyboard, display on LCD, and adjust the output sine wave amplitude. It also undertakes programming of the DDS device to produce the corresponding frequency signal, to control the relays and switch sine and square wave outputs.

3. Hardware Design of the Signal Generator

3.1. Design of the Power Supply Circuit. As shown in Figure 3, the inputs and outputs of AS1117-2.5, AS1117-3.3, and the DC-DC module use 330 μF and 0.1 μF capacitors for frequency compensation, to prevent the regulator from generating high-frequency self-excited oscillation and to suppress high-frequency interference in the circuit. The output terminals of the A05S05-2W and A05S15-1W power modules have resistors and capacitors to form a low-pass filter, which can smooth power output spikes and reduce interference to the subsequent circuit chip. The reverse protection design is also used in the power supply circuit, where a diode D3 is in cut-off state under normal power supply. If the input voltage is reversed, the diode conducts and the current from the external power supply flows through the FUSE and diode D3 into the negative terminal of the power supply. Since the diode forward resistance is very small, the current increases sharply and burns out the FUSE, disconnects the circuit, and thus protects the internal circuit from being damaged.

3.2. Design of LCD Module. The liquid crystal display module was built on LCM1602 that has 2 × 16 characters and internal font. The interface is shown in Figure 4, where R_{12}

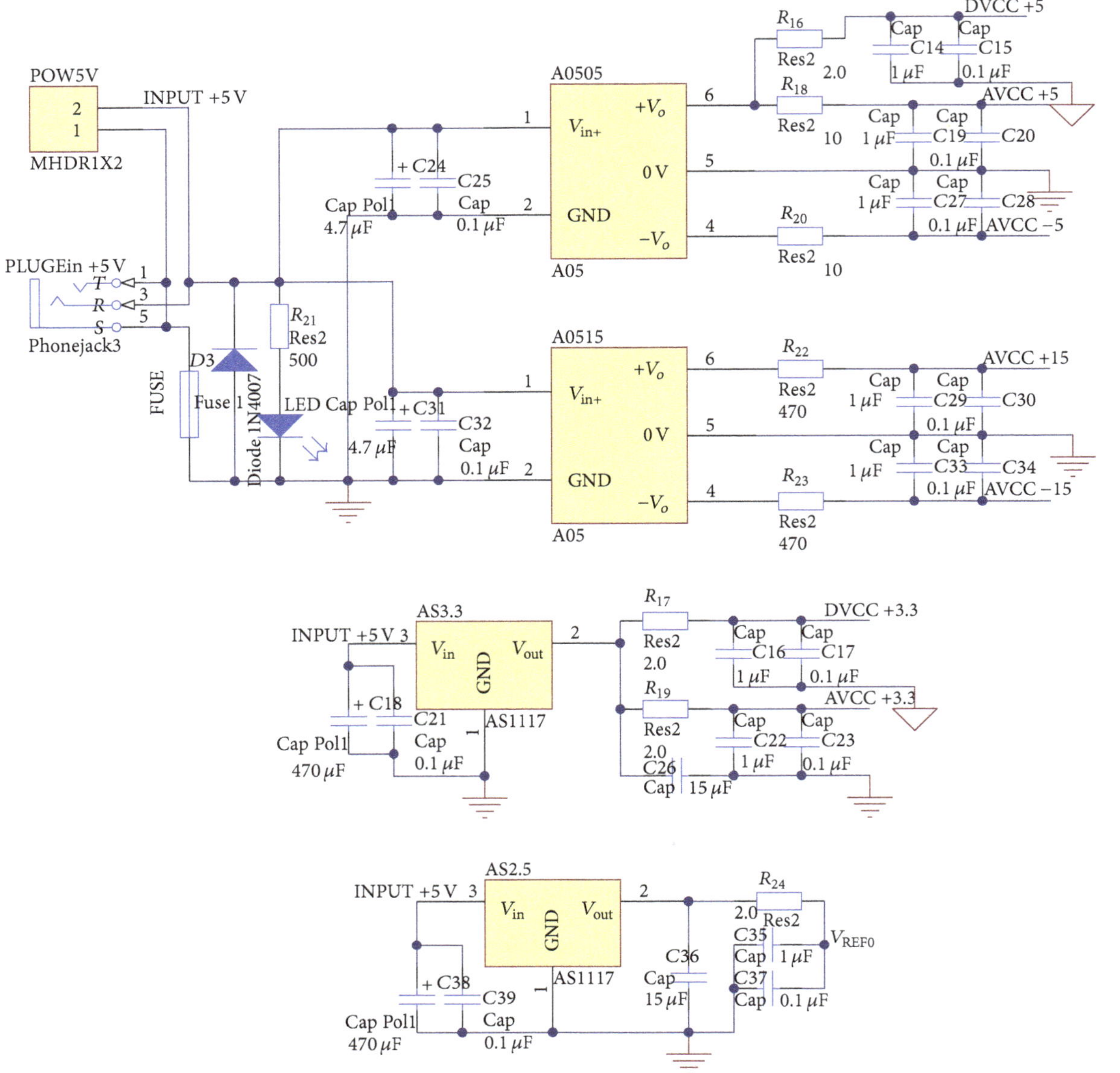

Figure 3: Circuit of the power supply.

potentiometer can be used to adjust the potential of feet 3 to change the contrast of the LCD display.

3.3. Design of Keyboard Circuit. The front panel of the signal generator is shown in Figure 5. By means of pressing the left or right buttons to select each bit of data to be transferred and then using add or subtract settings, the number of buttons can be reduced and remained the same when the data for transmission needs to be extended. This allows the circuit to be very flexible since appropriate changes can be made in the software. The circuit output functions can be selected by "sine/square wave" buttons. The output signal frequency can be configured under a square wave output state, whilst for a sine wave output state, the signal frequency or amplitude can be set or modified through the "Mode" button.

As shown in Figure 6, the "sine/square wave" is designed with a locking switch, whilst the "Add," "Reduce," "Shift left," "Shift right," and "Mode" buttons are designed without locking switch. The output is high level 3.3 V when SW1~SW5 are released and low level when pressing them. When Switch6 is released, PIN2 connects to PIN3 and PIN5 connects to PIN6, while PIN1 is hanging, SIN/REC is high level, and LED_SIN lights up, indicating the output is sine wave. When it is pressed, SIN/REC is low level, and PIN4 connects to PIN6, and LED_REC lights up, showing the output is squared wave signals.

3.4. The Circuit Design of Wave Generator

3.4.1. The Circuit Design of Sine Wave Generator. As is shown in Figure 7, the signal generator is based on a DDS chip

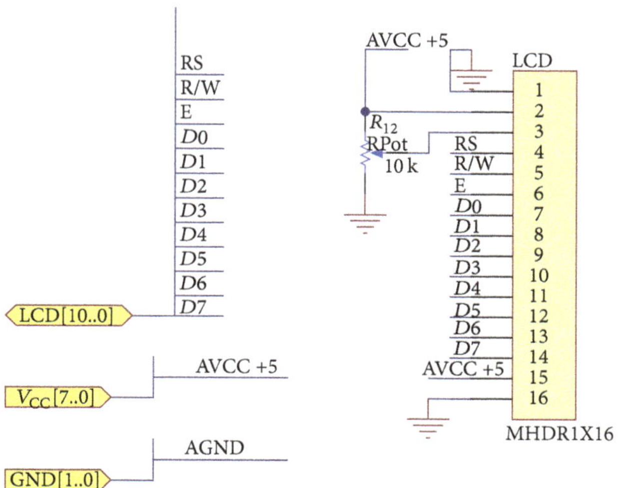

Figure 4: The interface circuit of LCD.

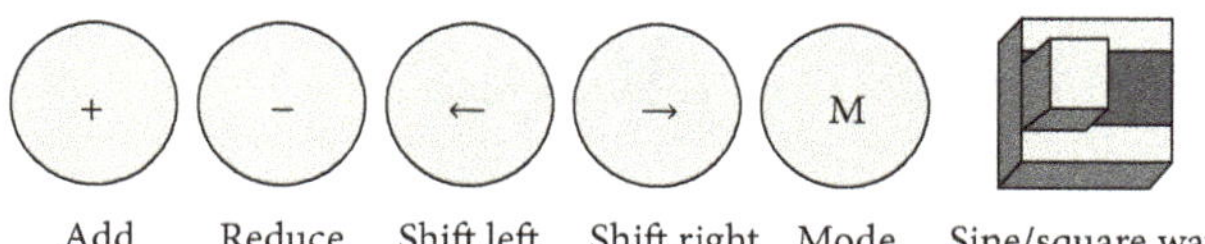

Figure 5: The shifting keyboard panel design.

AD9833 made by Analog Devices, Inc. (ADI), which is 25 MHz with active crystal as a reference clock.

3.4.2. Design of the Sine Wave Amplitude Modulation Circuit. According to the required sinusoidal amplitude which is input by the keyboard, the D/A port of the microcontroller produces a corresponding DC signal, then feeds it to the multiplier, and multiplies it with a fixed amplitude sinusoidal signal, to achieve the function of regulating the amplitude, as shown in Figure 8.

The multiplier is selected to be MPY634, which is powered by ±15 V voltage, and can adopt a single-ended input (±10 V) or a differential input (±11 V), while the output voltage can reach ±11 V. The relation of output voltage and input voltage is given by

$$V_{\text{out}} = A\left(\frac{(x_1 - x_2)(y_1 - y_2)}{\text{SF}} - (Z_1 - Z_2)\right). \quad (1)$$

Since the multiplier has internal negative feedback and V_{out} is limited, the amplification factor (A) is infinite in theory and is at least 85 dB in practice; therefore,

$$\frac{(x_1 - x_2)(y_1 - y_2)}{\text{SF}} - (Z_1 - Z_2) = 0. \quad (2)$$

The sinusoidal wave amplitude ranges from 0.038 V to 0.650 V; thus the DC component (V_o) is 0.344 V. The output from the multiplier must be the sine wave without DC component, so the DC component should be subtracted using circuit shown in Figure 8. The resistor R_{27} is linked between port Z_1 and port OUT, and resistor R_{26} is linked between ports Z_1 and Z_2.

Assuming the resistances of R_{26} and R_{27} are x and y, then

$$Z_1 - Z_2 = V'_{\text{out}} * \frac{y}{x + y}. \quad (3)$$

V'_{out} is the output voltage of multiplier, and the following can be concluded:

$$\frac{(x_1 - x_2)(y_1 - y_2)}{\text{SF}} = V'_{\text{out}} * \frac{y}{x + y}. \quad (4)$$

Therefore,

$$V'_{\text{out}} = \frac{x + y}{y} * \frac{(x_1 - x_2)(y_1 - y_2)}{\text{SF}}. \quad (5)$$

Assuming that

$$K_0 = \frac{x + y}{y} * \frac{1}{\text{SF}}. \quad (6)$$

K_0 is the gain of the multiplier of the two differential inputs,

$$V'_{\text{out}} = K_0 * (x_1 - x_2)(y_1 - y_2). \quad (7)$$

The current of the multiplier output cannot output directly because of its weakness of driving power, so a high speed operational amplifier must be set before the input of the multiplier to increase the output current.

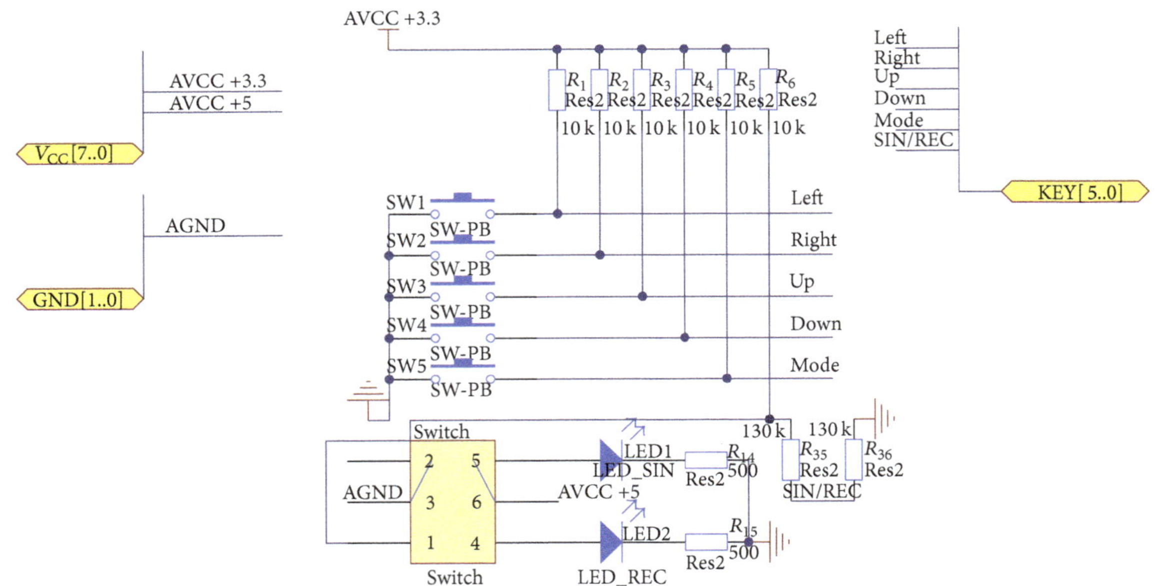

FIGURE 6: The keyboard circuit.

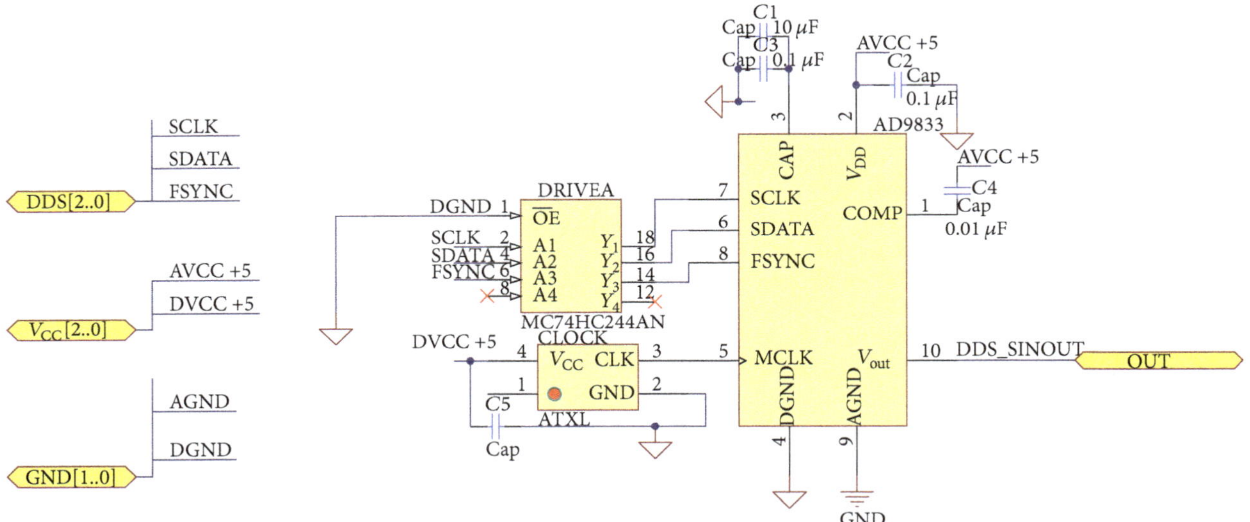

FIGURE 7: The circuit of wave generator.

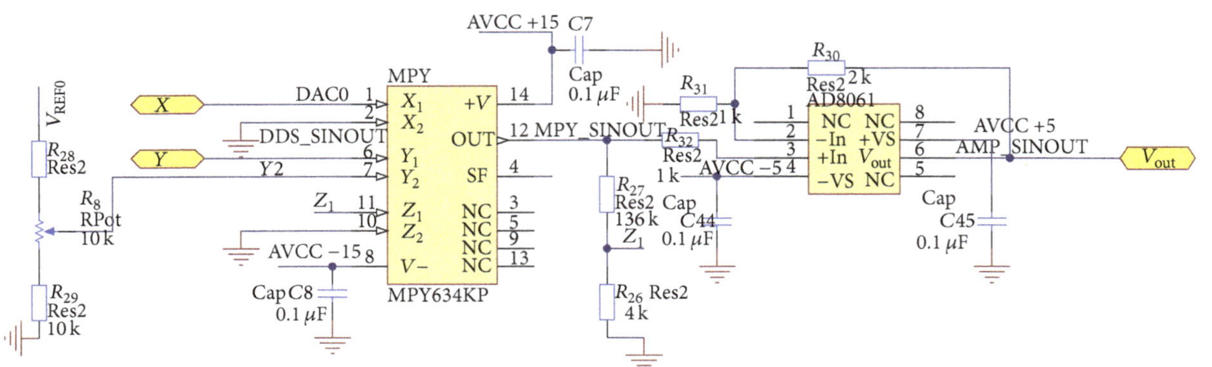

FIGURE 8: The circuit of regulating the sinusoidal amplitude.

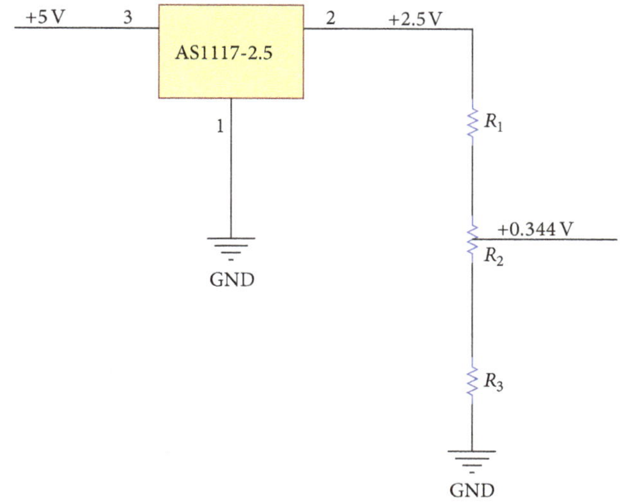

FIGURE 9: The circuit of offset voltage of the amplifier.

Taking $R_{30} = 2\,\text{K}\Omega$, $R_{31} = 1\,\text{K}\Omega$, $R_{32} = 1\,\text{K}\Omega$, the amplifier multiple is $A = 1 + R_{31}/R_{30} = 3$; then

$$V_{\text{out}} = 3 * V'_{\text{out}} = 3K_0 * (x_1 - x_2)(y_1 - y_2). \tag{8}$$

The output voltage of DAC of the MCU is ranging from 0 to 2.4 V and the sinusoidal waveform of DDS is ranging from 0 to 0.65 V, so the DC component of sinusoidal waveform (V_o) is 0.344 V.

Reducing the DC component at Y input of MPY634, the amplitude of sinusoidal waveform is

$$V_{\text{in}} = V_{\text{dds}} - V_o = \pm 0.306\,\text{V}. \tag{9}$$

Multiplying the two differential inputs X and Y gives $2.4 * 0.306 = 0.7344$, so it is required to get the amplitude of sinusoidal waveform more than 4 V; the gain must be chosen above $4/0.7344 = 5.4$.

The way of the circuit connection decides the value of $1/\text{SF}$.

When Port SF is hanging, the amplification factor is SF = 10 V which is accurately modified by laser in the integrated circuit and the error is 0.1% or less.

Through resistor R_{SF} linking between PIN SF and PIN $-V_S$, the value of SF can be changed:

$$R_{\text{SF}} = 5.4 * \left(\frac{\text{SF}}{10 - \text{SF}}\right). \tag{10}$$

Defining $x = 136\,\text{K}\Omega$, $y = 4\,\text{K}\Omega$, then

$$\begin{aligned} K &= 3 * K_0 = 3 * \frac{x + y}{y} * \frac{1}{\text{SF}} = 10.5 \\ V_{\text{out}} &= 10.3 * (x_1 - x_2)(y_1 - y_2) \\ V_{\text{outMax}} &= 10 * 2.4 * (\pm 0.306) = \pm 7.3\,\text{V}. \end{aligned} \tag{11}$$

Because of the amplifier providing ±5 V in practice, the output voltage can only reach about ±4.5 V.

The amplitude of sinusoidal waveform ranges from 0.038 V to 0.650 V and its DC component is 0.344 V, which is obtained by resistor divider. As shown in Figure 9, the AS1117-2.5 three-port voltage stabilizer provides the voltage of 2.5 V, providing the offset voltage of 0.344 V through resistor dividing.

Taking $R_1 = 60\,\text{K}\Omega$, $R_2 = 10\,\text{K}\Omega$, $R_3 = 10\,\text{K}\Omega$, the output scope of reference voltage is

$$\begin{aligned} V_{\text{ref0min}} &= 2.5 * \frac{R_3}{R_1 + R_2 + R_3} = 0.3125\,\text{V} \\ V_{\text{ref0max}} &= 2.5 * \frac{R_2 + R_3}{R_1 + R_2 + R_3} = 0.6250\,\text{V}. \end{aligned} \tag{12}$$

Therefore, the output scope of reference voltage ranges from 0.3125 V to 0.6250 V, and it is possible through adjusting potentiometer to get the offset voltage of 0.344 V. Therefore it is possible to eliminate the DC component of sinusoidal waveform from DDS output. Introducing a positive feedback to the circuit constitutes a hysteresis comparator, through adding a branch of voltage divider from comparator output to in-phase input. This can be seen from the schematic in Figure 10(a).

When input voltage V_I is gradually increased from zero but is less than or equal to V_T, which is called ceiling trigger level, $V_o = V_{\text{om}}^+$,

$$V_T = \frac{R_1 V_{\text{REF}}}{R_1 + R_2} + \frac{R_2}{R_1 + R_2} V_{\text{om}}^+. \tag{13}$$

When the input voltage V_I is greater than V_T, $V_o = V_{\text{om}}^-$, and the triggering level is changed into V'_T which is called the lower trigger level:

$$V'_T = \frac{R_1 V_{\text{REF}}}{R_1 + R_2} + \frac{R_2}{R_1 + R_2} V_{\text{om}}^-. \tag{14}$$

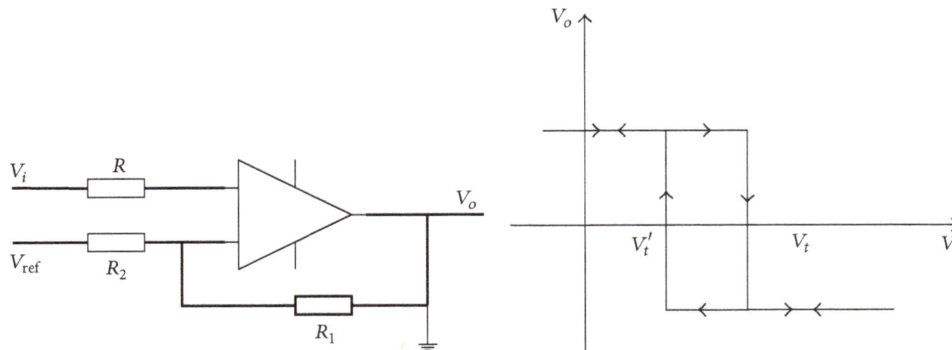

(a) The schematic diagram of hysteresis comparator (b) The voltage transmission characteristic of hysteresis comparator

FIGURE 10: Principle of the hysteresis comparator.

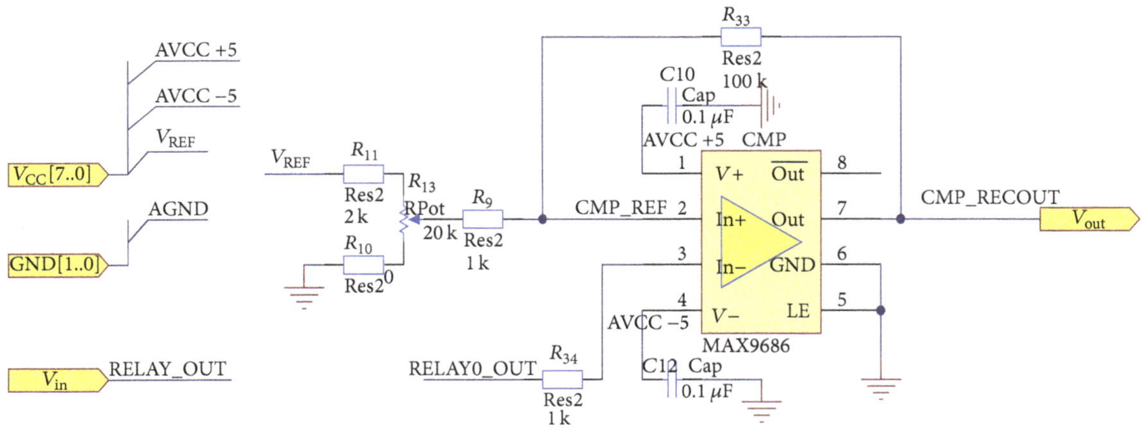

FIGURE 11: The circuit of generating square waveform.

The threshold voltage is

$$\Delta V = V_T - V_T' = \frac{R_2}{R_1 + R_2} V_{\text{om}}^{+} - \frac{R_2}{R_1 + R_2} V_{\text{om}}^{-} = \frac{R_2}{R_1 + R_2}\left(V_{\text{om}}^{+} - V_{\text{om}}^{-}\right). \tag{15}$$

From Figure 10(b) of voltage transmission characteristics of the hysteresis comparator, it can be seen that the comparison voltage is different between the direction of increment and the direction in input voltage, and the interference within the hysteresis range will not affect the output voltage.

The circuit of generating square waveform is shown in Figure 11. Port V_{REF} provides 2.5 V voltage for the voltage regulator module, the voltage can be regulated by adjusting a potentiometer, and the comparator outputs about 4 V of high level. So the backlash voltage of the circuit is

$$\Delta V = \frac{R_9}{R_{33} + R_9}\left(V_{\text{om}}^{+} - V_{\text{om}}^{-}\right) \approx 40\,\text{mV}. \tag{16}$$

3.5. The Design of Controlling Circuit of Relay Output. As shown in Figure 12, the relay module UA2 is equivalent to a double-pole double-throw switch. The control sides PIN1 and PIN8 are, respectively, connected to the +5 V supply and sine/square waveform changeover switch, PIN2, PIN3, PIN4 are the switch for final wave output switching, and PIN6, PIN7, PIN8 are the switch for shifting sine wave to hysteresis comparator to produce a square wave. Diode D1 is for current limiting to prevent excessive current impact on the circuit.

When the sine/square waveform switch is released, SIN/REC becomes high level, and there is no current through the relay coil or the current is very small, which means that it does not produce magnetic force. The connection of PIN6 and PIN7 will put the sinusoidal signal into the comparator to create square waveform. Then, the square signal will input into PIN2 of the decay, which would sent to external interface because of the connection of PIN2 and PIN3. When the sine/square waveform switch is pressed, SIN/REC changes to low level; thus the relay coil generates a magnetic force. The connection of PIN5 and PIN6 separates the sine wave signal from the comparator to prevent generating interference. The connected PIN4 and PIN3 deliver the sine signal generated from the amplitude to the external output interface.

3.6. The Design of MCU Controlling Circuit. As shown in Figure 13, a CYGNAL C8051F series MCU was chosen to

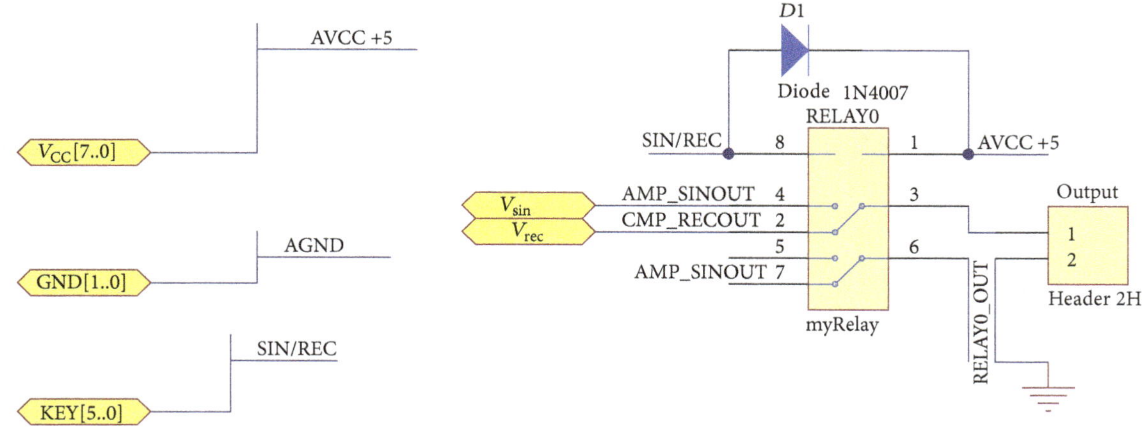

FIGURE 12: The circuit of relay controller.

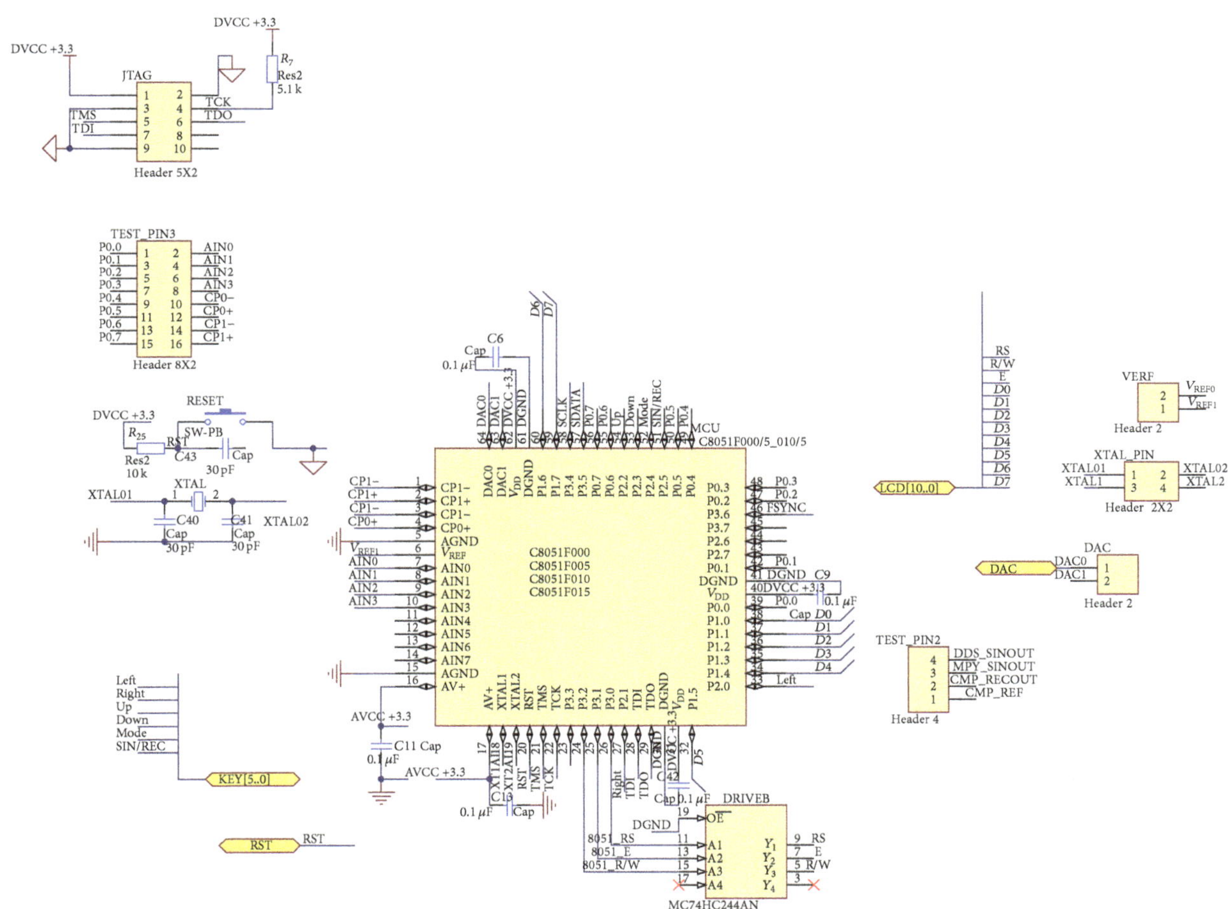

FIGURE 13: The controlling circuit of MCU.

implement the functions of reading keyboard, displaying LCD, programming DDS chip, and adjusting the amplitude of the sinusoidal signal. Considering the rules of microcontroller pin configuration, the internal digital resources start from P0 port; therefore when assigning peripheral device port, the MCU starts assigning from P3 port to reserve P0 port for the internal resources.

3.7. The Deployment of PCB and Anti-Interference Design. Because of the hybrid-system of digital circuit and analog circuit, and its high operating frequency, much attention should be put to the deployment of PCB and its anti-interference design. Considering the cost and the size, the PCB board is designed with double-layer plate and double wiring. Separate the digital circuit and analog circuit in PCB

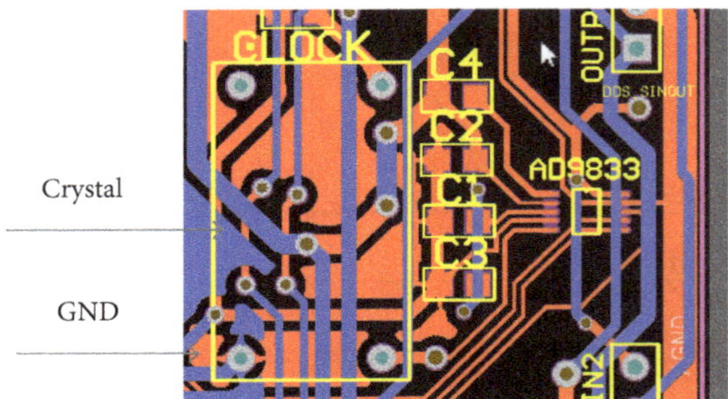

Figure 14: The PCB design diagram of AD9833.

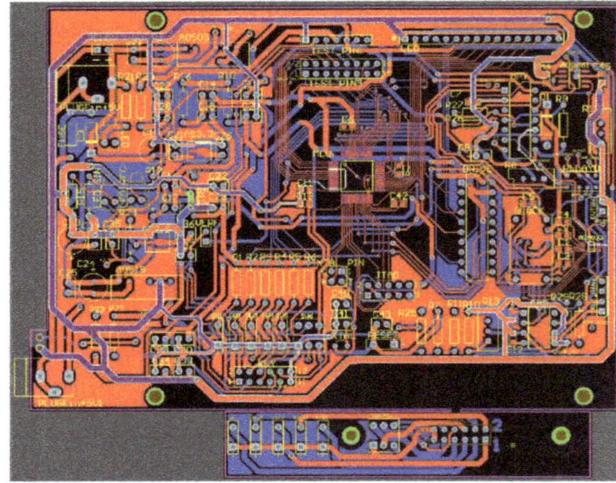

Figure 15: The whole PCB design diagram.

layout and wiring. In general, we should adopt the way of separating digital signal ground with analog signal ground and connecting them at a point. For the part of analog circuit, which includes DDS chip, multiplier, and relay, it should use the way of linking grounds, respectively, and linking the ground at a point, thicken the ground line at the same time.

The clock circuit of DDS wave generator is a critical part in the design, which can be easily interfered and have great influence on the quality of the output wave, so we should pay special attention to this part. In order to reach the purpose of isolation, the crystal oscillator should be close to the pin of DDS chip, thicken the line of crystal oscillator and the power, add cuprum to the shell of the crystal oscillator. The PCB design diagram of AD9833 is shown in Figure 14.

We should make the power line bold as large as possible because of its high current and take the impedance into consideration. The STAR structure is used in the power wiring. In fact, in this design, it is designed into the shortest structure by manual wiring at first, which must control the width of the conducting wire. Then, for each current channel of their device, the conducting wire must guarantee more than 20 mils. Finally, the circuit will be fulfilled into STAR structure. Finally fulfill the circuit into STAR structure. The whole PCB design diagram is shown in Figure 15.

4. The Software Design of Signal Generator

4.1. The Overall Design of Software. The main program diagram of SCM software is shown in Figure 16, which is based on the idea of structured and modular design. The initialization section mainly deals with writing operation on a few special function registers, to set the mode and initial value of each module and initialize variables used. Then, output the default waveform, and restore the last working state if reset source is the watchdog or the missing clock detector. Scanning of the keyboard is controlled by timer T0, implemented into different subfunctions according to different waveforms and waveform generating mode.

4.2. The Software Design of Signal Generator. AD9833 is programmable DDS signal generator with two 28-bit frequency registers inside and two 12-bit phase registers. The software block diagram is shown in Figure 17.

Firstly, write 16-bit operating mode command word to determine working conditions and select the frequency register and phase register, and, secondly, write one or two frequency control words to control the output frequency. Finally, write phase control word, so that the DDS signal generator can output waveforms corresponding to a frequency determined by the value of the frequency register, and the phase determined by the value in the mode register. The sequence chart of data writing is shown in Figure 18.

4.3. The Software Design of Keyboard Inputting. The block diagram of keyboard scanning is as shown in Figure 19. Use the 10 ms interrupt of timer T0 to implement keyboard scanning, and eliminate keyboard dither according to the number of interruption. In the timer several operations are governed such as mode switching, data adding, data reduction, and left or right shift of data to be configured. In different modes, the range of the input data can be limited to prevent the input out of range. If no keyboard action can be detected beyond seven seconds, it will automatically exit the FM or AM mode and return to normal waveform generation status.

4.4. The Software Design of LCD Displaying

4.4.1. The State of Sine Condition. In different modes, the LCD must display different interface for the user. If it is the normal sinusoidal waveform generation mode, as demonstrated in Figures 20(a) and 21(a), then turn off the LCD blinking of cursor and character, display the current amplitude in the first line, and show current frequency in the second line. The high bit 0 is not displayed. If it is sinusoidal FM mode, as shown in Figures 20(b) and 21(b), set the character and cursor to blink, and adjust the lowest bit by default and adjust the cursor by the left and right keys. The lowest bit to the right will move the cursor to the highest level, and the highest bit to the left will move the cursor to the lowest position. The data size is changed by the Add and Subtract keys; if the most significant bit to adjust the data size is limited to be only 0 or 1, the amplitude can not be higher

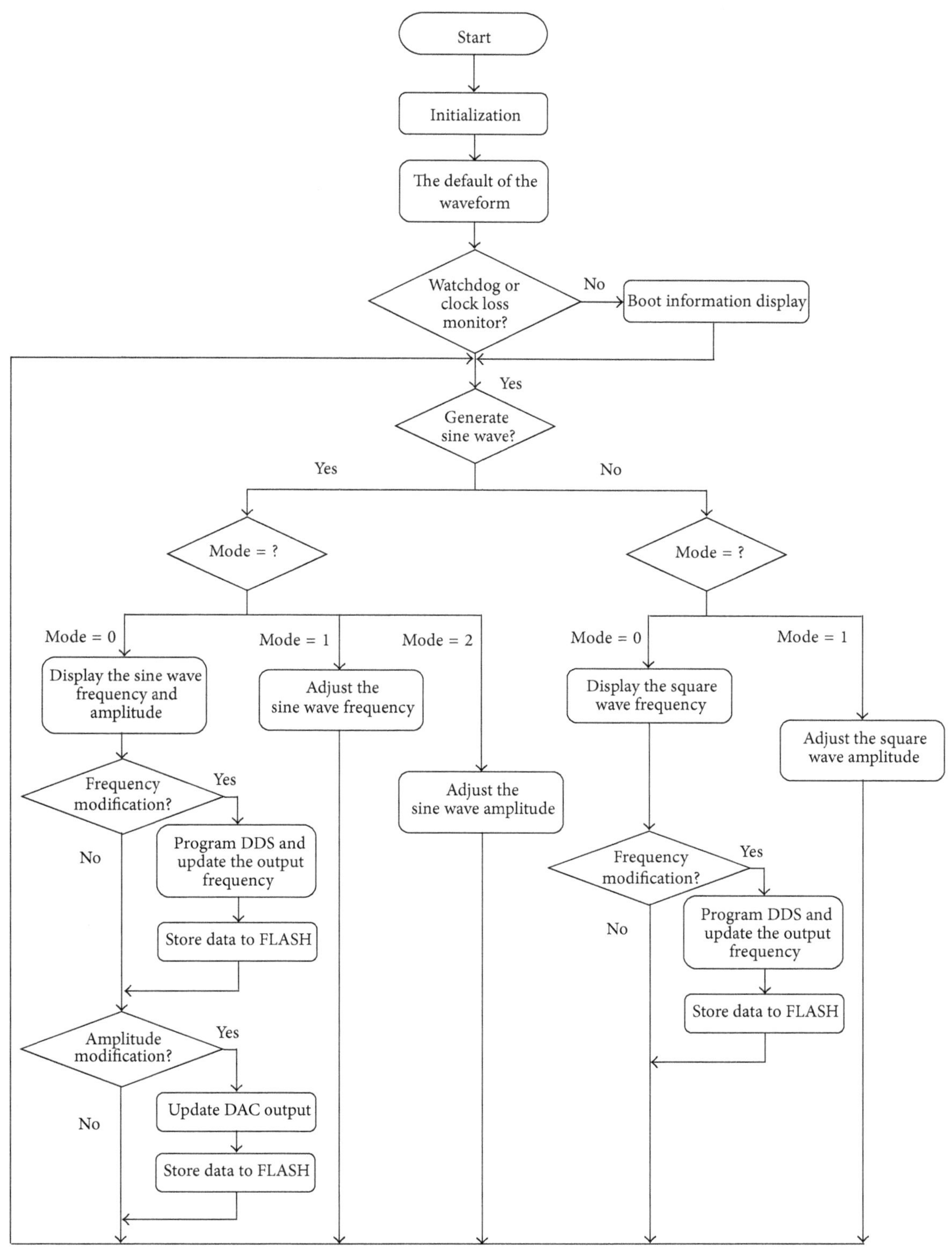

FIGURE 16: The main program diagram.

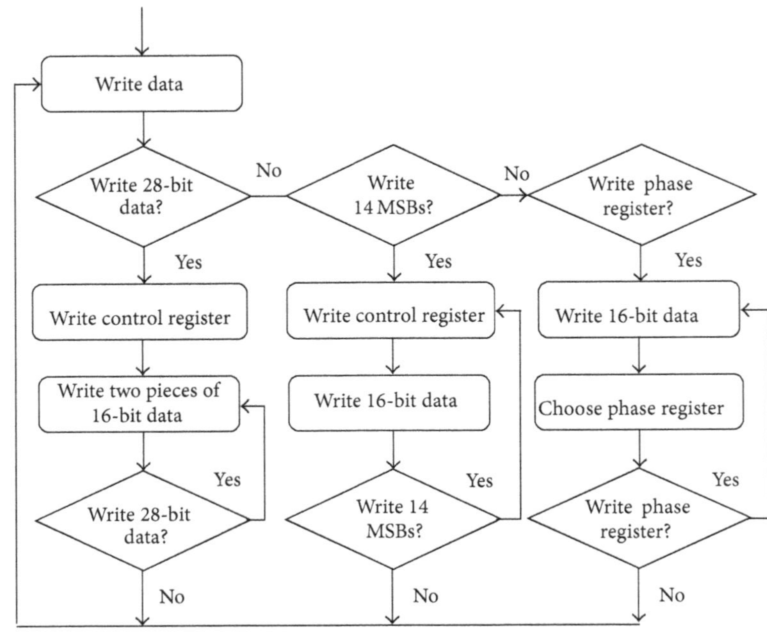

FIGURE 17: The block diagram of DDS writing data.

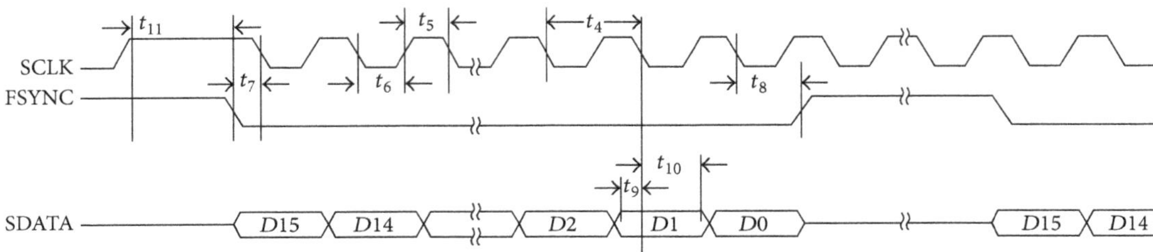

FIGURE 18: The sequence chart of DDS writing data.

than the magnitude of 10 V, and the frequency can not be set higher than the value of 20 MHz. Sinusoidal amplitude modulation is shown in Figures 20(c) and 21(c), the same case as the FM mode.

4.4.2. The State of Square Wave. Using the keys is the only way to adjust the frequency under the condition of the square wave, so there are only two modes in this state: normal mode and FM mode. If it is the normal mode as shown in Figure 22(a), then turn off the LCD cursor and character blinking, display "REC WAVE" in the first line, and show the current frequency in the second line, while the high bit 0 is not displayed. If it is the FM mode, as shown in Figure 22(b), the process is the same as the process occurring in the state of sinusoidal FM mode.

5. Experimental Data and Analysis

5.1. Frequency Characteristics. The control word ΔM generated from the microcontroller is an integer, so it is calculated using floating point operation and then transferred into an integer and written into the DDS chip. Ignoring the decimal part during integer conversion, therefore the error ΔM is ±1 and the frequency resolution is

$$\Delta F_{\text{out}} = \frac{\Delta M * f_{\text{MCLK}}}{2^{28}} = 0.093\text{ Hz}. \quad (17)$$

(1) Frequency Resolution. Inclusion of rounding algorithm can reduce the error caused by removing the decimal portion, and ΔM is 536.87; thus if the decimal portion is directly removed then the obtained result is 536 and the error is 0.87. After the improvement of the rounding algorithm, the result is 537 and the error is 0.13, which is significantly reduced. After rounding algorithm, the frequency control word M is

$$M \approx \left[\frac{F_{\text{out}} * 2^{28}}{f_{\text{MCLK}}} + 0.5 \right]. \quad (18)$$

Among the above, $f_{\text{MCLK}} = 25$ MHz.

The most significant deviation of the frequency control word is 0.5, so the frequency resolution is

$$\Delta F_{\text{out}} = \frac{\Delta M * f_{\text{MCLK}}}{2^{28}} = 0.046566\text{ Hz}. \quad (19)$$

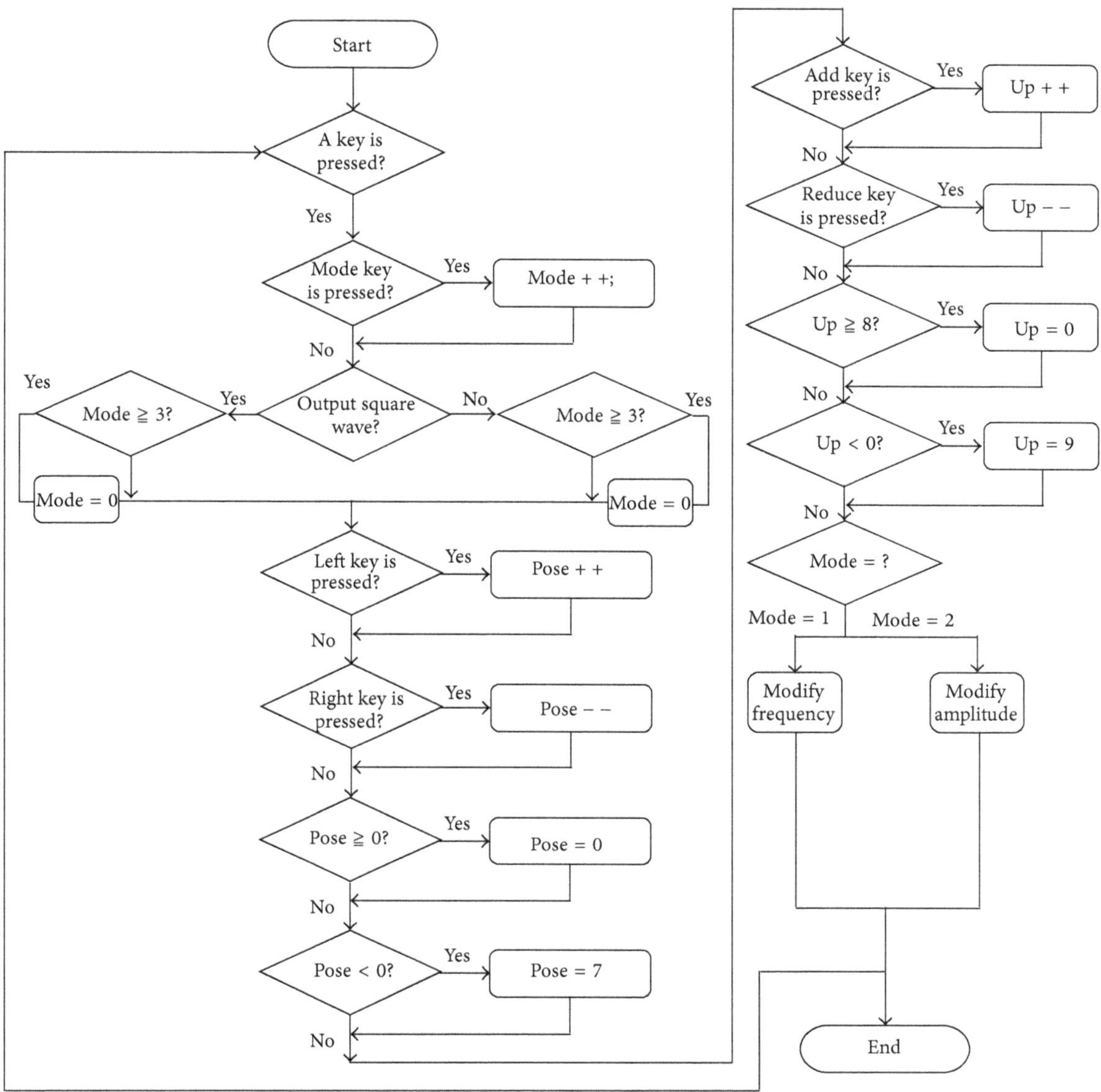

FIGURE 19: The block diagram of keyboard scanning.

The worst relative frequency accuracy is

$$\delta F_{\text{out}} = \frac{\Delta F_{\text{out}}}{F_{\text{out(min)}}} \times 100\% = 4.6566\%. \quad (20)$$

(2) Frequency Accuracy. The DDS working principle is based on digital sampling and the process of module recovery; therefore the number of sampling points will affect the frequency distortion and the accuracy of the composite signal. Through theoretical analysis, the sampling points and the frequency accuracy of output signal have the following mathematical relationship:

$$K_f = \frac{N}{N+1} \times 100\%. \quad (21)$$

In the formula, K_f is the frequency accuracy and N is the number of sampling points.

The more the sampling points, the higher the frequency accuracy; on the contrary, the fewer the number of samples, the lower the frequency accuracy. In this paper, the number of points varies with different output frequencies, which means the accuracy of frequency can be different:

$$K_{f\min} = 94.12\%$$
$$K_{f\max} > 99.99\%. \quad (22)$$

(3) Signal Distortion. The relationship between the signal distortion and sampling point is

$$R_{\text{THD}} = \sqrt{\left[\frac{\pi/N}{\sin(\pi/N)}\right]^2 - 1} \times 100\%. \quad (23)$$

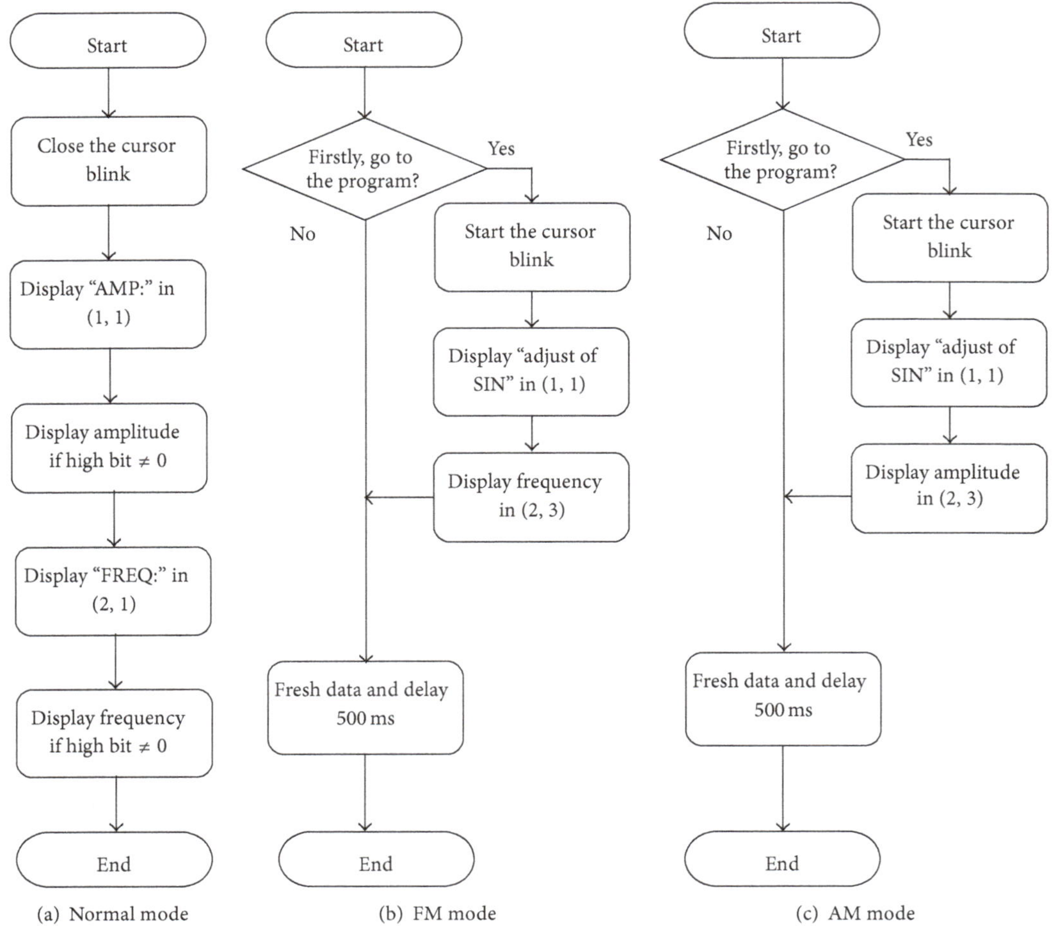

(a) Normal mode (b) FM mode (c) AM mode

FIGURE 20: The block diagram of LCD displaying under the state of sinusoidal wave.

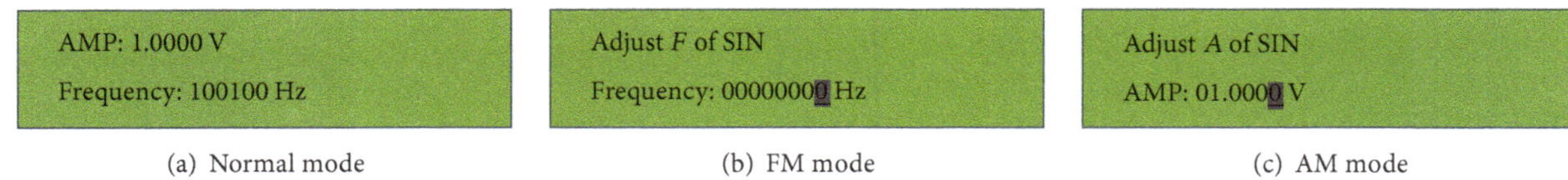

(a) Normal mode (b) FM mode (c) AM mode

FIGURE 21: The LCD working sketches under the state of sinusoidal wave.

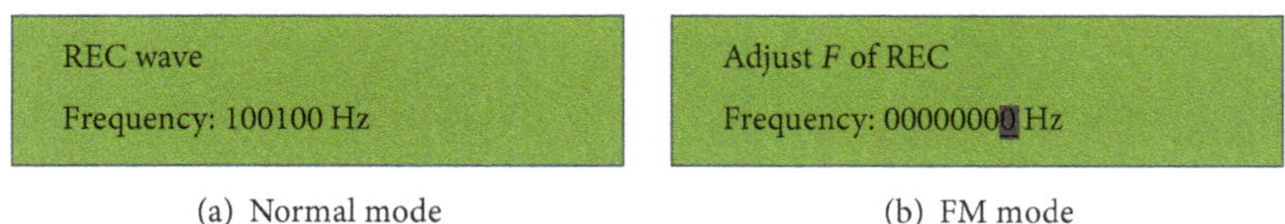

(a) Normal mode (b) FM mode

FIGURE 22: The LCD working sketches under the state of square wave.

When considering the impact of D/A converter on the accuracy of waveform distortion, the equation can be rewritten as

$$R_{\mathrm{THD}} = \sqrt{\left[1 + \frac{1}{6 \times 2^{D}}\right]\left[\frac{\pi/N}{\sin(\pi/N)}\right]^{2} - 1} \times 100\%. \quad (24)$$

In the formula, R_{THD} is waveform distortion, D is effective digit of the DAC converter, and N is the number of sampling points.

The frequency synthesizer in this paper is based on ROM look-up table. The principle suggests that the frequency control word not only determines the output frequency, but also determines the number of sampling points of the synthesized signal. The larger the frequency control word, the larger the output frequency and the smaller the number of sampling points; on the contrary, the smaller the frequency control word, the smaller the output frequency and the larger the number of sampling points. Along with the output

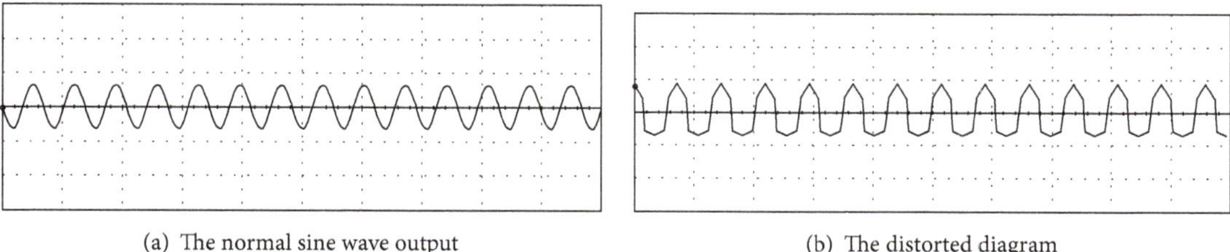

(a) The normal sine wave output (b) The distorted diagram

FIGURE 23: The waveform distortion contrast chart.

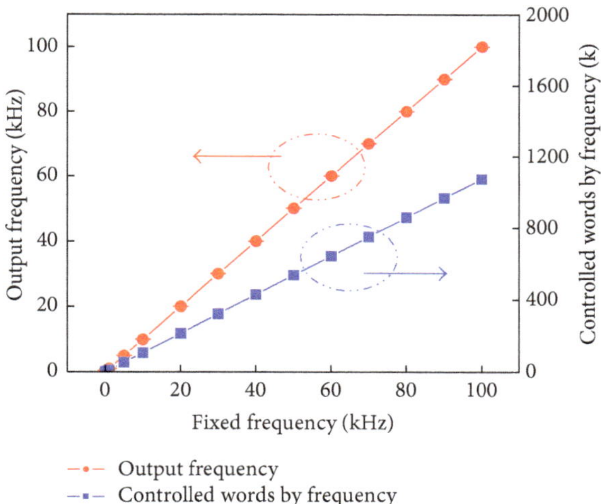

FIGURE 24: Experimental data of frequency testing.

frequency variation, the output signal waveform distortion also changes.

The waveform distortion contrast chart is shown in Figure 23, and the normal sine wave output is shown in Figure 22(a). When the frequency control word and the output frequency are increased, the sampling points will be reduced and the output waveform can be distorted, as shown in Figure 22(b).

This paper presents the maximum and minimum output signal waveform distortion:

$$R_{\text{THDmin}} = 0.638\%$$
$$R_{\text{THDmax}} = 11.67\%. \tag{25}$$

The frequency testing data curves are as shown in Figure 24.

5.1.1. Analysis of Frequency Error

(1) The Error of Phase Truncation. In order to obtain high-frequency resolution, the digit of phase accumulator N is generally rather large. However since ROM capacity is limited, the N-bit phase accumulator output only employs the high A bit for addressing ROM ($N > A$), and the low bit ($B - A$) is rounded, resulting in a phase truncation error. Spectrum emission caused by phase truncation is mainly about the spectrum of spurious signal staying at the spectrum on both sides of the output signal, and the spurious spectrum is the combination of the reference clock and output frequency; the spurious spectral amplitude varies with the function of $S_a(x)$. According to the theoretical analysis, the main spectrum S and the magnitude of the strongest stray spectrum S_{spur} satisfy the following relationship:

$$\frac{S}{S_{\text{spur}}} \geq 6\,(N - B)\ \text{dB}, \tag{26}$$

whereby N is the digit of phase accumulator and B is the discarded digit. The value of $(N - B)$ decides the level of the strongest truncated spectrum relative to the main spectrum which is caused by phase deduction. The design of the DDS module in this study uses a 28-bit phase accumulator and 12-line ROM address, and the rounded digit in the accumulator is $B = 16$. From (16) it can be calculated that the level of the strongest truncated spectrum relative to the main spectrum is more than −72 dB.

(2) Amplitude Quantization Error. Because the ROM stores the coding samples of waveform amplitude, these code words are represented by finite-bit binary data, which introduces amplitude quantization error. In general, the amplitude of the quantization noise signal is much smaller than the amplitude of the spurious signal caused by the phase truncation and

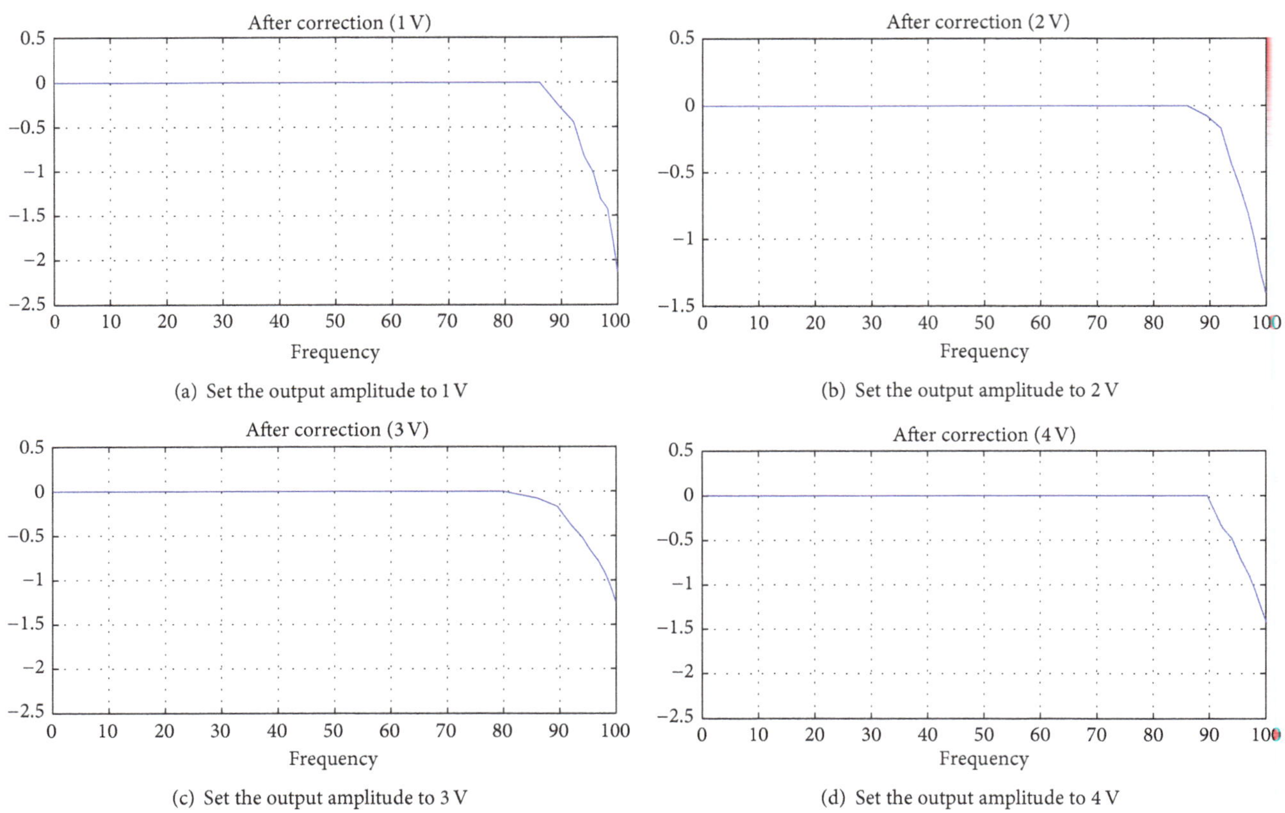

(a) Set the output amplitude to 1 V

(b) Set the output amplitude to 2 V

(c) Set the output amplitude to 3 V

(d) Set the output amplitude to 4 V

FIGURE 25: Amplitude characteristic.

DAC errors; within a certain range it is regarded as homogeneous distributed white noise. The total noise ratio can be obtained with statistical methods,

$$\text{SNR} = 6.02A + 1.76\,\text{dB}. \quad (27)$$

In the above A is effective addressing digit of phase accumulator output, and the total amplitude quantization SNR is 74 dB caused by spurious signals.

5.2. Amplitude Characteristics. As shown in Figure 25, when the input voltage is fixed at 1 V, 2 V, 3 V, and 4 V, within the low frequency range (0–10 kHz) the signal attenuation is 0 dB, which means the actual output amplitude is consistent with the given amplitude. When the frequency is over 10 kHz, the amplitude starts to decay; the higher the frequency, the greater the attenuation, and the system 3 dB cutoff frequency is at 80 kHz. The magnification of the multiplier will decrease with the increase of input signal frequency. While maintaining a constant amplitude sinusoidal signal input, we must change D/A output voltage. When the sine waveform amplitude is set to 1 V, the attenuation is less than −4 dB at the frequency of 100 kHz, when the amplitude is set to 2 V, the amplitude attenuation is close to −5 dB, and when the amplitude is set to 3 V, attenuation is greater than −5 dB.

5.3. Real Output Waveform. The amplitude and frequency are improved by the above method, and the output of the sine wave is measured as an example. Figure 26 shows the different frequency of sinusoidal signal output waveform when fixing the certain amplitude. Figure 27 shows the different amplitude of sinusoidal signal output waveform when fixing the certain frequency. The figures also show that the designed signal generator output is stability and has high accuracy.

6. Conclusions

A signal generator with integrated programmable DDS device is present. By DDS device and microcontroller, and it can change frequency and phase under the control of MCU. The amplitude of the output sinusoidal signal can be adjusted using the microcontroller; 12-bit D/A port was used to generate a variable voltage and then do the multiplication with the fixed amplitude of a sinusoidal signal in the multiplier. A sinusoidal signal with certain amplitude can be changed into a square wave signal through hysteresis comparator, changing comparison voltage to adjust the variable duty cycle of the square wave. Experimental results showed that the signal generator is of high resolution, high precision, small size, and light weight and is convenient and stable in use. According to the sampling theorem, the system DDS chip operates at 25 MHz reference clock, and the output frequency of the sinusoidal signal can theoretically reach 12.5 MHz. However, when the output frequency increases to 10 kHz the

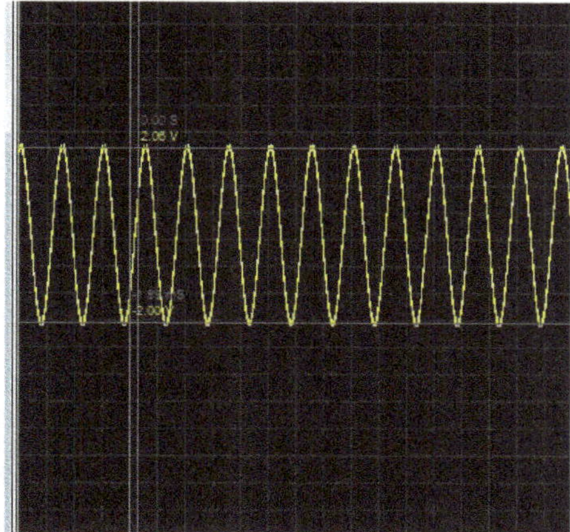

(a) The sinusoidal signal waveform output peak-peak voltage of 4 V and frequency of 6 Hz

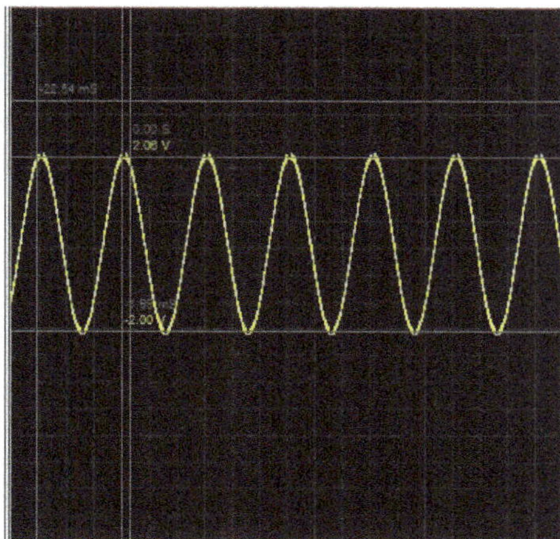

(b) The sinusoidal signal waveform output peak-peak voltage of 4 V and frequency of 3 Hz

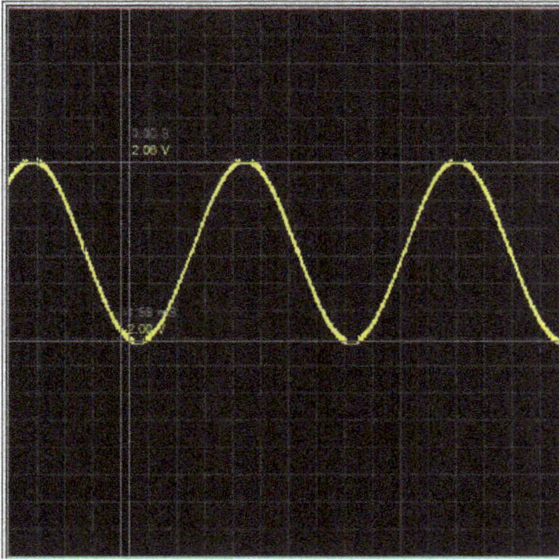

(c) The sinusoidal signal waveform output peak-peak voltage of 4 V and frequency of 1 Hz

FIGURE 26: Sinusoidal signal output waveform with different frequency and fixed amplitude.

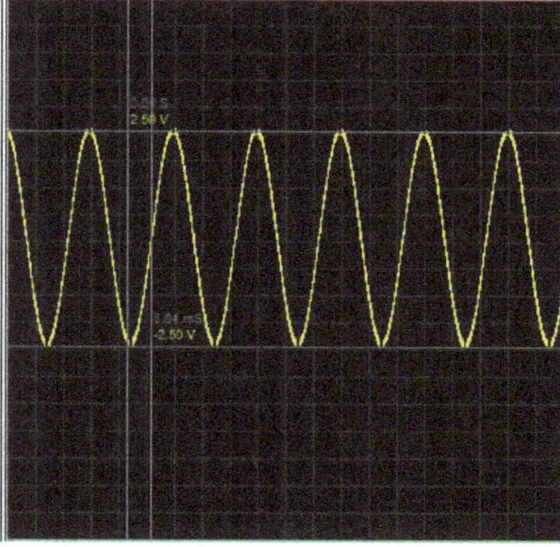

(a) The sinusoidal signal waveform output peak-peak voltage of 5 V and frequency of 3 Hz

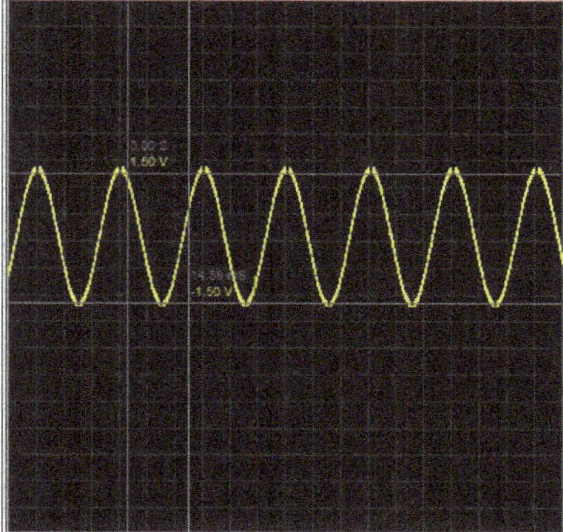

(b) The sinusoidal signal waveform output peak-peak voltage of 3 V and frequency of 3 Hz

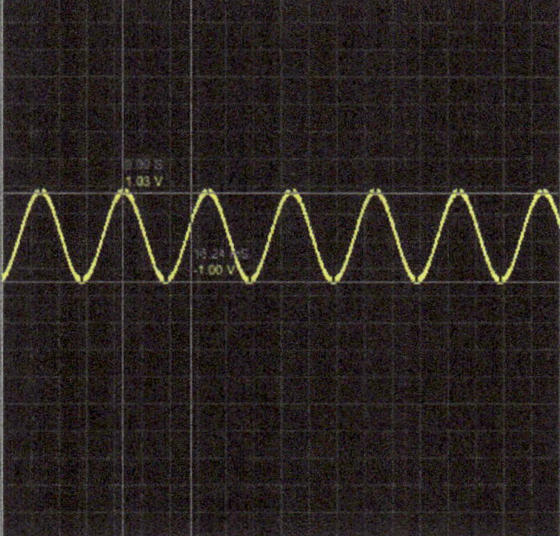

(c) The sinusoidal signal waveform output peak-peak voltage of 2 V and frequency of 3 Hz

FIGURE 27: Sinusoidal signal output waveform with different amplitude and fixed frequency.

amplitude begins to decay due to bandwidth limitations of the multiplier. Although nonlinear compensation algorithm has been used in the system software to enlarge bandwidth to some extent, further improvements are still needed.

Conflict of Interests

The authors declare that there is no conflict of interests regarding the publication of this paper.

Acknowledgment

This project is funded by Shandong Province special funding to upgrade technology research of large scientific instruments (ID: 2013SJGZ26).

References

[1] J. Si, H. Feng, P. Su, and L. Zhang, "Design and analysis of tubular permanent magnet linear wave generator," *The Scientific World Journal*, vol. 2014, Article ID 258109, 7 pages, 2014.

[2] R. Pandey, N. Pandey, S. K. Paul et al., "Voltage mode astable multivibrator using single CDBA," *ISRN Electronics*, vol. 2013, Article ID 390160, 8 pages, 2013.

[3] S. Minaei and E. Yuce, "A simple schmitt trigger circuit with grounded passive elements and its application to square/triangular wave generator," *Circuits, Systems, and Signal Processing*, vol. 31, no. 3, pp. 877–888, 2012.

[4] Y. Z. Shou, H. Zhang, and Y. H. Ge, "Design and implementation of DDS signal generator based on FPGA," *Journal of Jimei University (Natural Science)*, vol. 19, no. 5, pp. 393–400, 2014.

[5] N. Prashar, A. Singh, and B. Singh, "Design and analysis of digital wave generator using CORDIC algorithm with pipelining and angle recoding technique," *Computer Science & Engineering*, vol. 2, no. 3, pp. 123–132, 2012.

[6] G. Zhang, "Research of DDS-based high-precision multichannel signal generation systems," *Electronic Measurement Technology*, vol. 37, no. 4, pp. 125–129, 2014.

[7] Q. Sun and Q. Song, "Portable signal generator based on direct digital synthesis," *Instrument Technique and Sensor*, vol. 4, no. 4, pp. 67–70, 2009.

[8] Y. Sun, J. Lu, S. Liu, and H. Ben, "Design of sinusoidal signal generator based on AD9833 and potentiometer," *Electrical Measurement & Instrumentation*, vol. 7, pp. 93–96, 2012.

[9] D. Yang, X. Yang, and J. Chen, "Design and implementation of direct digital frequency synthesis multiple signal generator based on FPGA," *Journal of Xi'an University of Technology*, vol. 4, no. 29, pp. 439–443, 2013.

[10] Z. Cao, "Design of DDS signal generator based on FPGA," *Computer Measurement & Control*, vol. 19, no. 12, pp. 3175–3177, 3186, 2011.

[11] C. Zhang, "Visualization versatile waveform-generator based on S3C2440," *Chinese Journal of Liquid Crystals and Displays*, vol. 29, no. 6, pp. 939–943, 2014.

[12] J. Liao, "Design of phase adjustable signal generator based on DDS," *Journal of Luoyang Normal University*, vol. 33, no. 2, pp. 29–32, 2014.

4

Micro-Doppler Ambiguity Resolution based on Short-Time Compressed Sensing

Jing-bo Zhuang, Zhen-miao Deng, Yi-shan Ye, Yi-xiong Zhang, and Yan-yong Chen

School of Information Science and Engineering, Xiamen University, Xiamen 361005, China

Correspondence should be addressed to Zhen-miao Deng; dzm_ddb@xmu.edu.cn

Academic Editor: Igor Djurović

When using a long range radar (LRR) to track a target with micromotion, the micro-Doppler embodied in the radar echoes may suffer from ambiguity problem. In this paper, we propose a novel method based on compressed sensing (CS) to solve micro-Doppler ambiguity. According to the RIP requirement, a sparse probing pulse train with its transmitting time random is designed. After matched filtering, the slow-time echo signals of the micromotion target can be viewed as randomly sparse sampling of Doppler spectrum. Select several successive pulses to form a short-time window and the CS sensing matrix can be built according to the time stamps of these pulses. Then performing Orthogonal Matching Pursuit (OMP), the unambiguous micro-Doppler spectrum can be obtained. The proposed algorithm is verified using the echo signals generated according to the theoretical model and the signals with micro-Doppler signature produced using the commercial electromagnetic simulation software FEKO.

1. Introduction

Estimating and extracting micromotion information have attracted much attention in recent years [1–6]. Micromotion can be defined as the mechanical vibration, rotation, or other higher order motion components, excluding translational motion, of a target and will produce a frequency modulation on the returned signal that generates sidebands about the target's Doppler frequency. This is known as the micro-Doppler effect. This effect reflects the unique dynamic and structural characteristics of the target, which offers an approach for the recognition and identification of specific targets [7].

In Synthetic Aperture Radar (SAR)/Inverse Synthetic Aperture Radar (ISAR) imagery, micro-Doppler effect will introduce nonstationary phase modulation into returned signals, which will significantly decrease the readability of images [5, 8, 9]. In [8], an estimation method based on the discrete fractional Fourier transform is proposed to estimate the instantaneous vibration accelerations and frequencies. To separate the rigid body and the micro-Doppler parts, an L-statistics-based method for micro-Doppler effects removal is proposed in [9].

Most of the previous researches assume that the probing frequency to a micromotion target is large enough and thus there is no Doppler ambiguity and micro-Doppler ambiguity. However, a long range radar usually works in low PRF, which causes serious Doppler ambiguity and micro-Doppler ambiguity. In [10, 11], a CS-based Doppler ambiguity resolution method is proposed. However, this method cannot be applied to resolve micro-Doppler ambiguities. Therefore, it is necessary to study how to extract the unambiguous micro-Doppler time-frequency spectrum when the PRF is low. In [12], the CS is employed to remove undesirable cross terms in the Wigner-Ville distribution of the micro-Doppler radar signature. However, the micro-Doppler ambiguity problem is not discussed.

To solve micro-Doppler ambiguity problem, we propose a novel method based on CS in this paper. According to the RIP requirement of the sensing matrix, a sparse pulse train with random time stamps is designed based on the fixed-PRF pulses. The echo signals after matched filtering can be viewed as randomly sparse sampling of the micro-Doppler spectrum. To reconstruct the micro-Doppler signature from the sparse samples, we propose a short-time-compressed-sensing time-frequency analysis method. A short-time window slides along the slow-time domain echo signals and the reconstruction algorithm OMP is applied within this window to reconstruct the micro-Doppler spectrum. Two kinds of echo signals, one generated according to the theoretical model and one produced by the commercial electromagnetic simulation software FEKO, are used to verify the proposed algorithm.

2. Theory

2.1. Radar Echo Signal Model. The transmitted chirp signal can be modeled as

$$s_{\text{tran}}\left(\hat{t}, t_m\right) = \text{rect}\left(\frac{\hat{t}}{T_p}\right)\exp\left(j2\pi\left(f_c t + \frac{1}{2}\gamma\hat{t}^2\right)\right), \quad (1)$$

where $\text{rect}(u) = \begin{cases}1, & |u|\le 0.5\\ 0, & |u|>0.5\end{cases}$; f_c and γ are the center frequency and the chirp rate, respectively; T_p and $\hat{t}$ are the pulse width and fast time, respectively; $t = \hat{t} + t_m$ is the full time, where t_m denotes the slow time. The received echo signal $s_{\text{rec}}(\hat{t}, t_m)$ can be defined as

$$s_{\text{rec}}\left(\hat{t}, t_m\right) = A\text{rect}\left(\frac{\hat{t} - 2R_t/c}{T_p}\right)e^{j\pi\gamma(\hat{t}-2R_t/c)^2}e^{-j(4\pi f_c/c)R_t}, \quad (2)$$

where A is the backscattered field amplitude for the point scatter, c is the speed of electromagnetic wave propagation, and R_t is the instantaneous distance between the radar and the target. Performing matched filtering to the received signal and transforming the results into range-frequency domain yield

$$r\left(f_r, t_m\right) = A'\text{rect}\left(\frac{f_r}{B}\right)\exp\left(-j\frac{4\pi}{c}\left(f_r + f_c\right)R_t\right), \quad (3)$$

where B is the signal bandwidth. Ignoring the acceleration, jerk of the target, R_t can be approximated by

$$R_t \approx R_0 + vt_m + r_{\text{micro}}\left(t_m\right), \quad (4)$$

where R_0 is the distance between the target and radar at time t_0, v is the radial velocity, and $r_{\text{micro}}(t_m)$ is the micromotion of the target. Substituting (4) into (3) yields the slow-time domain echo signal:

$$r\left(f_d, t_m\right) = A''\exp\left[-j2\pi f_d\left(t_m\right)t_m\right], \quad (5)$$

where $A'' = A'\exp(2(f_r + f_c)R_0/c)$ and $f_d(t_m)$ is the Doppler frequency and can be written as

$$\begin{aligned} f_d\left(t_m\right) &= \frac{1}{2\pi}\frac{d\varphi\left(t_m\right)}{dt_m} = \frac{2}{\lambda}\frac{dr\left(t_m\right)}{dt_m} \\ &= \frac{2}{\lambda}\left[v + \frac{dr_{\text{micro}}\left(t_m\right)}{dt}\right] = f_{b\_d} + f_{m\_d}\left(t_m\right) \end{aligned} \quad (6)$$

which consists of two parts, that is, the Doppler frequency f_{b_d} caused by translational motion and the micro-Doppler frequency $f_{m\_d}(t_m)$ caused by micromotion.

2.2. Requirements of the PRF. According to the Nyquist-Shannon sampling theorem, for a band-limited baseband signal the sampling rate must be greater than or equal to two times the highest frequency of the signal. For a pulsed radar, the Doppler frequency $f_d(t_m)$ of the target is sampled with the sampling rate as the PRF. When the micro-Doppler frequency exceeds half of the PRF, the micro-Doppler ambiguity phenomenon will occur. Thus, the PRF must be greater than or equal to the highest micro-Doppler frequency shift.

In a short-time interval, $f_d(t_m)$ can be seen as a constant. If there are K scatterers each with different micro-Doppler frequencies, the echo signal can be rewritten as

$$r\left(f_d, t_m\right) = \sum_{k=1}^{K}A''\exp\left\{-j2\pi f_{dk}t_m\right\}. \quad (7)$$

If the PRF is smaller than $\max_{k=1,\dots,K}\{f_{dk}\}$, the spectrum aliases will occur in (7). According to the sub-Nyquist sampling theorem [13] or CS theory, the sparse signal can be sampled at a rate lower than the Nyquist. The slow-time domain radar echo signal can be viewed as the scalar sum of K sinusoidal signals, which is sparse in frequency domain [14]. Thus, the requirement for PRF can be loosened. Next we briefly review the compressing sensing theory.

2.3. Compressed Sensing. The theory of compressed sensing shows that when the signal is sparse or compressible, the signal can be reconstructed accurately or approximately by gathering very few projective values of the signal. Suppose x ($x \in R^N$) is a k-sparse (has k nonzero values) or compressive signal after orthogonal mapping projection and s is the sparse representation of x; then a measurement matrix $\mathbf{\Phi}$, which is irrelevant with the orthogonal map $\mathbf{\Psi}$, could be built to measure x linearly, resulting in only M ($M \ll N$) measured values y ($y \in R^M$):

$$y = \mathbf{\Phi}x. \quad (8)$$

The dimension of y is less than that of x, so the equation has infinity solutions. By solving the optimal problem above, signal x can be approximately reconstructed. x is sparse in $\mathbf{\Psi}$ domain; that is,

$$x = \mathbf{\Psi}s. \quad (9)$$

Substituting (9) into (8) yields

$$y = \mathbf{\Phi}x = \mathbf{\Phi\Psi}s = \mathbf{\Theta}s, \quad (10)$$

where $\mathbf{\Theta} = \mathbf{\Phi\Psi}$ is called sensing matrix with $M \times N$ dimension. As long as $\mathbf{\Theta}$ satisfies the Restricted Isometry Property (RIP), the sparse signal s could be recovered from the measured values y.

Signal reconstruction is the process of recovering x from the linear measured values y. The simplest method is to solve the ℓ_0 norm:

$$\begin{aligned} &\min_{x\in R^N} \quad \|\tilde{x}\|_0 \\ &\text{s.t.} \quad \mathbf{\Phi}\tilde{x} = y. \end{aligned} \quad (11)$$

It is optimal in theory but impractical in numerical computation, belonging to a nondeterministic polynomial (NP) hard problem. Donoho and Elad proved that ℓ_1 norm and ℓ_0 norm minimizations are equivalent if the solution is sufficiently sparse [15]. ℓ_1 norm

$$\begin{aligned} \min_{x \in R^N} \quad & \|\tilde{x}\|_1 \\ \text{s.t.} \quad & \Phi \tilde{x} = y \end{aligned} \tag{12}$$

is an optimization problem and could be solved using linear programming.

According to the description of Section 2.2, we know that the radar echo signals can be viewed as the scalar sum of sinusoidal signals and are sparse in frequency domain; thus, the compressed sensing can be utilized. The linear measurement y is a $M \times 1$ column vector, which is randomly extracted from the measurements of (7) with the corresponding random sampling time t_m. Since $r(f_d, t_m)$ is sparse in frequency domain, the Fourier basis is chosen as the basis matrix Ψ with its element defined by

$$\psi_{i,k} = \frac{1}{\sqrt{N}} \exp\left(-j2\pi \frac{ik}{N}\right), \quad 0 \le i,\ k \le N. \tag{13}$$

Extracting M rows corresponding to the random transmitted time t_m from the identity matrix $\mathbf{I}_{N\times N}$ yields the measurement matrix Φ. Thus, the sensing matrix Θ ($\Theta = \Phi\Psi$) is obtained by randomly extracting M row from the sparse matrix Ψ and meets the RIP property.

Generally, the value of M is related to the relevancy $u(\Phi_{N\times N}, \Psi)$ between the sparse matrix Ψ and the measurement matrix Φ and satisfies

$$M \ge C \cdot u^2 \left(\Phi_{N\times N}, \Psi\right) \cdot K \cdot \log N, \tag{14}$$

where C is a constant and $u(\Phi_{N\times N}, \Psi)$ is defined as [16]

$$u\left(\Phi_{N\times N}, \Psi\right) = \sqrt{N} \cdot \max_{1\le k, j\le N} \left|\left(\varphi_k, \psi_j\right)\right|. \tag{15}$$

The classical recovery algorithms of compressed sensing are Basis Pursuit and Greedy Matching Pursuit. BP algorithm is the global optimization algorithm which has several advantages including superresolution and stability, but it is also accompanied with high computational complexity. Greedy Matching Pursuit algorithm, such as Orthogonal Matching Pursuit (OMP), is a local optimization algorithm and has the low computational complexity and high level of localization accuracy. The characteristics of OMP, such as easy implementation and fast speed, make it a better choice than BP algorithm [16].

3. Short-Time Compressed Sensing

Doppler frequency shifts generally have the time-varying characteristic and should be analyzed via the joint time-frequency analysis technique. However, when the PRF is lower than the Nyquist, the traditional time-frequency analysis methods are invalid. According to the above theoretical analysis results and the time-variant properties of micro-Doppler, we design a window-weighted compressed sensing

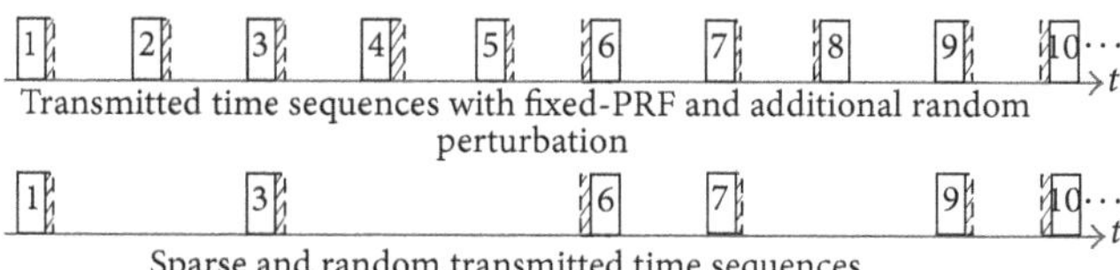

Figure 1: Illustration of radar dwell scheduling.

method. According to the RIP of the CS sensing matrix, a sparse pulse train is randomly extracted from traditional fixed repetition frequency pulses. By adding a random perturbation item to the transmitting time of each selected pulse, we can obtain a new transmitting time sequence, which is equivalent to those extracted from a high PRF pulse transmitting time set. Then the corresponding measurement matrix can be obtained according to the transmitting time sequence. After matched filtering to the radar echoes, a short-time window slides along the slow-time domain echo signals and the CS method is applied within this window to reconstruct the micro-Doppler spectrum. Similar to the short-time Fourier transform (STFT), we call this method as short-time compressed sensing.

The PRF of a LRR is usually low. For a LRR working in fixed PRF mode, the slow-time domain echoes suffer from serious micro-Doppler ambiguity. By modifying the ith pulse transmitting time t_i to $t_i' = t_i + \Delta_i$, where Δ_i is a random perturbation, we can obtain a new transmitting time sequence. Since Δ_i is random, the equivalent PRF for t_i' is larger than that for t_i. The unambiguous micro-Doppler frequency range for t_i' will be larger than that for t_i since higher PRF means wider unambiguous micro-Doppler range.

When the value of M meets (14), the designed sensing matrix can satisfy the RIP property and the slow-time echo signals can be reconstructed based on the sparse measurements. Therefore, we can reduce the number of transmitted pulses by randomly selecting M pulses from the traditional fixed-PRF transmitted time sequences and adding a random perturbation to them. The building of sparse and random transmitted time sequences is demonstrated in Figure 1.

Since $f_d(t_m)$ is relevant to the slow time t_m, there is no applicable sensing matrix to reconstruct the micro-Doppler spectrum via compressed sensing directly. However, if we multiplied a short-time window to the slow-time echoes, $f_d(t_m)$ within this window can be seen as a constant and then the sensing matrix can be generated. Assuming the time precision is Δt, the corresponding equivalent PRF is $f_s' = 1/\Delta t$ and the Doppler unambiguous range for t_i' is represented as $f_s' = (1/\Delta t) f_s$. Thus, the unambiguous micro-Doppler frequency range expands to $(-f_s'/2, f_s'/2)$.

The signal within the window is written as

$$\tilde{r}\left(f_d, t_m\right) = w\left(t_n, t_m\right) \sum_{k=1}^{K} A'' \exp\left\{-j2\pi f_{dk}\left(t_n\right) t_n\right\}, \tag{16}$$

where the sliding short-time window is defined as $w(t_n, t_m) = \begin{cases} 1, & t_m \le t_n \le t_{m+L-1} \\ 0, & \text{else} \end{cases}$. In the window, the transmitted time sequence is $t_n \in \{t_m, t_{m+1}, \ldots, t_{m+L-1}\}$. The length of the window is

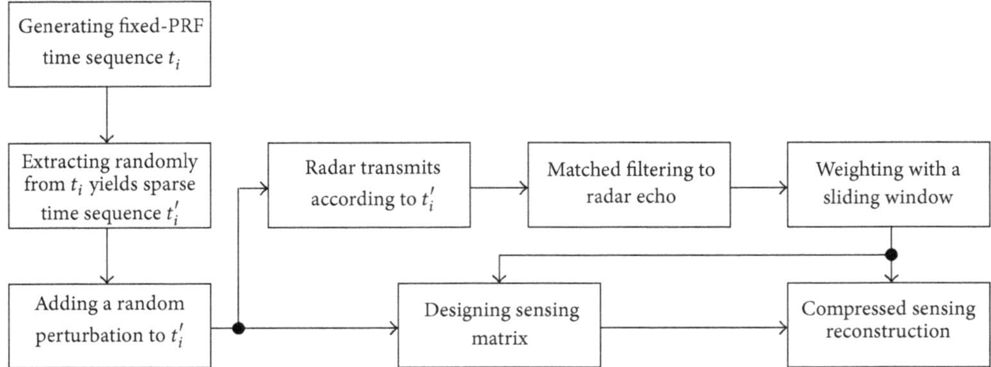

FIGURE 2: Flowchart of the proposed method.

L; that is, the length of the observed signal is $M = L$. In order to obtain more accurate Doppler estimate, an adaptive refinement algorithm of the sensing matrix proposed in [17] can be used to reconstruct the slow-time domain signal with higher precision. The column vector ψ_n of the basis matrix $\boldsymbol{\Psi} = [\psi_1, \psi_2, \ldots, \psi_N]$ is defined as $\psi_n(f) = \exp(j2\pi ft)$, and $\boldsymbol{\Psi}$ is an $N \times N$ matrix, where $N = 1/\Delta t$ and $\Delta f = (f_{\max} - f_{\min})/N$. According to t_n, extracting L rows from the basis matrix $\boldsymbol{\Psi}_{N\times N}$ yields the sensing matrix $\boldsymbol{\Theta}_{M\times N}$:

$$\boldsymbol{\Theta}_{M\times N} = \begin{bmatrix} e^{j2\pi f_0 t_0} & e^{j2\pi f_1 t_0} & \cdots & e^{j2\pi f_{N-1} t_0} \\ e^{j2\pi f_0 t_1} & e^{j2\pi f_1 t_1} & \cdots & e^{j2\pi f_{N-1} t_1} \\ \vdots & \vdots & \ddots & \vdots \\ e^{j2\pi f_0 t_{L-1}} & e^{j2\pi f_1 t_{L-1}} & \cdots & e^{j2\pi f_{N-1} t_{L-1}} \end{bmatrix}. \quad (17)$$

With the classical recovery algorithms, the unambiguous Doppler frequency $f_{dk}(t_m)$ can be reconstructed. Sliding the short-time window along the time axis and performing the CS operation, the unambiguous micro-Doppler time-frequency spectrum can be obtained.

To satisfy the RIP, the perturbation $\Delta_i = \varepsilon \cdot PRI$, where ε is a random variable uniformly distributed in $(0, 1)$; that is, $\varepsilon \sim U(0, 1)$. In practical application, there are two reasons causing that the term Δ_i could not be completely random: (1) the time sequence of radar is controlled by a reference clock. So the precision of Δ_i is limited to the precision of the clock; (2) theoretically, higher precision of Δ_i would yield larger unambiguous Doppler frequency range through refining the basis matrix. However, as pointed out in [14], the relationship between the DFT matrix refining factor, which is referred to the frequency redundancy factor, and the measurement number M should satisfy (20) in [14]. In other words, M would increase if the refining factor increases; therefore, the tradeoff between these two factors should be carefully considered. Furthermore, when the refining factor increases, the performance of recovery algorithm under noise environment will also degrade. Thus, we should carefully select proper Δt to ensure a good performance.

The process of the short-time compressed sensing on micro-Doppler spectrum reconstruction is summarized as follows:

(i) A new transmitting time sequence is generated by extracting randomly from the fixed-PRF transmitting time sequence. Then a perturbation is added to each element of the new transmitting time sequence. The radar transmits probing pulses according to the scheduled time sequence. Performing matched filtering to the echo pulses, the slow-time domain signals are obtained.

(ii) Slide a short-time window along the slow-time domain signals and construct the corresponding sensing matrix $\boldsymbol{\Psi}$ according to the time stamps of the window.

(iii) Performing the Orthogonal Matching Pursuit algorithm to the signals in the sliding window, the instantaneous micro-Doppler frequency is obtained.

The flowchart for the implementation of the algorithm is shown in Figure 2.

4. Simulation

In this section, simulations are performed to verify the proposed method. The proposed method can be applied to different kinds of micromotion. In the following, two micromotions, that is, spin and precession, are simulated.

Experiment 1 (one spin scatter case). The simulation parameters are shown in the list below. One scatter case is as follows.

Simulation Parameters

$f_c = 10\,\text{GHz}$,

$B = 5\,\text{MHz}$,

$f_r = 300\,\text{Hz}$, $t = 2\,\text{s}$,

$\omega = 2\pi\,\text{rad/s}$,

$r_1 = 0.6\,\text{m}$.

The corresponding slow-time echo signal for spin scatterers can be written as

$$r(f_d, t_m) = \sum_{k=1}^{1} A'' \exp\left\{-j2\pi \frac{2\omega_k r_k \cos(\omega_k t_m)}{\lambda} t_m\right\}. \quad (18)$$

The simulation results for spin motion are shown in Figure 3.

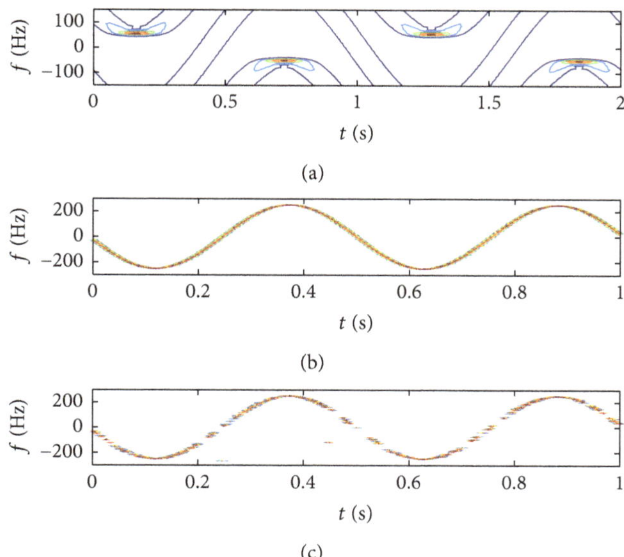

FIGURE 3: Micro-Doppler time-frequency spectrum: ambiguous time-frequency spectrum via STFT (a); unambiguous time-frequency spectrum of adding-perturbation fixed-PRF via CS (b); unambiguous time-frequency spectrum of adding-perturbation sparse-PRF via CS (c).

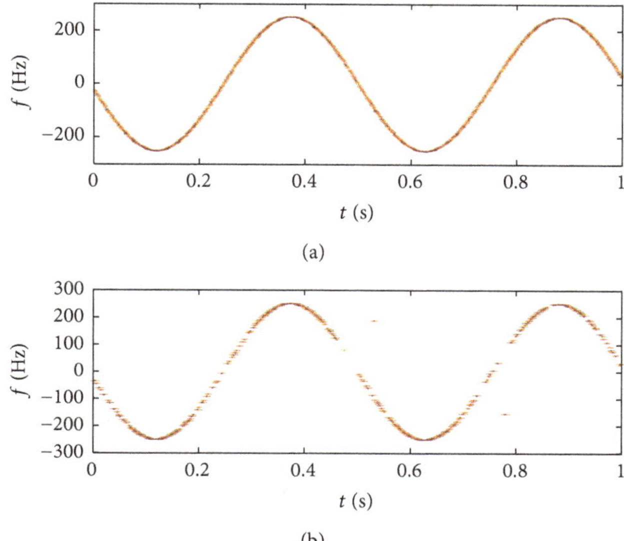

FIGURE 4: CS reconstruction quality with different measurement number. (a) $M = 10$; (b) $M = 6$.

According to the simulated parameters, the maximum micro-Doppler frequency shift is ±250 Hz. Since the PRF is not large enough, the micro-Doppler spectrum obtained from the STFT is ambiguous. Add a random perturbation Δ_i to each time stamp of the transmitting time sequence $\{t_1, t_2, \ldots, t_n\}$, respectively. The accuracy of the perturbation is $\Delta t = 1.667e-4$ and the corresponding $\text{PRF}' = 6000$. The length of the time window is $L = 10$; the number of measurements is $M = 10$; the length of the reconstructed signal is $N = 6000$; the sparse rate is $K = 2$; the sensing matrix $\mathbf{\Theta}$ is a 12×6000 matrix. The unambiguous micro-Doppler time-frequency spectrum can be recovered using the CS reconstruction algorithm, which is shown in (b) of Figure 3. Randomly selecting a portion of elements from $\{t_1, t_2, \ldots, t_n\}$ and adding a random perturbation Δ_i to each of the selected elements, we can generate sparse probing pulses. (c) of Figure 3 shows the unambiguous micro-Doppler spectrum reconstructed from these sparse pulses with good performance. The sparse rate, that is, the number of sparse probing pulses divided by the number of the original fixed-PRF pulses, is 0.6.

In order to investigate the effect of the number of measurements M on the reconstruction quality for a fixed sparse rate, we select different M values and compare the reconstruction results, which are shown in Figure 4. We can see that larger M corresponds to better reconstruction performance.

Experiment 2 (two spin scatterers' case). Two spin scatterers are simulated and the simulation parameters are shown in the list below. To demonstrate the ambiguity resolving ability of our method, we change the rotating radius $r_1 = 0.6$ m of Experiment 1 to $r_1 = 1.8$ m. With these simulation parameters, the micro-Doppler of this scatterer will be ambiguous

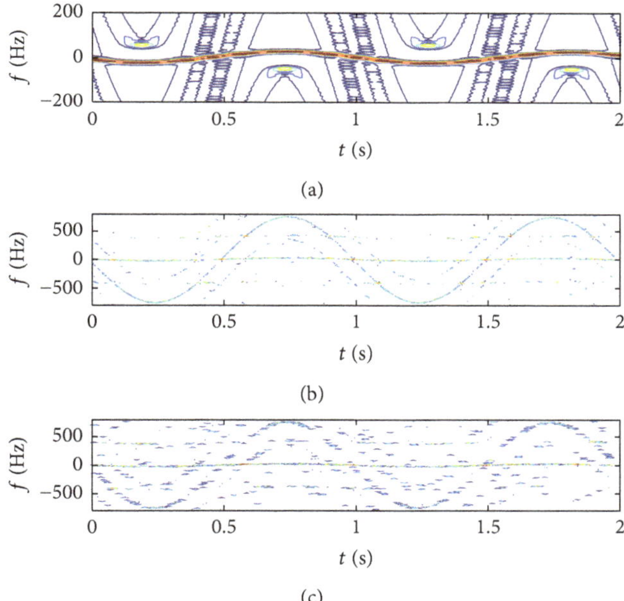

FIGURE 5: Micro-Doppler time-frequency spectrum: ambiguous time-frequency spectrum via STFT (a); unambiguous time-frequency spectrum of adding-perturbation fixed-PRF via CS (b); unambiguous time-frequency spectrum of adding-perturbation sparse-PRF via CS (c).

twice, which can be seen from (a) of Figure 5. Performing the short-time compressed sensing, the unambiguous micro-Doppler can be recovered successfully. The sparse rate of (c) of Figure 5 is 0.75.

Simulation Parameters

$$f_c = 10\,\text{GHz},$$

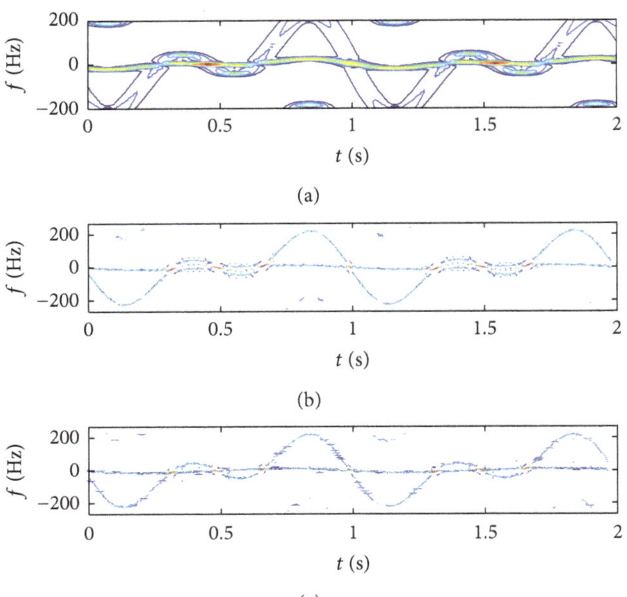

Figure 6: Micro-Doppler time-frequency spectrum: ambiguous time-frequency spectrum via STFT (a); unambiguous time-frequency spectrum of adding-perturbation fixed-PRF via CS (b); unambiguous time-frequency spectrum of adding-perturbation sparse-PRF via CS (c).

$$B = 5\,\text{MHz},$$
$$f_r = 400\,\text{Hz}, t = 2\,\text{s},$$
$$\omega = 2\pi\,\text{rad/s},$$
$$r_1 = 1.8\,\text{m}, r_2 = 0.06\,\text{m}.$$

Experiment 3 (two precession scatterers' case). Two precession scatterers are simulated and the simulation parameters are shown in the list below (19). The simulation results for precession motion can be shown in Figure 6. The corresponding slow-time echo signal for precession scatterers can be written as

$$r(f_d, t_m) = \sum_{k=1}^{2} A'' \cdot \exp\left\{-j2\pi\left[\frac{2\omega_1 r_k \cos(\omega_1 t_m)}{\lambda} + \frac{2\omega_2 R_k \cos(\omega_2 t)}{\lambda}\right] \cdot t_m\right\}. \tag{19}$$

Simulation Parameters

$$f_c = 10\,\text{GHz},$$
$$B = 5\,\text{MHz},$$
$$f_r = 400\,\text{Hz}, t = 2\,\text{s},$$
$$\omega_1 = 2\pi\,\text{rad/s}, \omega_2 = 4\pi\,\text{rad/s},$$
$$r_1 = 0.3\,\text{m}, R_1 = 0.15\,\text{m},$$
$$r_2 = 0.03, R_2 = 0.015.$$

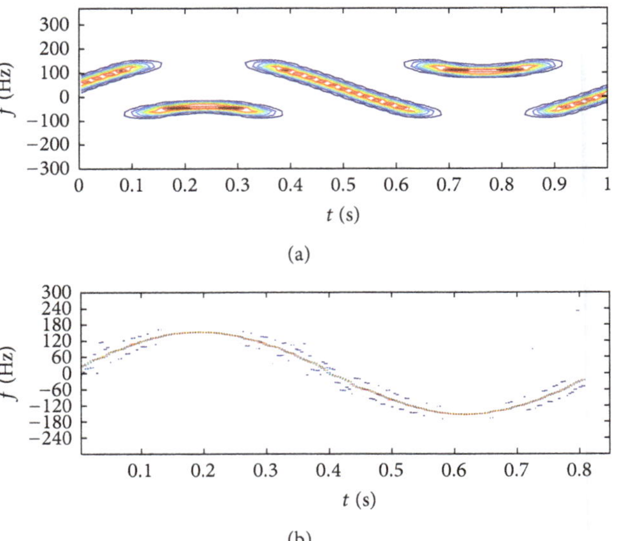

Figure 7: Micro-Doppler time-frequency spectrum: ambiguous time-frequency spectrum via STFT (a); unambiguous time-frequency spectrum of adding-perturbation sparse-PRF via CS (b).

The simulation results for the precession case are similar to those for the spin case. The sparse rate of (c) of Figure 6 is 0.5. The simulation results in Figures 3–6 show that the proposed method can expand the unambiguous micro-Doppler frequency range.

Experiment 4. In this experiment, we use the electromagnetic simulation software FEKO to generate sparse-sampled slow-time echo signals and reconstruct the unambiguous micro-Doppler spectrum via CS. A scatterer is rotating with a radius of 30 cm. The rate of angular motion is 2π rad/s. The carrier frequency is 10 GHz. The bandwidth is 500 MHz. The azimuth and pitch angle are 0° and 90°, respectively. The dynamic position of the scatterer at each sparse probing time stamp is calculated with MATLAB and then the corresponding slow-time echo is resolved with FEKO. The PRF is 200 Hz.

According to the configuration, the maximum micro-Doppler frequency is $f_{\max} = 40\pi$. Since the PRF $f_r < 2f_{\max}$, the micro-Doppler is ambiguous. The ambiguous micro-Doppler spectrum with fixed-PRF probing pulses is shown in (a) of Figure 7 and the unambiguous micro-Doppler spectrum with sparse-PRF probing pulses is shown in (b) of Figure 7. It can be seen that our proposed method is also suitable for the simulated echoes with FEKO.

5. Conclusion

The proposed short-time compressed sensing method can solve the micro-Doppler ambiguity problem. The sensing matrix built based on the sparse probing time stamps can meet the requirement of RIP property. Performing the OMP algorithm to the signals within the sliding short-time window, the ambiguous micro-Doppler spectrum can be reconstructed. The proposed method can work in low

PRF circumstances and requires transmitting fewer probing pulses while at the same time it can achieve larger unambiguous micro-Doppler range.

Conflict of Interests

The authors declare that there is no conflict of interests regarding the publication of this paper.

Acknowledgments

The research was supported by the National High-Tech R&D Program of China, the Open-End Fund National Laboratory of Automatic Target Recognition (ATR), the Open-End Fund of BITTT Key Laboratory of Space Object Measurement, and the National Natural Science Foundation of China (Grant no. 62101196).

References

[1] T. Thayaparan, L. J. Stanković, M. Daković, and V. Popović, "Micro-Doppler parameter estimation from a fraction of the period," *IET Signal Processing*, vol. 4, no. 3, pp. 201–212, 2010.

[2] K. Li, Y. Liu, K. Huo, W. Jiang, X. Li, and Z. Zhuang, "Estimation of micro-motion parameters based on cyclostationary analysis," *IET Signal Processing*, vol. 4, no. 3, pp. 218–223, 2010.

[3] T. Thayaparan, K. Suresh, S. Qian, K. Venkataramaniah, S. SivaSankaraSai, and K. S. Sridharan, "Micro-doppler analysis of a rotating target in synthetic aperture radar," *IET Signal Processing*, vol. 4, no. 3, Article ID ISPECX000004000003000245000001, pp. 245–255, 2010.

[4] T. Thayaparan, L. Stanković, and I. Djurović, "Micro-Doppler-based target detection and feature extraction in indoor and outdoor environments," *Journal of the Franklin Institute*, vol. 345, no. 6, pp. 700–722, 2008.

[5] Q. Wang, M. Pepin, A. Wright et al., "Reduction of vibration-induced artifacts in synthetic aperture radar imagery," *IEEE Transactions on Geoscience and Remote Sensing*, vol. 52, no. 6, pp. 3063–3073, 2014.

[6] P. Suresh, T. Thayaparan, T. Obulesu, and K. Venkataramaniah, "Extracting micro-doppler radar signatures from rotating targets using fourier-bessel transform and time-frequency analysis," *IEEE Transactions on Geoscience and Remote Sensing*, vol. 52, no. 6, pp. 3204–3210, 2014.

[7] L. Liu, D. McLernon, M. Ghogho, W. Hu, and J. Huang, "Ballistic missile detection via micro-Doppler frequency estimation from radar return," *Digital Signal Processing*, vol. 22, no. 1, pp. 87–95, 2012.

[8] S. B. Colegrove, S. J. Davey, and B. Cheung, "Separation of target rigid body and micro-Doppler effects in ISAR imaging," *IEEE Transactions on Aerospace and Electronic Systems*, vol. 42, no. 4, pp. 1496–1506, 2006.

[9] L. Stanković, T. Thayaparan, M. Daković, and V. Popović-Bugarin, "Micro-doppler removal in the radar imaging analysis," *IEEE Transactions on Aerospace and Electronic Systems*, vol. 49, no. 2, pp. 1234–1250, 2013.

[10] Y. H. Quan, L. Zhang, M. D. Xing, and Z. Bao, "Velocity ambiguity resolving for moving target indication by compressed sensing," *Electronics Letters*, vol. 47, no. 22, pp. 1249–1251, 2011.

[11] Y.-X. Zhang, J.-P. Sun, B.-C. Zhang, and W. Hong, "Doppler ambiguity resolution based on compressive sensing theory," *Journal of Electronics & Information Technology*, vol. 33, no. 9, pp. 2103–2107, 2011.

[12] N. Whitelonis and H. Ling, "Radar signature analysis using a joint time-frequency distribution based on compressed sensing," *IEEE Transactions on Antennas and Propagation*, vol. 62, no. 2, pp. 755–763, 2014.

[13] M. Mishali and Y. C. Eldar, "Sub-nyquist sampling," *IEEE Signal Processing Magazine*, vol. 28, no. 6, pp. 98–124, 2011.

[14] M. F. Duarte and R. G. Baraniuk, "Spectral compressive sensing," *Applied and Computational Harmonic Analysis*, vol. 35, no. 1, pp. 111–129, 2013.

[15] D. L. Donoho and M. Elad, "Optimally sparse representation in general (nonorthogonal) dictionaries via ℓ_1 minimization," *Proceedings of the National Academy of Sciences of the United States of America*, vol. 100, no. 5, pp. 2197–2202, 2003.

[16] J.-H. Wang, Z.-T. Huang, Y.-Y. Zhou et al., "Generalized incoherence principle in compressed sensing," *Signal Processing*, vol. 28, no. 5, pp. 675–679, 2012.

[17] L. Hu, Z. Shi, J. Zhou, and Q. Fu, "Compressed sensing of complex sinusoids: an approach based on dictionary refinement," *IEEE Transactions on Signal Processing*, vol. 60, no. 7, pp. 3809–3822, 2012.

5

Low-Complexity Detection Algorithms for Spatial Modulation MIMO Systems

Xinhe Zhang,[1] Yuehua Zhang,[1] Chang Liu,[2] and Hanzhong Jia[3]

[1]*School of Electronic and Information Engineering, University of Science and Technology Liaoning, Anshan 114051, China*
[2]*National Key Laboratory of Science and Technology on Communications, University of Electronic Science and Technology of China, Chengdu 611731, China*
[3]*State Grid Liaoning Information and Communication Company, Shenyang 110006, China*

Correspondence should be addressed to Xinhe Zhang; cdaszxh@sina.com

Academic Editor: Jit S. Mandeep

In this paper, the authors propose three low-complexity detection schemes for spatial modulation (SM) systems based on the modified beam search (MBS) detection. The MBS detector, which splits the search tree into some subtrees, can reduce the computational complexity by decreasing the nodes retained in each layer. However, the MBS detector does not take into account the effect of subtree search order on computational complexity, and it does not consider the effect of layers search order on the bit-error-rate (BER) performance. The ost-MBS detector starts the search from the subtree where the optimal solution is most likely to be located, which can reduce total searches of nodes in the subsequent subtrees. Thus, it can decrease the computational complexity. When the number of the retained nodes is fixed, which nodes are retained is very important. That is, the different search orders of layers have a direct influence on BER. Based on this, we propose the oy-MBS detector. The ost-oy-MBS detector combines the detection order of ost-MBS and oy-MBS together. The algorithm analysis and experimental results show that the proposed detectors outstrip MBS with respect to the BER performance and the computational complexity.

1. Introduction

To meet the demand of wireless communication systems for higher data transmission rate, multiple-input multiple-output (MIMO) technology has been adopted in mobile terminals. MIMO technology improves data throughput without increasing additional bandwidth and transmit power. Spatial modulation (SM) [1–3] is an emerging transmission scheme for MIMO systems. The main characteristic of SM is that only one transmit antenna is activated at one time slot, but simultaneously, the SM systems can use the original signal domain (signal constellation) and the transmit antenna (TA) indices (spatial constellation) to convey information. Compared to MIMO systems, SM systems can only equip one radio frequency (RF) chain, avoid interchannel interference (ICI) and interantenna synchronization (IAS), and also reduce the complexity of demodulation.

For the detection of SM signals, maximum ratio combining (MRC) algorithm was proposed in [4], in which the active-antenna index and the transmit symbol are separately estimated. The MRC detector has a low computational complexity and only performs well on the constrained channels. This detector was improved in [5], and it further can be applied in conventional channel conditions. The optimum maximum likelihood (ML) detector which involves joint detection of the TA index and of the transmit symbol was proposed in [6]. However, the computational complexity linearly grows as the number of TA (N_T), the number of receive antennas (N_R), and the size of the modulation scheme (N_M). In order to obtain the near-optimal solution with a lower computational complexity, several low-complexity detectors have been put forward [7–17]. In [7, 8], two low-complexity hard-limiter-based ML (HL-ML) detectors which have the same BER performance as the ML detector were proposed for M-PSK and square- or rectangular-QAM modulation. The computational complexity has nothing to do with the constellation size. In [9–11], sphere-decoding (SD) algorithms were put forward for SM systems, which are capable of achieving near-optimal performance with a lower computational complexity on average. At worst, the computational complexity is equivalent to that of the ML detector. However, its detection performance depends mainly on the initial search radius and the transmit parameters. Compared with SD detectors, SD aided by the ordering strategy proposed in [12] can greatly reduce the computational complexity. Two matched filter-(MF-) based detectors were proposed in [13]. In [14], Wang et al. proposed a novel signal vector-based detection (SVD) scheme. Tang et al. [15] presented a distanced-based ordered detection (DBD) algorithm to reduce the receiver complexity and achieve a near-maximum likelihood performance. To reduce the detection complexity of ML detection, Xu [16] presented simplified ML-based optimal detection (OD) and simplified multistage detection (MD). In the simplified ML-based detection and multistage detection schemes, the signal set is firstly partitioned into four "level-one subsets". Each level-one subset is further partitioned into four "level-two subsets" if each subset contains more than four signals. The simple low-complexity detection (SLCD) and adaptive simple low-complexity detection (ASLCD) were proposed in [17].

In [18, 19], the M-algorithm to maximum likelihood (MML) detector with prioritized tree-search structure was presented. The detection is considered as a breadth-first search tree with $N_T N_M$ branches and N_R layers, in which the ith layer corresponds to the ith receive antenna (RA). The MML detector only examines partial nodes in the tree, whereas the ML detector traverses all nodes. Compared with the ML detector, the MML detector can achieve a lower computational complexity. In [20], a low-complexity symbol detection based on modified beam search (MBS) was proposed. The detection process of the MBS algorithm can be represented by constructing a tree with N_T subtrees and $2N_R$ layers, where each subtree has N_M complete paths from the root node to the leaf nodes, and each of the paths stands for a candidate solution. The solution is found by performing modified beam search. Compared with the MML algorithm, the MBS algorithm reduces the computational complexity by discarding unpromising candidate solutions.

In the MBS detector, the detection sequence of different subtrees is confined to the ascending order of the subtree indices, whereas it ignores the influence of different search orders on the computational complexity. Moreover, the detection of all layers is confined to the ascending order of the layer indices, whereas it ignores the influence of different search orders on its bit-error-rate (BER) performance. That is to say, the influence of different search orders on the BER performance and the computational complexity is not considered in the MBS detector. In recent years, the sorting strategy has attracted more and more attention. To some degree, the sorting strategy can improve the algorithm detection performance. In [12, 21], different ordering strategies were proposed to improve the detection performance. In this paper, we proposed three MBS-based detectors with novel ordering strategies: (1) the ost-MBS detector rearranges the search order of subtrees; (2) the oy-MBS detector performs SM signal detection in a descending order of the received signal amplitude; (3) the detection orders of the abovementioned two detectors were jointly considered in the ost-oy-MBS detector.

The rest of this paper is organized as follows. In Section 2, the system model of SM systems is introduced. Section 3 gives a brief overview of the MBS detector. The ordering strategy is introduced to the MBS detector. Section 4 demonstrates ost-MBS, oy-MBS, and ost-oy-MBS detectors. Section 5 illustrates the simulation results. Finally, we conclude the paper with a summary in Section 5.

Notations. Boldface upper/lower case symbols denote matrices and column vectors; $\|\cdot\|_F$ is the Frobenius norm of a vector or a matrix; $|\cdot|$ is the amplitude of a complex quantity or the cardinality of a set; $\Re(\cdot)$ and $\Im(\cdot)$ are the real and imaginary parts of a complex-valued quantity; $(\cdot)^H$ is the conjugate transpose of a vector or a matrix; $\mathscr{CN}(\mu, \sigma^2)$ denotes a complex Gaussian random variable with mean μ and variance σ^2.

2. System Model

Consider an $N_T \times N_R$ SM system with constellation $S = \{s_1, s_2, \ldots, s_{N_M}\}$. In each time slot, the incoming data bits are rearranged into blocks of $\log_2 N_T N_M$ bits, in which $\log_2 N_T$ bits are used to select the activated TA and $\log_2 N_M$ bits are used to select the transmit symbol $s_m \in S$, $m \in \{1, 2, \ldots, N_M\}$. Hence, the system model for SM systems can be represented by

$$\mathbf{r} = \mathbf{G} \cdot \mathbf{s} + \mathbf{w}, \tag{1}$$

where $\mathbf{r} \in \mathbb{C}^{N_R}$ is the received signal vector; $\mathbf{s} \in \mathbb{C}^{N_T}$ is the transmit symbol vector, whose element is s_m at the lth position and zero at the other positions; $\mathbf{G} \in \mathbb{C}^{N_R \times N_T}$ and $\mathbf{w} \in \mathbb{C}^{N_R}$ are the channel matrix and the noise vector, whose elements follow the circularly symmetric complex Gaussian distributions with $\mathscr{CN}(0, 1)$ and $\mathscr{CN}(0, \sigma^2)$, respectively. The system model expressed in (1) can be reshaped as

$$\mathbf{y} = \begin{bmatrix} \Re(\mathbf{G}) & -\Im(\mathbf{G}) \\ \Im(\mathbf{G}) & \Re(\mathbf{G}) \end{bmatrix} \cdot \begin{bmatrix} \Re(\mathbf{s}) \\ \Im(\mathbf{s}) \end{bmatrix} + \begin{bmatrix} \Re(\mathbf{w}) \\ \Im(\mathbf{w}) \end{bmatrix} = \mathbf{H} \cdot \mathbf{x} + \mathbf{n}, \tag{2}$$

where $\mathbf{H} \in \mathbb{R}^{2N_R \times 2N_T}$, $\mathbf{x} \in \mathbb{R}^{2N_T}$, and $\mathbf{n} \in \mathbb{R}^{2N_R}$. Since only one TA is activated in each time slot, the system model expressed in (2) can be simplified as

$$\mathbf{y} = \begin{bmatrix} \mathbf{h}_l & \mathbf{h}_{l+N_T} \end{bmatrix} \cdot \begin{bmatrix} \Re(s_m) \\ \Im(s_m) \end{bmatrix} + \mathbf{n}, \tag{3}$$

where $\mathbf{h}_l$ is the lth column of $\mathbf{H}$.

It follows from (3) that the optimal ML-based demodulator can be formulated as

$$\begin{aligned}(\hat{l},\hat{s}_{\mathrm{m}}) &= \arg\min_{(l,s_{\mathrm{m}})\in\Lambda}\left\|\mathbf{y}-\left[\mathbf{h}_l\ \ \mathbf{h}_{l+N_{\mathrm{T}}}\right]\cdot\begin{bmatrix}\Re(s_{\mathrm{m}})\\ \Im(s_{\mathrm{m}})\end{bmatrix}\right\|_F^2 \\ &= \arg\min_{l,s_{\mathrm{m}}\in\Lambda} A^{(l,s_{\mathrm{m}})},\end{aligned} \tag{4}$$

where Λ denotes the set containing all possible transmit antenna indices and complex constellation points, $\Lambda = \{(l,s_{\mathrm{m}}) \mid l \in \{1,2,\ldots,N_{\mathrm{T}}\}, s_{\mathrm{m}} \in \{s_1,s_2,\ldots,s_{N_{\mathrm{M}}}\}\}$, $A^{(l,s_{\mathrm{m}})} = \sum_{k=1}^{2N_R} |y_k - h_{k,l}\cdot\Re(s_{\mathrm{m}}) - h_{k,l+N_{\mathrm{T}}}\cdot\Im(s_{\mathrm{m}})|^2$, and $h_{k,l}$ is the $(k,l)^{th}$ entry of matrix $\mathbf{H}$.

3. MBS Detector

According to Kim and Yi [20], the detection of the SM signal can be regarded as a tree with N_{T} subtrees and $2N_{\mathrm{R}}$ layers, where each subtree has N_{M} complete paths from the root node to the leaf nodes. For ease of understanding, we give an illustration (Figure 1) for the idea of the MBS detector. Suppose we have a 2×2 SM system with 4QAM modulation; thus the search tree has 4 layers and 8 branches. 4 branches of each subtree (TA) correspond to 4 symbols from the 4QAM constellation. We define the branch metric of node (l,s_{m}) at the kth layer as the squared Euclidean distance between the received and the transmit signals, which can be denoted as $B_k^{(l,s_{\mathrm{m}})} = |y_k - h_{k,l}\cdot\Re(s_{\mathrm{m}}) - h_{k,l+N_{\mathrm{T}}}\cdot\Im(s_{\mathrm{m}})|^2$. The accumulated metric of node (l,s_{m}) at the kth layer is the summation of the branch metric at the kth layer and the accumulated metric at $(k-1)$th layer, which can be expressed as $A_k^{(l,s_{\mathrm{m}})} = B_k^{(l,s_{\mathrm{m}})} + A_{k-1}^{(l,s_{\mathrm{m}})}$.

In the first subtree, the M_k nodes in the kth layer with the smallest accumulated metrics are kept as the candidate nodes for the next layer. At the last layer, the node with the smallest accumulated metric is considered as the solution in the first subtree. The smallest accumulated metric and its corresponding TA index and the transmit symbol are represented by ρ_Ω and $\Omega = \{\hat{l},\hat{s}_{\mathrm{m}}\}$, respectively. In the subsequent subtrees, ρ_Ω and Ω are gradually updated. In the kth layer, at most M_k nodes with the accumulated metrics smaller than the threshold value, ρ_Ω are selected as the survival branches for the next layer. If the accumulated metric is not less than ρ_Ω, the search of the current branch is terminated. Otherwise, we should continue to search the next branch. If the accumulated metric is smaller than ρ_Ω at the last layer, ρ_Ω and Ω are updated. Repeat the search process until all subtrees are checked. The cross symbol in Figure 1 shows that the branch with the accumulated metric not less than the threshold value ρ_Ω is pruned.

4. Proposed Ordering MBS-Based Detectors

In MBS detector, the detection of the SM signal is in the ascending order of the subtree indices and the RA indices. Essentially, the MBS detector is used to calculate (4) and select partial reserved nodes. When the number of the reserved nodes is a constant, it is of great importance to choose which nodes. In other words, the computation order of (4) can directly affect the BER performance and the computational complexity. In this section, in order to investigate the influence of different detection orders on the BER performance and computational complexity, we propose three ordering MBS-based detectors by adjusting the search orders.

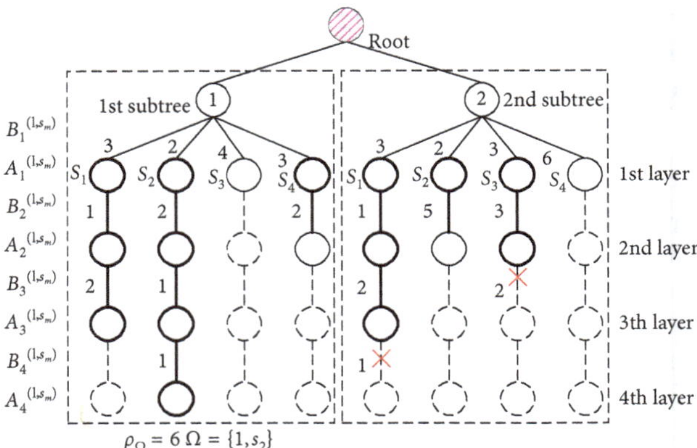

Figure 1: The tree structure of 2×2 SM with 4QAM and $M = [3,2,2,1]$.

4.1. Ost-MBS Detector. The computational complexity of the MBS detector is reduced by pruning the branches, whose accumulated metric is larger than or equal to ρ_Ω. However, the MBS algorithm does not take into account the influence of subtree search order on the computational complexity. If the optimal solution stands in the first subtree, only fewer nodes are searched in the subsequent subtrees. But, if the optimal solution exists in the last subtree, we need to search more nodes in the top $N_{\mathrm{T}}-1$ subtrees. In other words, searching firstly from the subtree where the optimal solution most probably belongs to can reduce total searches of nodes in the subsequent subtrees. Thus, it can decrease the computational complexity. That is, the searching order of subtrees directly affects the computational complexity. We propose the ost-MBS detector, which estimates the optimal solution by rearranging the order of subtrees. In the first stage, we order the TA indices based on the suboptimal antenna detection algorithm. In the second stage, we first search from the most probable TA index, the estimated solution is obtained by searching all N_{T} subtrees. The proposed ost-MBS detector works as follows.

Stage 1. Reorder the TA indices in descending order using the suboptimal modified maximum ratio combining (MMRC) algorithm proposed in [5]. The MMRC filter outputs are obtained by

$$t_j = \frac{\left|\mathbf{h}_j^H\mathbf{y}\right|}{\left\|\mathbf{h}_j\right\|_F}, \quad j = 1,\ldots,N_{\mathrm{T}}. \tag{5}$$

The higher the value of t_j, the more likely it is that the jth TA was the one activated. The TA indices are sorted in descending order by t_j. The set of ordered TA indices is denoted by $\mathbf{u}$. Suppose $\mathbf{T} = [t_1,t_2,\ldots,t_{N_{\mathrm{T}}}]^{\mathrm{T}}$, the ordered TA indices can be obtained as

$$\mathbf{u} = \left(u_1, u_2, \ldots, u_{N_T}\right) = \text{argsort}\,(\mathbf{T}), \quad (6)$$

where sort$(\cdot)$ denotes a descending order function and u_1 and u_{N_T} are the indices of the maximum and minimum elements of $\mathbf{T}$, respectively. In other words, u_1 and u_{N_T} are the most likely and the least likely estimates of the TA index, respectively.

Stage 2. Determine the TA index and the transmit symbol using the MBS detector.

4.2. Oy-MBS Detector. Since only one TA is activated at one time slot, assuming that the lth antenna sends symbol s_m, the received signal can be expressed as

$$r_i = g_{l,i} \cdot s_m + w_i, \quad i = 1, \ldots, N_R. \quad (7)$$

Each signal at the receiver is related to the channel gain, the transmit symbol, and the white Gaussian noise. Due to the difference of the channel gain and noise in each channel, the channel gain and noise together determine the amplitude of the received signal. Generally speaking, a strong received signal contributes to the demodulation. In the MBS detector, the TA index and the transmit symbol are estimated in ascending order of the RA index from the root node to the leaf nodes. However, the effect of search order on the BER is not taken into account. For this reason, we propose the oy-MBS detector, which detects in descending order of the amplitude of the received signal. The oy-MBS detector is described in detail as follows.

Stage 1. Sort the RA indices in descending order by $|r_i|$. Let $\mathbf{Z} = [z_1, z_2, \ldots, z_{N_R}]^T$, where $z_i = |r_i|$, and the set of the ordered RA indices can be obtained as

$$\widetilde{v} = \left(\widetilde{v}_1, \widetilde{v}_2, \ldots, \widetilde{v}_{N_R}\right) = \text{argsort}\,(\mathbf{Z}), \quad (8)$$

where $\widetilde{v}_1$ and $\widetilde{v}_{N_R}$ are the indices of the maximum and minimum values in $\mathbf{Z}$. The layer search order set $\mathbf{v}$ can be obtained as

$$\begin{aligned} v_{2i-1} &= \widetilde{v}_i, \\ v_{2i} &= N_R + \widetilde{v}_i, \quad i = 1, \ldots, N_R. \end{aligned} \quad (9)$$

The new search tree, whose $(2i-1)^{th}$ and $(2i)^{th}$ layers correspond to the $\widetilde{v}_i^{th}$ RA, can be built by exchanging the layers of Figure 1.

Stage 2. Determine the TA index and the transmit symbol using the MBS detector.

4.3. Ost-Oy-MBS Detector. In ost-MBS detector, all N_T subtrees are searched in descending order of $|\mathbf{h}_j^H \mathbf{y}|/\|\mathbf{h}_j\|_F$. That is to say, we first detect the most probable subtree and then detect the most impossible subtree at last. To some extent, the computational complexity can be reduced. The oy-MBS detector performs the MBS detection in descending order of the received signals amplitude, which can improve the BER performance. In this subsection, we combine the detection orders of ost-MBS and oy-MBS detectors together. We propose the ost-oy-MBS detector whose detection order is based on the subtrees and the received signals. We detect all subtrees in descending order of $|\mathbf{h}_j^H \mathbf{y}|/\|\mathbf{h}_j\|_F$ and detect each layer of subtrees in descending order of $|r_i|$.

The detection process of the proposed ordering MBS-based detectors is summarized in Algorithm 1. In Algorithm 1, lines 2–6 and 7–11 correspond to subtree-ordering and receiver-ordering strategies, respectively, whereas lines 12–28 describe the detection process of the MBS detector.

5. Simulation Results

In this section, the computational complexity and the BER performance of the proposed detectors and the MML, MBS, ML, simplfied OD, simplified MD, SLCD, and ASLCD detectors are compared. The label, (N_1, N_2) OD, denotes the simplified OD detector with N_1 level-one subsets and N_2 level-two subsets. The label, $(N, N_1$, and $N_2)$ MD, stands for the simplified MD detector with N estimated transmit antennas, N_1 level-one subsets, and N_2 level-two subsets. The label, (N) SLCD, denotes the simplified low-complexity detection with N most probable estimates. The label, (N, α) ASLCD, stands for the adaptive low-complexity detection with N most probable estimates and threshold coefficient α. The ideal channel state information (CSI) is assumed available at the receiver. In the simulation, the signal-to-noise ratio (SNR) is the ratio of the signal power to the noise power, i.e., $\rho = (\sum_{m=1}^{N_M} s_m^2/N_M)/\sigma^2$.

To validate the BER performance of the abovementioned detectors, the theoretical bound [17] is drawn in BER simulation figures. Figures 2–3 compare the BER performance and the computational complexity of the proposed detectors and existing detectors for 4×4 64QAM SM systems with $M = [64, 26, 26, 8, 8, 2, 2, 1]$ in MBS and the proposed detectors and $M = [256, 104, 104, 32, 32, 8, 8, 1]$ in the MML detector. The BER performance and computational complexity of the proposed detectors are shown in Figures 4–5 with $M = [16, 10, 8, 4, 4, 2, 1, 1]$ in MBS and the proposed detectors and $M = [128, 80, 64, 32, 32, 16, 8, 1]$ in the MML detector for 8×4 16QAM SM systems. Since each level-two subset must contain more than four signals (i.e., the modulation order $N_M > 16$) in OD and MD detectors, the simulation curves of OD and MD detectors are not listed in Figures 4–5.

To estimate the computational complexity of an algorithm, we define the computational complexity as the total number of the real-valued multiplications/divisions required in the detection process.

In the MML algorithm, we need to compute the accumulated metrics of all $N_T N_M$ nodes in the first layer and compute the accumulated metrics of M_{k-1} nodes in the kth $(2 \le k \le 2N_R)$ layer. Since computing the accumulated metrics of one node needs 3 real multiplications, the computational complexity of the MML detector is $3N_T N_M + \sum_{i=1}^{2N_R-1} 3M_i$.

Since the ML detector computes the accumulated metrics of all nodes in the tree, the computational complexity of the ML detector is $6N_T N_R N_M$.

According to Section 3.2 and 3.3 in [16], we can obtain the computational complexity of OD and MD. The

Initialization: $\Lambda_l = \{(l, s_{\mathrm{m}}) | s_{\mathrm{m}} \in \{s_1, \cdots, s_{N_{\mathrm{M}}}\}\}$, for each $l \in \{1, \cdots, N_{\mathrm{T}}\}$.
if subtree-ordering is used **then**
    $\mathbf{u} = (u_1, \ldots, u_{N_{\mathrm{T}}}) = \mathrm{argsort}(|\mathbf{h}_j^H \mathbf{y}| / \|\mathbf{h}_j\|_{\mathrm{F}})$
else
    $\mathbf{u} = (1, \cdots, N_{\mathrm{T}})$
end if
if receiver-ordering is used **then**
    $\overline{v} = (\overline{v}_1, \ldots, \overline{v}_{N_{\mathrm{R}}}) = \mathrm{argsort}(|r_i|)$, obtain layer search order $\mathbf{v}$ by Equation (9).
else
$\overline{v} = (1, \cdots, N_R)$, obtain layer search order $\mathbf{v}$ by Equation (9).
end if
$\rho_\Omega = 0, \Omega = \phi$.
for $i = 1 N_{\mathrm{T}}$
    $\Psi = \Lambda_{u_i}$, $[(p,q), \mathrm{value}] =$ search subtree($\Psi \mathbf{v}$)
    $\Omega = (p,q)$ and $\rho_\Omega = \mathrm{value}$, if (p,q) is not null and value $< \rho_\Omega$.
end for
End the algorithm by returning Ω corresponding to ρ_Ω.
function search subtree($\Psi, \mathbf{v}$)
    For each $(p,q) \in \Psi$, $A^{(p,q)} = 0$.
    for $i = 1 2 N_{\mathrm{R}}$
        $k = v_i$, for each $(p, q \in \Psi)$, $A^{(p,q)} = A^{(p,q)} + B_k^{(p,q)}$.
        while $i < 2N_R$ and $|\Psi| > M_i$
            $\Psi = \Psi - \{(p,q)\}$, where $(p,q) = \underset{(p,q)\in\Psi}{\mathrm{argmax}}\, A^{(p,q)}$.
        end while
        For each $(p,q) \in \Psi$, $\Psi = \Psi - \{(p,q)\}$, if $A^{(p,q)} \geq \rho_\Omega$.
    end for
    return $[(p,q), \mathrm{value}] = \underset{(p,q)\in\Psi}{\mathrm{argmin}}\, A^{(p,q)}$, if Ψ not empty; otherwise return null.
end function

ALGORITHM 1: Ordering-aided MBS-based detectors.

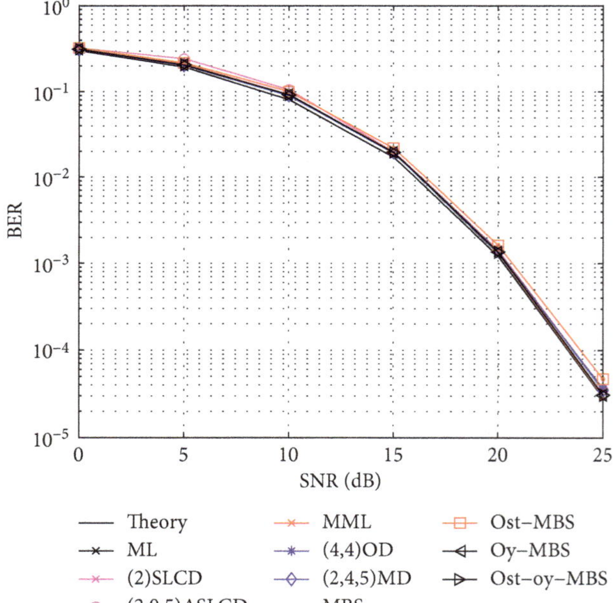

FIGURE 2: The BER performance of the proposed detectors with $N_{\mathrm{T}} = 4, N_{\mathrm{R}} = 4$, and 64QAM modulation.

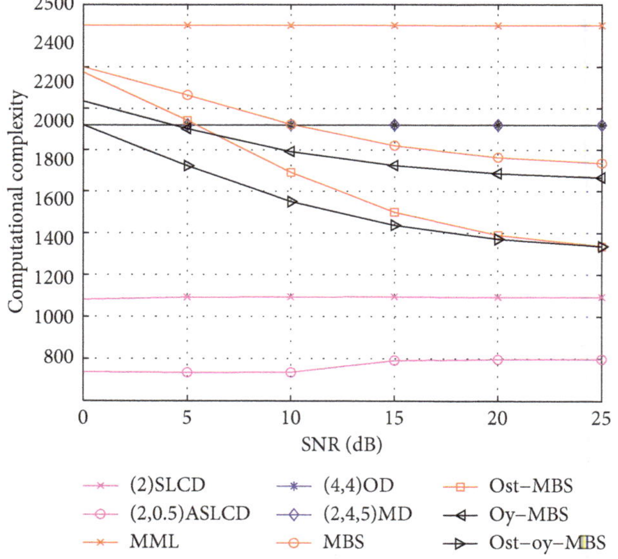

FIGURE 3: The computational complexity of the proposed detectors with $N_{\mathrm{T}} = 4, N_{\mathrm{R}} = 4$, and 64QAM modulation.

computational complexity of the (N_1, N_2) OD detector and (N, N_1, N_2) MD detector is $10N_R(4N_T + 4N_1 + N_{\mathrm{M}}N_2/16)$ and $10N_{\mathrm{R}}(N_{\mathrm{T}} + 4N + 4N_1 + N_{\mathrm{M}}N_2/16)$, respectively.

According to Section 3.5 and 3.6 in [17], the computational complexity of SLCD and ASLCD depends on the parameter N and the size of estimated transmit symbol set. Since the size of estimated transmit symbol set is not constant, the computational complexity can only be obtained by simulation.

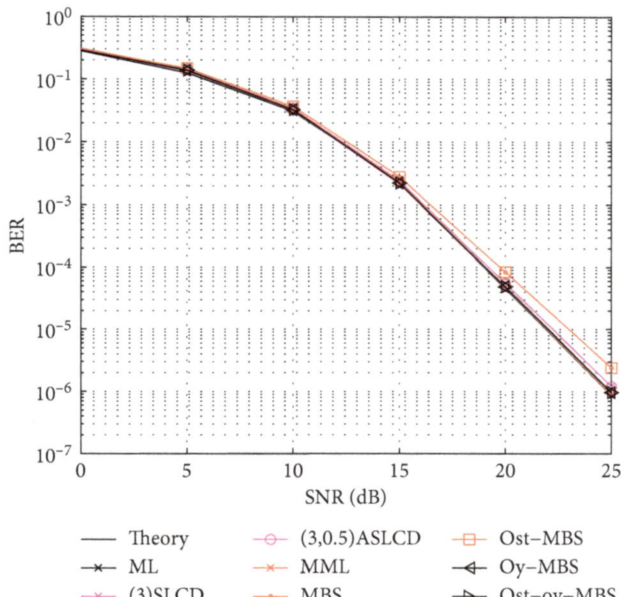

FIGURE 4: The BER performance of the proposed detectors with $N_T = 8, N_R = 4$, and 16QAM modulation.

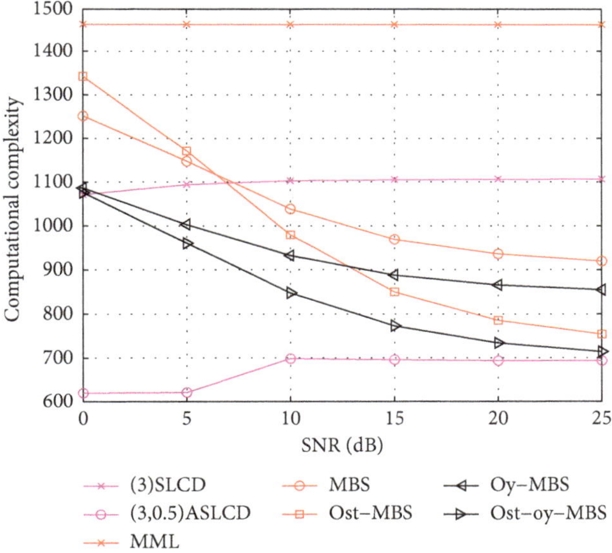

FIGURE 5: The computational complexity of the proposed detectors with $N_T = 8, N_R = 4$, and 16QAM modulation.

From the above analysis, we can conclude that the computational complexity of MML, OD, and MD detectors is lower than that of the ML detector but with a BER performance loss. Meanwhile, the complexity of MML, OD, MD, SLCD, and ASLCD depends on the preset parameters.

The computational complexity of the proposed MBS-based detectors includes the number of real-valued multiplications of computing $|\mathbf{h}_j^H\mathbf{y}|/||\mathbf{h}_j||_F$ and $|r_i|$ in stage 1, and the number of real-valued multiplications of MBS detector in stage 2. The computational complexity of stage 1 can be easily obtained by calculation. The computational complexity of stage 2 depends on the number of retained nodes. In MBS and the proposed MBS-based detectors, the parameter M_k is at most the number of retained nodes in each layer. That is, the number of retained nodes is not fixed. Therefore, the computational complexity of proposed MBS-based detectors can only be obtained by simulation.

From the above simulation curves, we can draw the following conclusions:

(1) The computational complexity of MBS, OD, MD, SLCD, ASLCD, MML, and the proposed detectors is lower than that of the ML detector. Since the number of retained nodes of MML, OD, and MD detectors is fixed under different SNRs, the computational complexity does not change with the SNR. The computational complexity of SLCD, ASLCD, MBS, ost-MBS, oy-MBS, and ost-oy-MBS detectors changes with the SNR.

(2) The BER performance of ost-MBS detector is the same as that of the MBS detector, and the complexity is lower than that of the MBS detector. The ost-MBS detector only changes the search order of subtrees and does not affect the detection performance. Therefore, the ost-MBS and MBS detectors have the same BER performance. The ost-MBS detector searches the subtrees in the descending order of $|\mathbf{h}_j^H\mathbf{y}|/\|\mathbf{h}_j\|_F$, which increases the probability of the optimal solution in the first subtree. Since the accumulated metric of the optimal solution is minimal, the number of retained nodes can be reduced in the subsequent subtrees, thus reducing the total computational complexity.

(3) The BER performance of the oy-MBS detector is superior to that of the MBS detector. The oy-MBS detector estimates the solution in the descending order of the received signals amplitude. To some degree, the strong received signal contributes to the demodulation. Therefore, compared with the MBS detector, the oy-MBS detector has better BER performance. Meanwhile, we also notice that the sorting strategy of oy-MBS also reduces the computational complexity.

(4) The ost-oy-MBS detector and oy-MBS detector have the same BER performance, which is superior to the MBS detector. The ost-oy-MBS detector has the advantages of both ost-MBS and oy-MBS detectors. That is, the ost-oy-MBS detector has the best BER performance and the lowest computational complexity among the proposed MBS-based detectors.

(5) Under the current simulation conditions, the BER performance of OD, MD, SLCD, ASLCD, MML, and ost-oy-MBS detectors is almost the same as the ML detector. Compared to OD and MD detectors, MBS-based detectors have a lower computational complexity in moderate-to-high SNRs.

6. Conclusion

In this paper, novel ordering MBS-based detectors for SM systems are proposed to improve the BER performance and reduce the computational complexity. The ost-MBS detector first searches each subtree from the most probable TA. Compared to the MBS detector, it has the lower computational complexity. The oy-MBS algorithm detects each subtree in the descending order of the received signal amplitude. The BER performance of the oy-MBS detector is superior to that of the MBS detector. The ost-oy-MBS detector combined the orders of ost-MBS and oy-MBS detectors. Among all proposed MBS-based methods, the ost-oy-MBS detector has the best BER performance and the lowest computational complexity. Meanwhile, we notice that the computational complexity and the BER performance of OD, MD, SCLD, ASLCD, MML, and MBS-based detectors depend on the preset parameters. Regrettably, how to select the parameters in the proposed MBS-based detectors can only be obtained by simulation. Next, we will study how to select parameters and try to give a theoretical derivation.

Conflicts of Interest

The authors declare that they have no conflicts of interest.

References

[1] R. Mesleh, H. Haas, C. W. Ahn, and S. Yun, "Spatial modulation-a new low complexity spectral efficiency enhancing technique," in *Proceedings of 2006 First International Conference on Communications and Networking in China*, pp. 1–5, Beijing, China, October 2006.

[2] M. D. Renzo, H. Haas, A. Ghrayeb, S. Sugiura, and L. Hanzo, "Spatail modulation for generalized MIMO: challenges, opportunities and implementation," *Proceedings of the IEEE*, vol. 102, no. 1, pp. 56–103, 2014.

[3] P. Yang, M. D. Renzo, Y. Xiao, S. Li, and L. Hanzo, "Design guidelines for spatial modulation," *IEEE Communications Surveys and Tutorials*, vol. 17, no. 1, pp. 6–26, 2015.

[4] R. Mesleh, H. Haas, S. Sinanovic, C. W. Ahn, and S. Yun, "Spatial modulation," *IEEE Transactions on Vehicular Technology*, vol. 57, no. 4, pp. 2228–2241, 2008.

[5] N. R. Naidoo, H. Xu, and T. Quazi, "Spatial modulation: optimal detector asymptotic performance and multiple-stage detection," *IET Communications*, vol. 5, no. 10, pp. 1368–1376, 2011.

[6] J. Jeganathan, A. Ghrayeb, and L. Szczecinski, "Spatial modulation: optimal detection and performance analysis," *IEEE Communications Letters*, vol. 12, no. 8, pp. 545–547, 2008.

[7] R. Rajashekar, K. V. S. Hari, and L. Hanzo, "Reduced-complexity ML detection and capacity-optimized training for spatial modulation systems," *IEEE Transactions on Communications*, vol. 62, no. 1, pp. 112–125, 2014.

[8] H. Men and M. Jin, "A low-complexity ML detection algorithm for spatial modulation system with MPSK constellation," *IEEE Communications Letters*, vol. 18, no. 8, pp. 1375–1378, 2014.

[9] A. Younis, R. Mesleh, H. Haas, and P. Grant, "Reduced complexity sphere decoder for spatial modulation detection receivers," in *Proceedings of 2010 IEEE Global Telecommunications Conference GLOBECOM 2010*, pp. 1–5, Miami, FL. USA, December 2010.

[10] A. Younis, M. D. Renzo, R. Mesleh, and H. Hass, "Sphere decoding for spatial modulation," in *Proceedings of 2011 IEEE International Conference on Communications (ICC)*, pp. 1–6, Kyoto, Japan, June 2011.

[11] A. Younis, S. Sinanovic, M. D. Renzo, R. Mesleh, and H. Haas, "Generalised sphere decoding for spatial modulation," *IEEE Transactions on Communications*, vol. 61, no. 7, pp. 2805–2815, 2013.

[12] K. Lee, "Doubly ordered sphere decoding for spatial modulation," *IIEEE Communications Letters*, vol. 19, no. 5, pp. 795–798, 2015.

[13] S. Sugiura, C. Xu, S. X. Ng, and L. Hanzo, "Reduced-complexity coherent versus non-coherent QAM-aided space-time shift keying," *IEEE Transactions on Communications*, vol. 59, no. 11, pp. 3090–3101, 2011.

[14] J. Wang, S. Jia, and J. Song, "Signal vector based detection scheme for spatial modulation," *IEEE Communications Letters*, vol. 16, no. 1, pp. 19–21, 2012.

[15] Q. Tang, Y. Xiao, P. Yang, Q. Yu, and S. Li, "A new low-complexity near-ML detection algorithm for spatial modulation," *IEEE Wireless Communications Letters*, vol. 2, no. 1, pp. 90–93, 2013.

[16] H. Xu, "Simplified maximum likelihood-based detection schemes for M-ary quadrature amplitude modulation spatial modulation," *IET Communications*, vol. 6, no. 11, pp. 1356–1363, 2012.

[17] H. Xu, "Simple low-complexity detection schemes for M-ary quadrature amplitude modulation spatial modulation," *IET Communications*, vol. 6, no. 17, pp. 2840–2847, 2012.

[18] J. Zheng, X. Yang, and Z. Li, "Low-complexity detection method for spatial modulation based on M-algorithm," *Electronics Letters*, vol. 52, no. 21, pp. 1552–1554, 2014.

[19] Z. Tian, J. Yang, and Z. Li, "M-algorithm-based optimal detectors for spatial modulation," *Journal of Communications*, vol. 10, no. 4, pp. 245–251, 2015.

[20] T. Kim and K. Yi, "Low-complexity symbol detection based on modified beam search for spatial modulation MIMO systems," *Electronics Letters*, vol. 51, no. 19, pp. 1546–1548, 2015.

[21] X. Zhang, G. Zhao, Q. Liu, N. Zhao, and M. Jin, "Enhanced M-algorithm-based maximum likelihood detectors for spatial modulation," *AEU - International Journal of Electronics and Communications*, vol. 70, no. 9, pp. 1361–1366, 2016.

6

Doppler Ambiguity Resolution based on Random Sparse Probing Pulses

Yunjian Zhang,[1] Zhenmiao Deng,[1] Jianghong Shi,[1] Linmei Ye,[1] Maozhong Fu,[1] and Chen Zhao[2]

[1]*School of Information Science and Engineering, Xiamen University, Fujian 361005, China*
[2]*Faculty of Science, University of Auckland, Auckland 1052, New Zealand*

Correspondence should be addressed to Zhenmiao Deng; dzm_ddb@xmu.edu.cn

Academic Editor: Igor Djurović

A novel method for solving Doppler ambiguous problem based on compressed sensing (CS) theory is proposed in this paper. A pulse train with the random and sparse transmitting time is transmitted. The received signals after matched filtering can be viewed as randomly sparse sampling from the traditional fixed-pulse repetition frequency (PRF) echo signals. The whole target echo could be reconstructed via CS recovery algorithms. Through refining the sensing matrix, which is equivalent to increase the sampling frequency of target characteristic, the Doppler unambiguous range is enlarged. In particular, Complex Approximate Message Passing (CAMP) algorithm is developed to estimate the unambiguity Doppler frequency. Cramer-Rao lower bound expressions are derived for the frequency. Numerical simulations validate the effectiveness of the proposed method. Finally, compared with traditional methods, the proposed method only requires transmitting a few sparse probing pulses to achieve a larger Doppler frequency unambiguous range and can also reduce the consumption of the radar time resources.

1. Introduction

Long range radars (LRRs) usually work in low Pulse Repetition Frequency (PRF). Although there is no or low range ambiguity, it suffers from serious Doppler or velocity ambiguity. Therefore, Doppler ambiguity resolution is necessarily required to ensure accurate measurement of the radial velocity of a target. One resolution introduced from pulse Doppler technology is adopted by LRRs, which measure the velocity by transmitting high PRF impulse train. However, the high PRF mode requires more radar time resources, which are very precious to phased array radars when tracking multiple targets at the same time. On the other hand, the low signal pulse width may lead to a weak corresponding echo signal and hence a small operational range of the LRR, due to the high repetition frequency.

Many Doppler ambiguity resolution methods have been proposed in [1–7]. The so-called robust refinement Chinese Remainder Theorem (CRT) algorithm under noisy environment is proposed in [1–4]. The staggered PRF is proposed in [5] for overcoming Doppler ambiguity. Clearly, by exploiting the difference of the measured Doppler frequencies between two staggered PRF pulses, the blind frequency could be expanded by ten times or more [5]. In [6], a CS-based Doppler ambiguity resolution method is proposed, which can resolve Doppler ambiguities of several targets. The clutter suppression and velocity resolution in moving target indication are investigated in [7].

In this paper, we proposed a novel method to solve Doppler ambiguous problem, which tremendously increases fixed-PRF and randomly determines the small amount of measurement corresponding to the time sequence of transmitted probing pulses. Explicitly, a sparse pulse train is randomly selected from traditional fixed repetition frequency pulses, according to the Restricted Isometry Property (RIP) of the CS sensing matrix, and a random perturbation item is added to the transmitting time of each selected pulse before transmitting. The received signals after Matched Filtering (MF) can be considered as randomly sparse sampling of Doppler. By using the CS recovery algorithms, the sensing matrix can be built based on the transmitting time sequence and recover the whole Doppler spectrum. As a result, the number of columns of sensing matrix becomes quite large such that many reconstruction algorithms of CS are not adequate, namely, l_1-minimization [8], Orthogonal Matching Pursuit (OMP) [9], Spectral Compressive Sensing (SCS) [10], and so forth, since these algorithms require explicit operations on the complete sensing matrix, which introduces extremely high computational complexity, especially in the scenario of large-scale applications.

To address this problem, the proposed method for enlarging the Doppler unambiguous range in this paper is based on Complex Approximate Message Passing (CAMP) algorithm [11], which is an extension of the original Approximate Message Passing (AMP) algorithm [12] from real number field to complex domain. Moreover, the process of message passing algorithm only involves the sensing matrix or transposition of the sensing matrix, and hence, it is especially suitable for radar applications in which the echo signals are often complex [13] and the range of Doppler frequencies for unknown targets is large.

The performance of the CAMP algorithm can be accurately predicted by State Evolution (SE) formalism introduced in [12]. In this formalism, the Mean Square Error (MSE) of reconstruction is a state variable; its change from iteration to iteration is modeled by a scalar function. Therefore, the variance caused by each iteration can be obtained accurately. Combining CRLB results of frequency estimation with nonuniform sampling, the CRLB of the proposed method can be derived.

The rest of the paper is organized as follows. In Section 2, we present the radar echo signal model utilized for this work and the brief revision of CS theory. Section 3 discusses the proposed Doppler ambiguity resolution method based on the adaptive CAMP algorithm in detail. In Section 4, sensing matrix design and radar dwell scheduling scheme are explained. We also derive the Cramer-Rao Lower Bound (CRLB) for the proposed method. We then present numerical simulation results to demonstrate the improvement of the proposed method in Section 5. Finally, we conclude the paper in Section 6.

2. Model and the CS Theory

2.1. Radar Echo Signal Model. Let the center frequency to be denoted by f_c and the chirp rate to be denoted by γ; the transmitted chirp signal can be modeled as

$$s(\hat{t}, t_m) = \operatorname{rect}\left(\frac{\hat{t}}{T_p}\right) \exp\left(j2\pi\left(f_c t + \frac{1}{2}\gamma \hat{t}^2\right)\right), \quad (1)$$

where $\operatorname{rect}(u) = \begin{cases} 1, & |u| \le 0.5 \\ 0, & |u| > 0.5 \end{cases}$ and T_p and $\hat{t}$ are the pulse width and the fast time, respectively; $t = \hat{t} + t_m$ is the full time, where t_m denotes the slow time. When the distance between the target and the radar is R_t, the received echo signal $s_r(\hat{t}, t_m)$ can be defined as

$$\begin{aligned} s_r(\hat{t}, t_m) &= A\operatorname{rect}\left(\frac{(\hat{t} - 2R_t/c)}{T_p}\right) \\ &\cdot \exp\left(j\pi\gamma\left(\hat{t} - \frac{2R_t}{c}\right)^2\right) \exp\left(-\frac{j4\pi f_c}{cR_t}\right), \end{aligned} \quad (2)$$

where A is the coefficient scattering and c is the propagation speed of electromagnetic wave. Performing MF to the received signal and transforming the results into range-frequency domain yield

$$r(f_r, t_m) = A'\operatorname{rect}\left(\frac{f_r}{B}\right) \exp\left(-\frac{j4\pi(f_r + f_c) R_t}{c}\right), \quad (3)$$

where B is the signal bandwidth. Here, R_t could be approximated by ignoring the acceleration, jerk, and rotation velocity of the object, leading to

$$R_t \approx R_0 + vt_m, \quad (4)$$

where R_0 is the distance between the target and radar at time t_0 and v is the radial velocity. By substituting (4) into (3), the slow-time domain echo signal is obtained as

$$r(f_d, t_m) = A'' \exp(-j2\pi f_d t_m), \quad (5)$$

where $f_d = 2(f_r + f_c)v/c$, $A'' = A' \exp(2(f_r + f_c)R_0/c)$, and f_d is the Doppler frequency.

As can be seen from (5), the slow-time domain radar echo signal could be viewed as a sinusoidal signal, which is sparse in frequency domain. Therefore the Doppler frequency estimation problem could be solved via CS theory. Furthermore, if there are K targets with different Doppler frequencies, the echo signal can be rewritten as

$$r(f_{d_i}, t_m) = \sum_{i=1}^{K} A_i'' \exp(-j2\pi f_{d_i} t_m), \quad (6)$$

where f_{d_i} notes the ith Doppler frequency.

2.2. Compressed Sensing Theory. The CS theory is able to reconstruct the original signal from a set of nonadaptive measurements sampled at a much lower rate than required by the Nyquist sampling theorem by simultaneously sensing and compressing sparse or compressible signals. The original signal denoted by $x \in C^N$ and the sampled linear measurements $y \in C^M$ can be written as

$$y = Ax + n, \quad (7)$$

where $A \in C^{M\times N}$ denotes the sensing matrix with $M < N$ and $n \in C^M$ is the measurement noise. Note that any signal x can be represented in terms of a basis of $N \times 1$ vectors $\{\psi_i\}_{i=1}^N$. According to the $N \times N$ basis matrix $\Psi = [\psi_1|\psi_2|, \ldots, \psi_N]$ with the vectors $\{\psi_i\}$ as columns, the original signal x can be expressed as $x = \sum_{i=1}^N s_i\psi_i = \Psi s$ where s_i represents

Input: y, A, τ, x
Initialization: $\tilde{x}^0 = 0$, $\tilde{z}^0 = y$
for $t = 1$: maxiter
$\tilde{x}^t = A^H z^{t-1} + \tilde{x}^{t-1}$
$\sigma_t = \text{std}(\tilde{x}^t - x)$
$z^t = y - A\tilde{x}^{t-1} + z^{t-1}\frac{1}{2\delta}\left(\left\langle \frac{\partial \eta^R}{\partial x_R}(\tilde{x}^t; \tau\sigma_t)\right\rangle + \left\langle \frac{\partial \eta^I}{\partial x_I}(\tilde{x}^t; \tau\sigma_t)\right\rangle\right)$
$\hat{x}^t = \eta(\tilde{x}^t; \tau\sigma_t)$
end
Output: $\tilde{x}$, $\hat{x}$, σ_*

Algorithm 1: Ideal CAMP algorithm.

weighting coefficient. If x is sparse or compressible, and Ψ is chosen properly, the vector s can be computed directly from y provided that the Restricted Isometry Property (RIP) condition holds. The undersampling ratio, that is, the aspect ratio of the sensing matrix, is defined as $\delta \triangleq M/N$ and the sparsity rate is defined as $\rho \triangleq k/M$, where k is the number of nonzero entries. By using the convex optimization, a close approximation of x is expressed as

$$\hat{x} = \arg\min_{x'} \left\|x'\right\|_1 \quad \text{s.t.} \quad \left\|Ax' - y\right\|_2^2 \le \varepsilon, \tag{8}$$

where ε indicates the noise level. However, in some interesting large-scale applications, since the matrix A and vector x may contain millions of entries, it may become significantly slow to solve the standard convex optimization problem as the Doppler ambiguity problem we addressed previously. Furthermore, the original signal needed to be reconstructed contains enormous number of entries, due to the fact that the practical Doppler frequency of target may be much higher than the PRF of radar. Therefore, it becomes the interest of CS research community to solve the l_1-penalized least squares problem, expressed as

$$\hat{x} = \min_{x'} \frac{1}{2}\left\|y - Ax'\right\|_2^2 + \lambda\left\|x'\right\|_1, \tag{9}$$

where λ is a positive regularization parameter, which represents a tradeoff between sparsity and reconstruction error; that is greater λ denotes greater sparsity and greater reconstruction error and vice versa. The complex l_1-norm is defined as $\|x\|_1 \triangleq \sum_i |x_i| = \sum_i \sqrt{(x_i^R)^2 + (x_i^I)^2}$. Note that this is also referred to as Complex Least Absolute Shrinkage and Selection Operator (c-LASSO) problem [14] in other research fields.

3. The CAMP Algorithm

3.1. Adaptive CAMP Algorithm. The CAMP is a fast and efficient iterative algorithm for solving the problem with complex vectors and matrices involved. In the real number field, the algorithm is referred to as AMP algorithm, which only requires the transpose operation on the sensing matrix A [12]. Therefore, it indicates that the analogous method can be applied in complex domain, namely, replacing the matrix A^T by A^H, where H denotes the complex conjugate transpose. As reported in [11], the performance of CAMP has been asymptotically analyzed in the case of noise-free and noisy measurements; here the brief revision of the basics of CAMP is discussed as follows.

In the ideal scenario, the complex original signal x that we intend to reconstruct is assumed to be known at the receiver; thus the exact current standard deviation of the noise is available during the iterative processing. In this paper, this is referred to as Algorithm 1.

Clearly, in Algorithm 1, the complex soft thresholding is the function of u and λ, as expressed as $\eta(u;\lambda) \triangleq (|u| - \lambda)e^{j\angle u}\mathbf{1}(|u| > \lambda)$, where $\mathbf{1}$ is the so-called indicator function. The main purpose of the thresholding function is to impose the sparsity at each iteration. $\partial\eta^R/\partial x_R$ is the partial derivative of η^R with respect to the real part of the input and $\partial\eta^I/\partial x_I$ is the partial derivative of η^I with respect to the imaginary part of the input. The predefined maximum number of iterations is named maxiter. The estimation of x at iteration t is labeled as $\tilde{x}^t$, and $\tilde{x}^t$ is a nonsparse, noisy estimate of x, where σ_t is the standard deviation and $\langle\cdot\rangle$ denotes arithmetic average.

Algorithm 1 first searches for a noisy estimation of the original signal $\tilde{x}^t$, which is not sparse. In order to obtain the sparse estimation $\hat{x}^t$, the soft thresholding function is subsequently adapted. Here the assumption is made that the algorithm uses the known x for calculating the noise standard deviation σ_t. However, in fact, the original signal x we intend to reconstruct is unknown. An alternative approach of estimating σ_t is adopted to tackle it, which utilizes median estimation shown as

$$\hat{\sigma}_t = \sqrt{\frac{1}{\ln 2}}\text{median}\left(\left|\tilde{x}^t\right|\right). \tag{10}$$

Furthermore, the optimal CAMP thresholding control parameter $\hat{\tau}_o$ (typically in the range $2 < \tau_o < 4$ [15]) can be derived based on the estimation of σ_t, which minimizes $\hat{\sigma}_*^2$. Note that it is assumed that we know $\tau_{\max}$ such that $\tau_o < \tau_{\max}$. Given a step δ_τ, we define a sequence of thresholds $\tau = \{\tau_l\}_{l=1}^L$ such that $\tau_1 = \tau_{\max}$ and $\tau_l = \tau_{l-1} - \delta_\tau$. Starting from $\tau_{\max}$, at the first iteration l CAMP is initialized with $\hat{x}_l^0 = \hat{x}_{l-1}$ and $z_l^0 = z_{l-1}$. Using the solution of CAMP at the previous iteration $l-1$ as initial value for $\tau = \tau_l$.

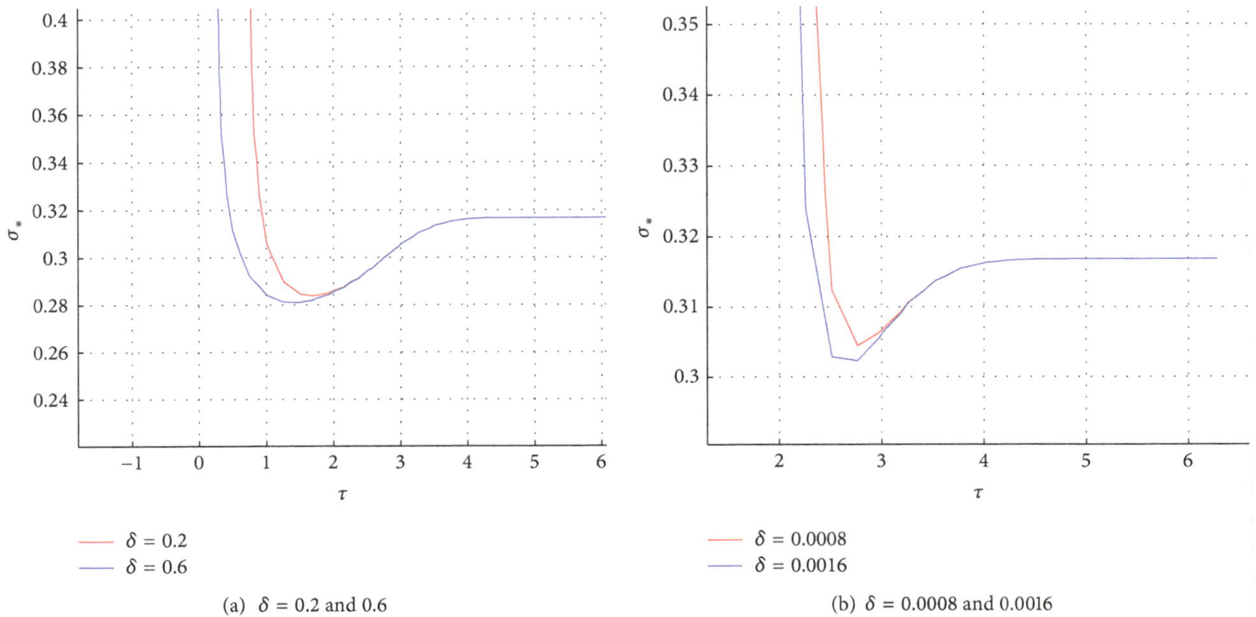

FIGURE 1: Fixed point σ_* versus optimal threshold τ_o for ideal CAMP with $\sigma = 0.23$; undersampling ratios δ equal (a) 0.2, 0.6 and (b) 0.0008, 0.0016, respectively. Sparsity rate $\rho = 0.1$ and the nonzero entries of vector x have all amplitudes equal to 1 and phase uniformly distributed between $-\pi$ and π.

After L iterations, the resultant matrix $\widehat{X} = [\widehat{x}_1, \widehat{x}_2, \ldots, \widehat{x}_L]$ of size $N \times L$ contains the CAMP solution for a given τ_l in each column. Also, we have L estimates $\{\widehat{\sigma}_*^l\}_{l=1}^L$. The optimal threshold $\widehat{\tau}_o$ is chosen as the one that minimizes the estimated CAMP output noise variance $\widehat{\sigma}_*^2$. The value of $\tau_{\max}$ can be determined with assist of the c-LASSO problem. It is known that for $\lambda > \lambda_{\max} = \|A^H y\|_\infty$ the only solution is zero solution. One uses the calibration equation [11], which denotes the relationship between λ and τ, to evaluate the $\widehat{\tau}_{\max}$. The calibration equation is given by

$$\lambda \triangleq \tau\sigma_* \left(1 - \frac{1}{2\delta} E\left(\frac{\partial \eta^R}{\partial x_R}(X + \sigma_* Z; \tau\sigma_*) + \frac{\partial \eta^I}{\partial x_I}(X + \sigma_* Z; \tau\sigma_*)\right)\right), \tag{11}$$

where E is with respect to the two independent random variables $Z \sim CN(0,1)$ and $X \sim p_X$. With $\lambda = \lambda_{\max}$ and $\sigma_* = \widehat{\sigma}_0$, the value of the estimation $\widehat{\tau}_{\max}$ can be evaluated. The expectation in (11) equals the MSE of the estimate $\widehat{x}^t$ after applying the soft thresholding function. The details in evaluating the function with a given p_X are shown in [13, Appendix A]. This algorithm is called Adaptive CAMP, since both the noise variance $\widehat{\sigma}_t$ and the threshold $\widehat{\tau}_o$ are adaptively estimated inside the algorithm itself, and the only input variables are y and A. The relationship of fixed-point σ_* and the optimal threshold τ_o can be expressed as

$$\sigma_{t+1}^2 = \sigma^2 + \frac{1}{\delta} E\left(\left|\eta\left(X + \sigma_t Z; \tau\sigma_t\right) - X\right|^2\right) \tag{12}$$

by using SE formalism. Figure 1 demonstrates the relation when different values are chosen for the undersampling ratios. Equation (12) exhibits that the time $t+1$ deviation is constituted by the input noise variance σ^2 and the MSE of the estimation after applying the soft thresholding function. Then we can analytically calculate the time $t+1$ deviation caused by CAMP. It has been proved that the right-hand side of (12) is concave [11]; thus there exists at most one stable fixed point σ_*^2. Moreover, for $\tau = \tau_o$, the value of σ_* for DFT frame is not very different from the value for Gaussian matrix. For extremely sparse situation where the number of Doppler frequencies is much lower than the number of frequency points, for example, the Doppler ambiguity problem, the relation between σ_* and τ_o is shown in Figure 1(b). Furthermore, from Figure 1 we can find that both the optimal threshold τ_o and the corresponding output noise standard deviation increase as the undersampling ratio δ decreases; namely, the number of measurements decreases.

4. CS Sensing Matrix Design and Radar Dwell Scheduling

4.1. Sensing Matrix Design. To construct the measurement matrix Φ, the M number of rows form an $N \times N$ identity matrix is selected, which corresponds to the time sequence of transmitted probing pulses. In order to increase unambiguous Doppler range, M pulses from the traditional fixed-PRF pulse train are selected randomly and a random perturbation is added to the transmitting time of every selected pulse.

Since $r(f_d, t_m)$ in (5) is sparse in frequency domain, the basis matrix Ψ can be defined as Discrete Fourier Transform (DFT) of x:

$$\Psi_{i,k} = \exp\left(-\frac{j2\pi ik}{N}\right), \quad 0 \le i,\ k \le N. \tag{13}$$

The sensing matrix A, also known as "randomly extracting matrix," is evaluated by the measurements matrix Φ multiplying the basis matrix Ψ. Note that the sensing matrix A is equivalent to an oversampled DFT matrix, that is, DFT frame.

4.2. Radar Dwell Scheduling. The PRF of LRR is usually low and the typical value of f is $10\,\text{Hz} < f < 1000\,\text{Hz}$. When radar works in fixed PRF mode, the pulse transmitting time sequences are $t_i = i/f_s$, $i = 1, \ldots, n$. In this case, the echoes in the slow-time domain may suffer from severe Doppler ambiguity. By modifying t_i as $t_i' = t_i + \Delta_i$, where Δ_i is a random perturbation, we can obtain a new transmitting time sequence, which is equivalent to those extracted from a high PRF pulse transmitting time set. It is obvious that the precision of perturbation determines the equivalent value of PRF. The slow-time domain echo signal could be recovered by CAMP algorithms from the measurements y as discussed before.

Clearly, M elements are randomly selected from the transmitting time set $\{t_1, t_2, \ldots, t_n\}$ and each of them is added by a perturbation Δ_k, respectively, where Δ_k is a random variable with its mean and variance to be 0 and ξ^2, labeled as $\Delta_k \sim N(0, \xi^2)$. The pulse transmitting time can be expressed as

$$t_k' = t_k + \Delta_k, \tag{14}$$

where k is the time index and is a subset of $\{1, 2, \ldots n\}$; that is, $k = \{k_1, k_2, \ldots, k_M\} \subset \{1, 2, \ldots, n\}$.

In practical, the term Δ_k is not completely random, which is caused by the factor that the time sequence of a radar controlled by a reference clock, which indicates that the precision of Δ_k should be lower than that of clock. Actually, high precision of Δ_k means large unambiguous Doppler range; meanwhile, the refining factor is enlarged, which introduces high computational burden. Therefore, to ensure enough Doppler unambiguous range, a proper value of Δ_k is necessarily required. Here, assuming the time precision is Δt, the corresponding equivalent PRF is $f_s' = 1/\Delta t$, and the Doppler unambiguous range for t_i' is represented as

$$f_s' = \frac{1}{\Delta t} \cdot f_s. \tag{15}$$

The unambiguous Doppler range is extended to f_s'. The purpose of enlarging unambiguous Doppler range is to widen the measure velocity range, which is equal to

$$v = \frac{c f_s'}{2\left(f_c + f_r\right)} = \frac{c f_s'}{2\left(f_c + \gamma t_m\right)}. \tag{16}$$

It is obvious that the velocity of resolution Δv depends on precision of f_s', according to (16).

After determining the dwell time $t_1, t_2, \ldots, t_M$, radar transmits pulses at these moments. The measurements $r_1, r_2, \ldots, r_M$ in slow-time domain are obtained by matched filtering to the echo signals. By using the CAMP algorithm, the slow-time domain echo signals can be recovered and the Doppler frequency under expanded unambiguous Doppler range can be estimated. The estimation accuracy is determined by the precision of the basis matrix. In order to obtain more accurate Doppler estimate, an adaptive refinement algorithm of the sensing matrix proposed in [16] can be used to reconstruct the slow-time domain signal with higher precision.

4.3. SNR and CRLB. The definition of SNR for the CAMP-based CS radar system is provided by [13, Ch. 3.4]. The SNR at the input (SNR_{in}) and output (SNR) of the MF are defined by

$$\begin{aligned}
\text{SNR}_{\text{in,MF}} &= \frac{a^2}{N\sigma^2}, \\
\text{SNR}_{\text{in,CAMP}} &= \frac{a^2}{M\sigma^2}, \\
\text{SNR}_{\text{MF}} &= \frac{a^2}{\sigma^2}, \\
\text{SNR}_{\text{CAMP}} &= \frac{a^2}{\sigma_*^2},
\end{aligned} \tag{17}$$

where a^2 is the received power of a target at bin i and σ_* is the minimized standard deviation defined in Section 3. Based on previous analysis, the SNR_{CAMP} is lower than SNR_{MF} due to the increase of noise standard deviation imposed by the iterative CAMP algorithm.

Consider the case of M samplings of a single complex sinusoid corrupted by Additive White Gaussian Noise (AWGN):

$$\begin{aligned}
x_{t_m} &= A e^{j(\omega t_m + \theta)} + n_{t_m}, \\
\mathbf{x} &= \left[x_{t_0}, x_{t_1}, \ldots, x_{t_{M-1}}\right]^T,
\end{aligned} \tag{18}$$

where $\{t_m \mid 0 \le m \le M-1\}$ is an index set for the sampling time, ω is the unknown frequency in radians, θ is a random phase which is uniformly distributed over $[-\pi, \pi]$, n_t is zero mean complex AWGN with variance σ^2, and A is the energy per sampling.

The conditional Probability Density Function (PDF) of $\mathbf{x}$ given ω and θ is

$$\begin{aligned}
p(\mathbf{x} \mid \omega, \theta) = &\frac{\exp\left\{-\left(\|\mathbf{x}\|^2 + MA\right)/\sigma^2\right\}}{\left(\pi\sigma^2\right)^M} \\
&\cdot \exp\left(\frac{2A}{\sigma^2} \operatorname{Re}\left\{\sum_{m=0}^{M-1} x_{t_m} e^{-j(\omega t_m + \theta)}\right\}\right).
\end{aligned} \tag{19}$$

Taking the log of (19) and neglecting constants yield the log likelihood function for samplings in AWGN

$$L(\mathbf{x} \mid \omega, \theta) = \frac{2A}{\sigma^2} \operatorname{Re}\left\{\sum_{m=0}^{M-1} x_{t_m} e^{-j(\omega t_m + \theta)}\right\}. \tag{20}$$

Fisher's information matrix (FIM) is the expected value of the second order derivatives

$$\mathbf{J} = \frac{2A^2}{\sigma^2} \begin{bmatrix} \sum_{m=0}^{M-1} t_m^2 & \sum_{m=0}^{M-1} t_m \\ \sum_{m=0}^{M-1} t_m & M \end{bmatrix}. \tag{21}$$

Inverting $\mathbf{J}$ yields the CRLB for the estimation variance of the frequency in AWGN. In our proposed method, when the CAMP reconstruction algorithm is applied, the variance σ^2 becomes σ_*^2 and the CRLB is given by

$$\mathrm{CRLB}(\widehat{w}) = \frac{1}{A^2/\sigma_*^2} \frac{M/2}{M\sum_{m=0}^{M-1} t_m^2 - \left(\sum_{m=0}^{M-1} t_m\right)^2}. \tag{22}$$

Note that the variance σ_*^2 is larger than the variance σ^2, indicating that the CRLB of our proposed method is higher than that of traditional nonuniform sampling methods [17].

5. Numerical Simulations

In this section, the performance of proposed method is studied with the aid of numerical simulations. We commence the section by investigating the noise-free scenario, where the unambiguous Doppler frequencies are 345.482 Hz and 347.158 Hz. Since radar transmits a fixed-PRF (PRF = 100 Hz) pulse train. It is worth pointing out that the pulse train can be used in some LRRs for the velocity measurement or the normal waveform for tracking targets. M is set to 12, and the pulse transmitting time set t_k is obtained by randomly extracting M elements from the fixed-PRF transmitting time set with the perturbation added. The precision of pulse transmitting time is $\Delta t = 0.001$, and the corresponding dimension of basis matrix is $N = 1/\Delta t = 1000$. Doppler frequencies are determined corresponding to the peaks of frequency response. The estimated Doppler frequencies are 345 Hz and 347 Hz, closing to the true values 345.482 Hz and 347.158 Hz. In order to obtain higher estimate accuracy, redundancy sensing matrix can be refined by multiple. As shown in Figure 2, the Doppler frequencies are estimated as 345.3369 Hz and 347.3057 Hz with higher accuracy, when the refining factor q is set as 100, with SNR equal to 20 dB.

Since the tradeoff between success rate of reconstruction and calculation amount is determined by the value of the refining factor, the influence of refining factor for reconstruction performance is investigated as follows. Monte Carlo (MC) simulation results under different refining factors are depicted in Figure 3. The number of iterations is set to 1 and 1000 MC simulations are implemented. The success rate of reconstruction is related to the precision of true frequency Δf of the unknown target. As it is clear from Figure 3 as Δf increases the curves go up. Higher precision of the frequency of the target requires larger refining factor and *vice versa*. On the other hand, the significantly low refining factor, namely, the significantly large distance between adjacent columns in the DFT frame, introduces the spectrum leakage, as discussed in [10].

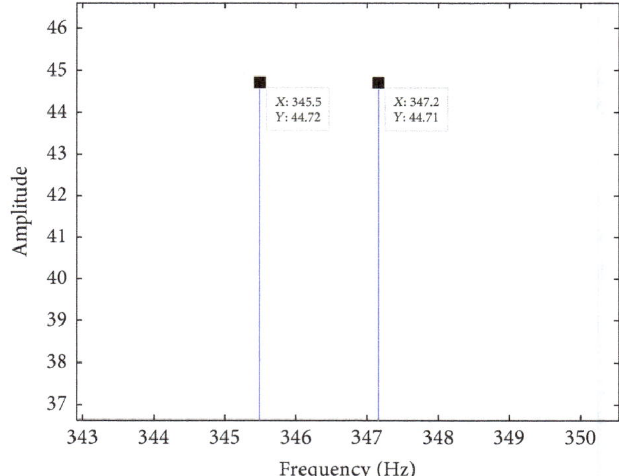

FIGURE 2: Spectrum of recovered slow-time echo signal.

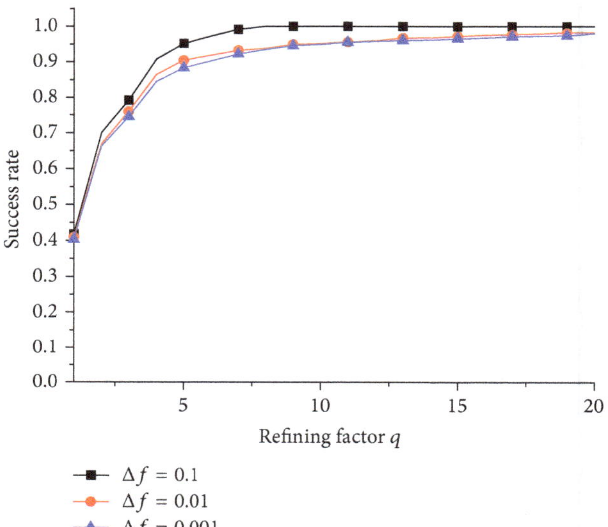

FIGURE 3: The success rate of reconstruction with different refining factor.

TABLE 1: Comparison of simulation times versus different refining factors.

Refining factor	10	100	500	1000
Time (s)	0.044040	0.417513	2.064184	3.909073

The success rate of reconstruction estimated at different SNR is shown in Figure 4, which demonstrates the performance with the presence of noise. Here the number of iterations is set to 31.

As shown in Figure 4, more than 95% trials succeed provided that the SNR is greater than 10 dB. Running times under different refining factors are recorded in seconds using the tic/toc functions of Matlab. The frequency span of the sensing matrix is set to 10 kHz. Simulation results shown in Table 1 indicate that the proposed method owns good real time response.

TABLE 2: Compared with CRT, Staggered PRF, and nonuniform sample.

Method	CS	CRT	Staggered frequency	Nonuniform sampling
Sample property	Randomly probing pulses	Relatively prime	Close to each other $\lvert f_{s1} - f_{s2}\rvert \ll f_{s1}, f_{s2}$	$f_s \pm \Delta f_i$ $(\Delta f_i \ll f_s)$
Pulse number	$M \ll f_s$	$f_{s1} + f_{s2}$	$f_{s1} + f_{s2}$	f_s
Unambiguous range	$q \cdot f_s$	$f_{s1} \cdot f_{s2}$	$\lvert f_{s1} \cdot f_{s2} / (f_{s1} - f_{s2})\rvert$	f_s

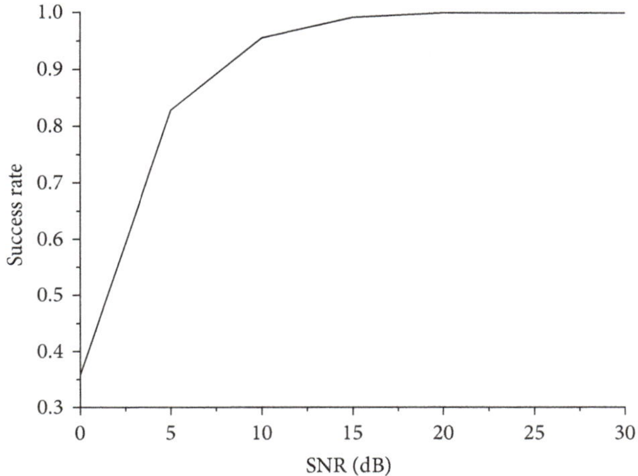

FIGURE 4: The success rate of reconstruction at different SNR.

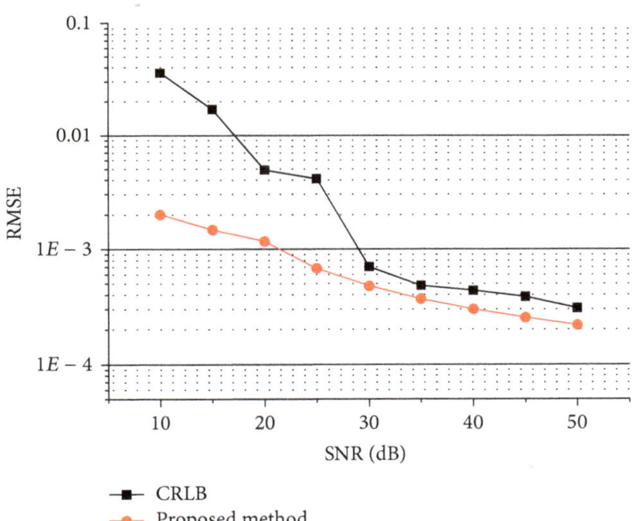

FIGURE 5: Performance of the proposed method.

The comparison between the Root Mean Square Error (RMSE) and the CRLB for frequency estimation of a sinusoid with the presence of noise is investigated, as shown in Figure 5. The refining factor q is set to 1000 and the number of samplings M is set to 12. The simulation is repeated 1000 times. It is implied from Figure 5 that the Doppler frequency can be obtained from recovered slow-time domain echo signal. The curve of RMSE approaches that of the CRLB due to the error caused by reconstruction which decreases slowly

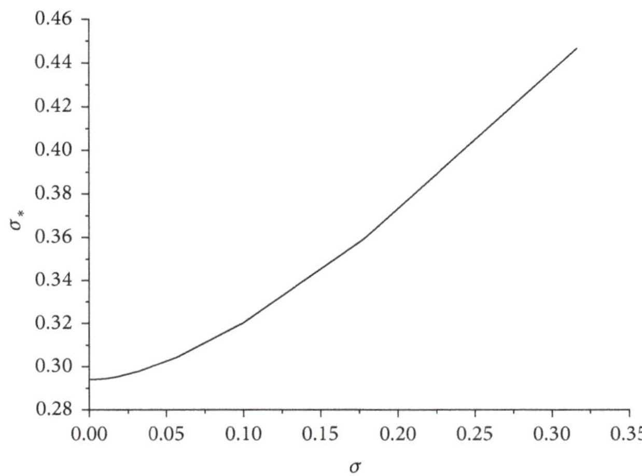

FIGURE 6: The relationship between σ and σ_*. The signal amplitude is 1 and the corresponding SNR is set from 10 dB to 70 dB.

as the SNR increases; thus the SNR defined in (17) increases slowly. Figure 6 demonstrates the relation between σ and σ_*, and it is observed that the deviation σ_* converges to 0.2942 when σ decreases to 0.

In the following, our proposed method is compared with traditional methods, including CRT, Staggered PRF, and nonuniform sample [18], and the comparison results are presented in Table 2. Seen from Table 2, our proposed method transmits the least transmitting pulses. Furthermore, the Doppler frequencies are obtained from recovered Doppler spectrum via CAMP in our proposed method, while unambiguous Doppler frequencies are obtained from ambiguous Doppler measurements in those traditional algorithms.

6. Conclusion

Relied on sparse property of radar echo and CS theory, a novel Doppler ambiguous resolution utilizing sparse probing pulses is proposed. The whole slow-time domain echo signals can be recovered via the adaptive CAMP reconstruction algorithm. By refining the basis matrix, the equivalent high PRF echoes can be obtained and therefore, the Doppler unambiguous range can be enlarged. Using the SE formalism, we have derived the CRLB of the proposed method. The accuracy of estimation is demonstrated by theoretical inference and simulations. Compared with traditional methods, our novel method only transmits a few sparse probing pulses and reduces more than 80% of the time resource consumption of radar. Furthermore, due to the fact that the pulse transmitting time can be controlled by the radar clock, which generally has a frequency of 10 MHz or integral multiple of 10 MHz, the precision of pulse transmitting time adjustment needed in the paper is achievable. Therefore, our proposed method is adequate to be implemented in practical, since only slight adjustment is required on the radar dwell scheduling.

Conflict of Interests

The authors declare that there is no conflict of interests regarding the publication of this paper.

Acknowledgments

The research was supported by the National High-tech R&D Program of China and the Open-End Fund National Laboratory of Automatic Target Recognition (ATR). The authors would like to thank Laura Anitori and Zai Yang for sharing the CAMP Matlab package.

References

[1] A. Ferrari, C. Bérenguer, and G. Alengrin, "Doppler ambiguity resolution using multiple PRF," *IEEE Transactions on Aerospace and Electronic Systems*, vol. 33, no. 3, pp. 738–751, 1997.

[2] X. Li, H. Liang, and X.-G. Xia, "A robust Chinese remainder theorem with its applications in frequency estimation from undersampled waveforms," *IEEE Transactions on Signal Processing*, vol. 57, no. 11, pp. 4314–4322, 2009.

[3] W.-J. Wang and X.-G. Xia, "A closed-form robust Chinese remainder theorem and its performance analysis," *IEEE Transactions on Signal Processing*, vol. 58, no. 11, pp. 5655–5666, 2010.

[4] A. Maroosi and H. K. Bizaki, "Digital frequency determination of real waveforms based on multiple sensors with low sampling rates," *IEEE Sensors Journal*, vol. 12, no. 5, pp. 1483–1495, 2012.

[5] T. Liu and Y.-H. Gong, "Staggered RPF analysis and design," *Journal of University of Electronic Science and Technology of China*, vol. 38, no. 2, pp. 97–101, 2009.

[6] Y.-X. Zhang, J.-P. Sun, B.-C. Zhang, and W. Hong, "Doppler ambiguity resolution based on compressive sensing theory," *Journal of Electronics and Information Technology*, vol. 33, no. 9, pp. 2103–2107, 2011 (Chinese).

[7] Y. H. Quan, L. Zhang, M. D. Xing, and Z. Bao, "Velocity ambiguity resolving for moving target indication by compressed sensing," *Electronics Letters*, vol. 47, no. 22, pp. 1249–1251, 2011.

[8] S. S. Chen, D. L. Donoho, and M. A. Saunders, "Atomic decomposition by basis pursuit," *SIAM Journal on Scientific Computing*, vol. 20, no. 1, pp. 33–61, 1998.

[9] Y. C. Pati, R. Rezaiifar, and P. S. Krishnaprasad, "Orthogonal matching pursuit: recursive function approximation with applications to wavelet decomposition," in *Proceedings of the 27th Asilomar Conference on Signals, Systems & Computers*, pp. 40–44, November 1993.

[10] M. F. Duarte and R. G. Baraniuk, "Spectral compressive sensing," *Applied and Computational Harmonic Analysis*, vol. 35, no. 1, pp. 111–129, 2013.

[11] A. Maleki, L. Anitori, Z. Yang, and R. G. Baraniuk, "Asymptotic analysis of complex LASSO via complex approximate message passing (CAMP)," *IEEE Transactions on Information Theory*, vol. 59, no. 7, pp. 4290–4308, 2013.

[12] D. L. Donoho, A. Maleki, and A. Montanari, "Message-passing algorithms for compressed sensing," *Proceedings of the National Academy of Sciences of the United States of America*, vol. 106, no. 45, pp. 18914–18919, 2009.

[13] L. Anitori, *Compressive sensing and fast simulations: applications to radar detection [Dissertation]*, Delft University of Technology, Delft, The Netherlands, 2012.

[14] R. Tibshirani, "Regression shrinkage and selection via the lasso," *Journal of the Royal Statistical Society, Series B: Methodological*, vol. 58, no. 1, pp. 267–288, 1996.

[15] A. Maleki and D. L. Donoho, "Optimally tuned iterative reconstruction algorithms for compressed sensing," *IEEE Journal on Selected Topics in Signal Processing*, vol. 4, no. 2, pp. 330–341, 2010.

[16] L. Hu, Z. Shi, J. Zhou, and Q. Fu, "Compressed sensing of complex sinusoids: an approach based on dictionary refinement," *IEEE Transactions on Signal Processing*, vol. 60, no. 7, pp. 3809–3822, 2012.

[17] J. A. Gansman, J. V. Krogmeier, and M. P. Fitz, "Single frequency estimation with non-uniform sampling," in *Proceedings of the 30th Asilomar Conference on Signals, Systems and Computers*, vol. 1, pp. 399–403, IEEE, Pacific Grove, Calif, USA, November 1996.

[18] S.-W. Park, W.-D. Hao, and C. S. Leung, "Reconstruction of uniformly sampled sequence from nonuniformly sampled transient sequence using symmetric extension," *IEEE Transactions on Signal Processing*, vol. 60, no. 3, pp. 1498–1501, 2012.

7

A Novel Denoising Algorithm of Electromagnetic Ultrasonic Detection Signal based on Improved EEMD Method

Wenkang Gong, Qi Liu, Wenhao Du, Weichen Xu, and Gang Wang

School of Information Science and Engineering, Northeastern University, Shenyang, Liaoning 110003, China

Correspondence should be addressed to Wenkang Gong; 906833158@qq.com

Academic Editor: Sos S. Agaian

In this paper, we propose a new denoising algorithm for electromagnetic ultrasonic signals based on the improved EEMD method, which can adaptively adjust for added noise and average times in different noisy environments, so that the effect of the residual difference of white noise on the results can be eliminated as far as possible. First, the way to add white noise in the EEMD method is processed, and then the permutation entropy algorithm is used to identify the nature of the components obtained during the decomposition. Then the wavelet transform modulus maximum denoising method is used to deal with the IMF components of the high-frequency part obtained before. Finally, the processed IMF results and residual difference are summed up. The results show that after processing, the noise component in the signal is less and the original information is more reserved, which prevents the signal distortion to a great extent and provides more effective data for subsequent processing. In the experiment, the crack defect data collected by the electromagnetic ultrasonic experiment system were processed by the improved EEMD method. Compared with the traditional EEMD method, it can retain the information of crack location more accurately, which proves the effectiveness of the proposed method.

1. Introduction

In recent years, the electromagnetic ultrasonic nondestructive testing technology for pipeline defect detection has been paid more and more attention. Compared with the traditional ultrasonic testing technology, it is simpler and more effective, having a variety of different detection modes. However, due to the influence of environment, human operation, and other factors, there are some singularities in the data collected by the receiving end of the electromagnetic ultrasonic transducer. What is more, it contains a certain degree of noise interference, so that it may cause great disturbance to the identification of signal position and feature in the later period.

Literature [1] proposed an improved denoising algorithm based on wavelet transform modulus maxima reconstruction. This method had a good approximation to the original wavelet transform coefficients of the signal. However, the wavelet denoising method was limited by both the time domain and the frequency domain, and it could not meet the analysis requirements of high resolution in the time domain and frequency domain. In document [2], authors utilized special symmetric matrices to construct the new nontensor product wavelet filter banks, which could capture the singularities in all directions. In document [3], authors proposed an image denoising method based on nonseparable wavelet filter banks and two-dimensional principal component analysis (2D-PCA). This method could achieve both good visual quality and a high peak signal-to-noise ratio for the denoised images. In document [4], the permutation entropy was introduced into the threshold function as the representation parameter of signal denoising, and the permutation entropy of the wavelet packet coefficients of the signal was calculated. Literature [5] proposed a new EEMD (Ensemble Empirical Model Decomposition) harmonic detection method based on new wavelet threshold denoising preprocessing to effectively eliminate the effect of random noise on harmonic detection. However, the EEMD denoising method was to reject the high-frequency partial components directly, which would result in the loss of valid information in the high-frequency components.

The core of the improved algorithm in this paper is to adaptively adjust the added noise and the average times under different noise environments, so that the effect of the residual difference of white noise on the results can be eliminated as far as possible.

In this paper, we first introduce the conception of modulus maxima denoising method based on wavelet transform and EEMD denoising method. Then, the differential threshold method is used to remove the singularities in the data. Next, we make innovative improvements to the EEMD algorithm so that it can adaptively get the ratio coefficient, and the useless residual difference of the added white noise can be reduced to the maximum extent at the same time. Aiming at the added white noise in the EEMD method, we use the permutation entropy algorithm to identify the nature of the components obtained during the decomposition.

For the remaining signal of the low-frequency stationary part, the EMD (Empirical Mode Decomposition) is directly used in the processing, while the other high-frequency IMF (Intrinsic Mode Function) components are continuously obtained by the EEMD decomposition, thereby reducing the influence of noise on the effective part.

Afterwards, the wavelet transform modulus maximum denoising method is used to deal with the IMF components of the high-frequency part obtained before. Finally, the processed IMF components and residual difference are summed up. The results show that after processing, the noise component in the signal is less and the original information is more reserved, which can prevent the signal distortion to a great extent and provide more effective data for subsequent processing.

2. Preliminaries

2.1. Preparation. There are many traditional discrete data denoising methods. At the following, the applicable characteristics of various methods will be combined to explain the relevant knowledge involved in this article algorithm. And then the method proposed in this paper is applied to the processing of electromagnetic ultrasonic nondestructive testing signal. By comparison, the advantages of this article algorithm are highlighted.

In order to compare the denoising effects of several methods, we used the ETG-100 ultrasonic thickness gauge to test three steel plates of the same material as X56 in the laboratory environment. Their length is 50 cm, the width is 30 cm, and the thickness is 12.37 mm, 13.35 mm, and 15.21 mm. Three sets of clean thickness echo signals are obtained. Noise is added to the first set of signals, as shown in Figure 1.

2.2. Introduction to EMD Method. EMD is suitable for the analysis of nonlinear, unsteady signals. The core of the method is to decompose the more complex signals and get the IMF components of the signals. The IMF components obtained by this method represent the characteristics of the data series at different time scales, respectively. In this way, the fluctuation trend of the signal under different scales of the original signal can be decomposed and refined and then analyzed.

For the IMF, Huang et al., had given the qualified conditions [6]:

(1) In the whole data set, the number of extrema and zero-crossings must either be equal or differ at most by one

(2) The average value of the envelope of the maximum and the minimum value of a data sequence is zero

For the signal sequence X_i that needs to be processed, the interpolation function is, respectively, used to obtain the envelopes of the maximum $X_{\max}$ and the minimum point $X_{\min}$ of the signal. Then the average of two envelopes is obtained:

$$X_{\text{mid}} = \frac{(X_{\min} + X_{\max})}{2}. \tag{1}$$

After using the original signal and the average line signal to deal with the difference component, we have

$$H_1 = X_i - X_{\text{mid}}. \tag{2}$$

After getting the component, we first judge whether it is IMF. According to the two principles mentioned above, if the conditions are met, we define $C_{1i} = H_1$. Otherwise, the original sequence is replaced with H_1. The above operation is repeated until a satisfactory sequence function H_{1-k} is obtained, denoted as C_{1i}. The calculation will be stopped, when the following cutoff condition is met:

$$\text{sd} = \sum_{i=0}^{T} \frac{\left|H_{(1-k)i} - H_{(1-(k-1))i}\right|}{H^2_{(1-(k-1))i}}, \tag{3}$$

where sd is generally based on experience value of 0.2–0.3. When meeting the above requirements, we note $C_{1i} = H_{1-k}$.

The C_{1i} is stripped from the original sequence, X_i, and repeated for the rest of the sequence to obtain C_{2i}, and so on, to obtain C_{ji}. The residual sequence obtained by separating all the eigenfunctions from the original data sequence is defined as R_i. Overall expressed as

$$X_i = \sum_{j=1}^{k} C_{ji} + R_i. \tag{4}$$

For the EMD method, it is difficult to ensure that the local mean value limited by condition (2) is equal to zero during the screening process because of the complexity of the electrical signals collected by electromagnetic ultrasound. When the signal is abnormal, it will affect the signal envelope, and the IMF component, resulting in model aliasing, which may lead to the loss of the original physical meaning of the component.

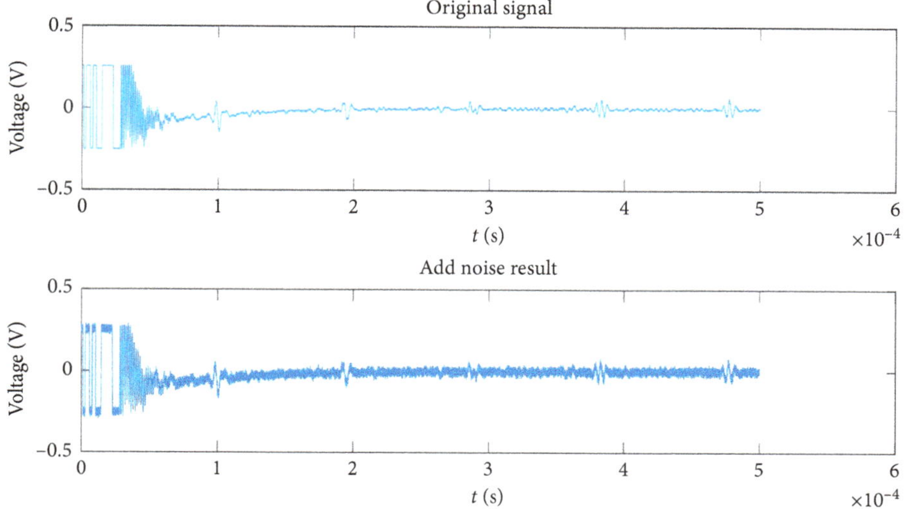

Figure 1: Adding noise to the original signal.

2.3. EEMD Denoising Method. The improvement proposed by Norden. E. Huang to the EMD method in solving the problem of model aliasing is called EEMD [7]. The steps of the EEMD method are as follows:

(1) Add a white noise sequence in the original signal.

(2) Get IMF components with EMD decomposition method.

(3) Repeat the above two steps, and the added white noise is different each time. When a signal is applied to a uniformly distributed white noise background, the signal regions at different scales are automatically mapped to the appropriate scales associated with the background white noise. The decomposed IMF components are shown in Figure 2.

(4) Integrate and average the IMF components obtained each time. Since the noise is different in each individual test, the noise will be removed when the overall mean at a sufficient number of tests is used. After that, the overall mean will eventually be considered as the true result. With more and more repetitions of the above steps, additional noise can be eliminated, and the only permanent part is the signal itself. The general EEMD decomposition flowchart is shown in Figure 3.

The traditional EEMD algorithm is based on the principle of noise-assisted signal processing; the mode aliasing phenomenon is effectively solved by adding a small amplitude of white noise to equalize the signal. The real signal is retained by using the zero-mean characteristic of Gaussian white noise, which is a great improvement to the traditional EMD analysis method.

But the disadvantage of the traditional EEMD method is that the added white noise can not be completely offset from each other in practical application, so the signal is still affected by noise to a certain extent. In the decomposed component, the high-frequency part contains a lot of noise, which is usually removed directly, and then the signal with a large correlation is reconstructed to get the denoised signal.

Because the high-frequency IMF component which is removed directly contains effective information, it will affect the original signal to some extent. In addition, the added white noise and the number of processing have a greater impact on the decomposition results, so that mode aliasing cannot be completely eliminated and may produce more useless components. Therefore, EEMD cannot adjust these decomposition parameters according to the actual situation, especially when the noise is changeable.

3. Abnormal Data Removed

Before the postprocessing of the data, some "damage data" of the data collected by electromagnetic ultrasound needs to be checked and removed. In this paper, the differential threshold method is used to distinguish the numerical changes between the sampling points of the collected data. When the absolute value of the difference between two adjacent points is greater than the set threshold, it is regarded as the wrong data and will be replaced. The principle of differential threshold method is as follows:

$$X_i = \begin{cases} \dfrac{x_{i-1} + x_{i+1}}{2}, & |x_i - x_{i-1}| > T, \\ x_i, & |x_i - x_{i-1}| < T, \end{cases} \quad (5)$$

where x_i is discrete data obtained after normalization, X_i is data obtained through algorithm detection and processing, and T is the selected differential threshold and is the maximum value of the difference between adjacent sampling points in the ideal signal.

When the data difference between two adjacent points is less than the selected threshold, the original data will

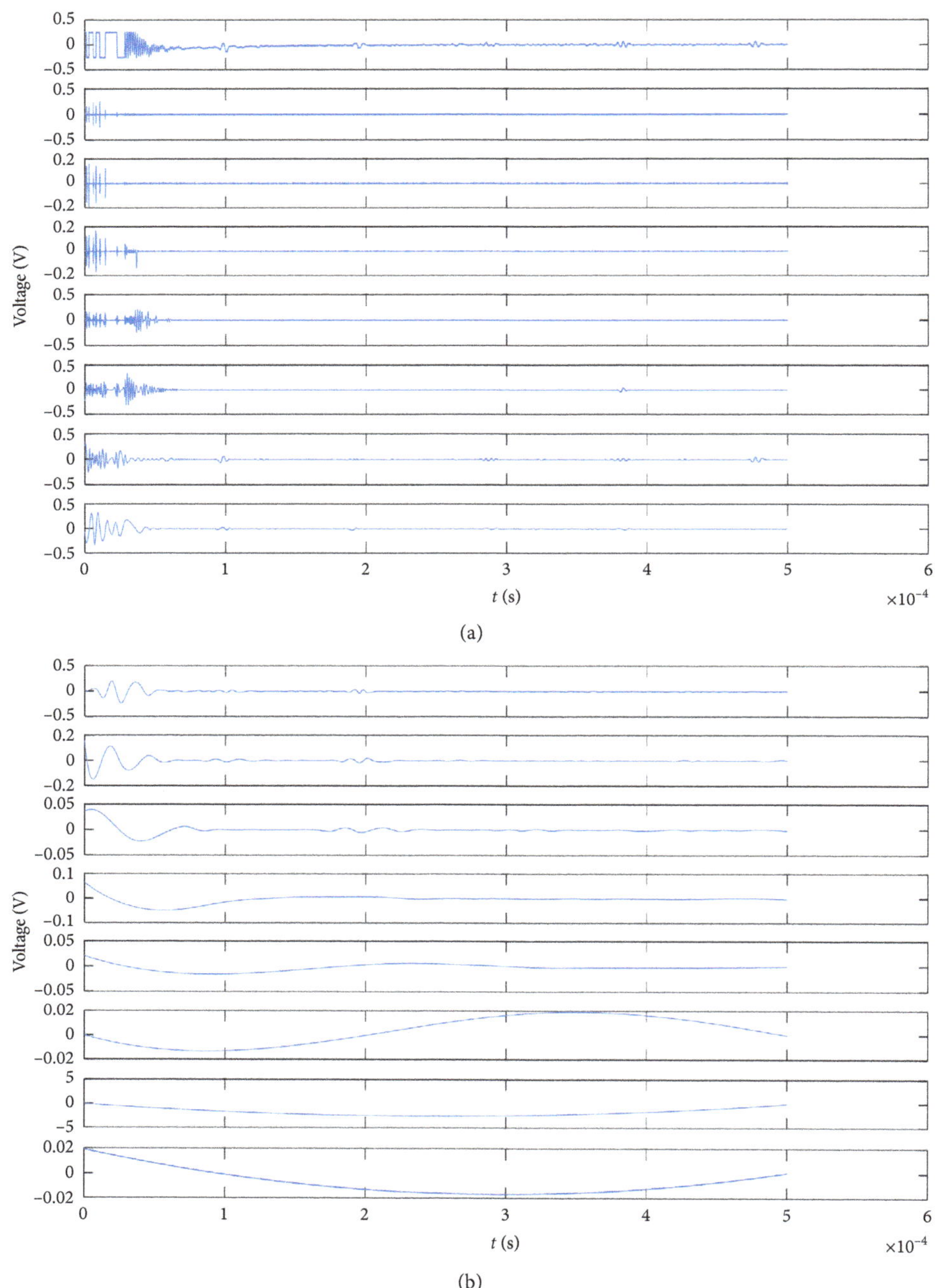

FIGURE 2: Decomposition results of EEMD method. (a) IMF 1–8. (b) IMF 9–16.

continue to be used. For the mutation data, the data of the two sample points before and after the mutation sample data can be used to supplement.

4. The Improved EEMD Method

In view of the lack of traditional EEMD in electromagnetic ultrasonic testing data processing, for the first time, a new denoising algorithm based on the improved EEMD method, which can adaptively adjust for added noise and integrated average times in different noisy environments, is proposed in this paper.

4.1. Permutation Entropy Algorithm. Permutation entropy [8] is an algorithm used to describe the complexity of time series signals. The algorithm is simple and efficient. What is more, it can be used to analyze the correlation of nonlinear and nonstationary complex signals. In this paper, it is used to identify the properties of the components obtained during the decomposition.

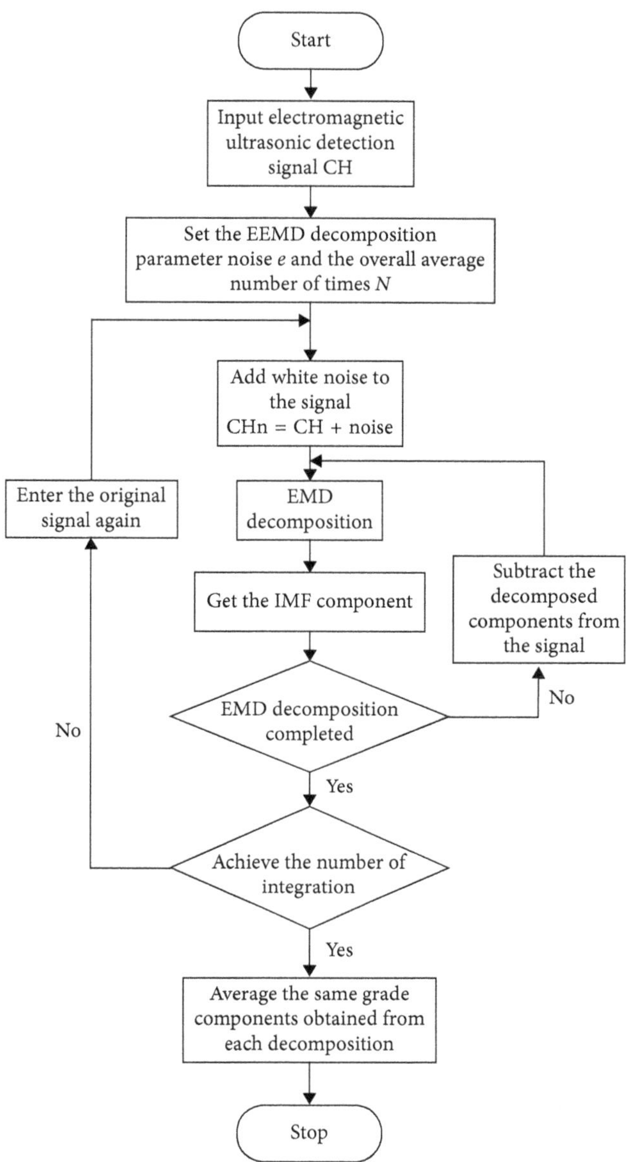

FIGURE 3: Processing flow of EEMD.

For a time series $S(i)$, the signal is first reconstructed to obtain the following:

$$\begin{bmatrix} s(1) & s(1+\tau) & \cdots & s(1+(n-1)\tau) \\ s(2) & s(2+\tau) & \cdots & s(2+(n-1)\tau) \\ \vdots & \vdots & \ddots & \vdots \\ s(N) & s(N+\tau) & \cdots & s(N+(n-1)\tau) \end{bmatrix}, \quad (6)$$

where τ is the delay time, n is the number of dimensions embedded, and N is the difference between the signal length and $(n-1)$.

The reconstructed data are sorted in the ascending order of magnitude. Then the position index of each element in the reconstruction component is labeled as $[j_1, j_2, \ldots, j_n]$, respectively. By this way, we get N sets of different labels such as $[j_1, j_2, \ldots, j_n]_{1,\ldots,N}$. According to the embedding dimension, the symbol sequence has a total of $m!$ kinds. The probability that we get each sequence of sequence numbers is $P_1, \ldots, P_N$. The form as defined by Shannon entropy is shown in Equation (7), and the normalized method is shown in Equation (8), so that the value of entropy is between 0 and 1:

$$H_p(m) = -\sum_{g=1}^{N} P_g \ln P_g, \quad (7)$$

$$H_p = \frac{H_p(m)}{\ln(m!)}. \quad (8)$$

After that, we process the simulation signal for the combined sequence of noise and related sinusoidal signals. According to relevant research experience, we set the parameters $\tau = 1$ and $n = 5$. The signal sequence s is as follows:

(i) $s(1)$: white noise with a signal sequence length of 98000 (with late detection data length)
(ii) $s(2)$: Gaussian random noise and random signal with signal sequence length of 98000
(iii) $s(3)$: mixing of random noise and white noise with a signal sequence length of 98000
(iv) $s(4)$: $\sin(2\pi \cdot 500 \cdot t)$, $t = 0 : 1/97999 : 1$
(v) $s(5)$: $\sin(2\pi \cdot 10 \cdot t)\sin(2\pi \cdot 100 \cdot t)$, $t = 0 : 1/97999 : 1$
(vi) $s(6)$: $[1 + \sin(2\pi \cdot 5 \cdot t)]\sin(2\pi \cdot 50 \cdot t^2 + 2\pi \cdot 10 \cdot t)$, $t = 0 : 1/97999 : 1$

The sequence is processed to obtain the permutation entropy values, which are 0.9897, 0.9722, 0.9815, 0.2443, 0.1159, and 0.2105. We can see that the entropy of the noise is large and irregular, while the entropy of the sinusoidal composite signal is low. We can set a threshold value of 0.58 to provide the parameter support for the improved follow-up study of the EEMD algorithm mentioned below.

4.2. Wavelet Transform Modulus Maxima. Wavelet transform is used to decompose the original signal into high-frequency part and low-frequency part. The low-frequency parameters are retained while the high-frequency part is decomposed again, followed by progress [9].

The modulo-maximum method is a typical method in the wavelet denoising method. Wavelet coefficients can reflect the transient characteristics of the original signal at different scales. Modular-maximum denoising based on wavelet transform is to process the modulus maxima of wavelet decomposition coefficients. Since the modulus maximum point of the signal will increase with the expansion of the scale, the noise maximum modulus point will be opposite, and the signal will be reconstructed from the modulus maxima at different scales by the processed wavelet coefficient, which is the basic idea of WTMM (wavelet transform modulus maxima).

Because of the complexity of the signal processing, the extreme point of the wavelet decomposition coefficient usually corresponds to the abrupt point of the signal, and the singularity of the signal corresponds to the variation rule of the modulus of the wavelet coefficients. Therefore, the paper incorporates the WTMM method into the EEMD denoising algorithm, and a new improved EEMD denoising algorithm is proposed. The following describes the specific implementation process.

4.3. Parameter Selection Criteria. The improved method first determines the principle of adding noise and the average number of times. Different from the traditional empirical judgment, through a considerable number of experimental studies, the specification of adding white noise in the EEMD method has been derived:

$$0 \le \beta \le \frac{\rho}{2}, \tag{9}$$

where β is the ratio of the standard deviation σ_{noise} of artificially added white noise to the standard deviation σ_{ch} of the original signal, and ρ is the ratio of the standard deviation σ_h of the high-frequency component of the signal to the standard deviation σ_{ch} of the original signal.

So, Equation (9) can be equivalent to Equation (10):

$$0 \le \sigma_{\text{noise}} \le \frac{\sigma_h}{2}. \tag{10}$$

In normal conditions, we choose $\sigma_{\text{noise}} = \sigma_h/4$, that $\beta = \rho/4$.

Another important parameter is the average number of times. Empirical studies have shown that the formula is chosen as shown in

$$e = \frac{\beta}{\sqrt{N}}, \tag{11}$$

where N is the average times, and e is the relative decomposition error, the general value is 1%.

According to the above formula, the average number of integration N is obtained as shown in the following equation:

$$N = \left(\frac{\beta}{e}\right)^2 = \left(\frac{\sigma_{\text{noise}}}{e \times \sigma_{\text{ch}}}\right)^2 = \left(\frac{\sigma_h}{4\sigma_{\text{ch}} \times e}\right)^2. \tag{12}$$

Obviously, the standard deviation of artificially added white noise and the average number of integration are all related to the ratio coefficient β.

4.4. Denoising with the Improved EEMD Method. The core of the improved algorithm is to adaptively adjust the added noise and the average times under different noise environments, so that the effect of the residual difference of white noise on the results can be eliminated as far as possible. Research shows that the white noise added to the high-frequency part of the EEMD method has negligible influence on the mode aliasing, while the white noise added to the low-frequency part has a greater influence factor, so the low-frequency part is directly decomposed by the EMD method to eliminate the influence of mode aliasing.

Firstly, we make improvements to the EEMD algorithm so that it can adaptively get the ratio coefficient. At the same time, it can minimize the useless residual difference caused by added white noise in the result. Secondly, aiming at the way to add white noise to the EEMD method, we use the permutation entropy algorithm to identify the nature of the components obtained during the decomposition. As for the remaining signal of the low-frequency stationary part, the EMD decomposition is directly used in the processing, while the others are continuously obtained by the EEMD decomposition, thereby reducing the influence of noise on the effective part. Afterwards, the wavelet transform modulus maximum denoising method is used to deal with the IMF components of the high-frequency part obtained before. Finally, the processed IMF results and residual difference are summed up. The results show that after processing, the noise content in the signal is less and the original information is more reserved, which prevents the signal distortion to a great extent and provides effective data for subsequent processing.

The flowchart of improvement is shown in Figure 4. The specific steps are as follows:

(1) First of all, the original signal is decomposed by the EMD method to obtain the eigenfunction group, and the first set of high-frequency IMF components in the decomposition result is recorded as the high-frequency component of the original signal. Then, the ratio of the standard deviation of the original signal and the high-frequency IMF components, combined with the previous formula, is used to find the value of β. In this way, we can find out the deviation standard of artificially added noise in the EEMD decomposition operation and the average integral times of this state.

(2) Firstly, the original signal is decomposed into IMF components by EEMD algorithm, and then, all IMF components are sequentially calculated by using the permutation entropy algorithm. If the entropy value is greater than the set threshold, the next component will continue to be calculated.

(3) The WTMM method is used to denoise the components whose entropy values are greater than the threshold. Then, the IMF components and the residuals of the high-frequency part can be obtained.

(4) These high-frequency components whose entropy values are greater than the threshold are removed from the original signal, the remaining part of the signal is decomposed by the EMD method to get the low-frequency partial IMF components.

(5) By summing the IMF components and the residuals obtained in the above two steps, the processed result is as shown in the following equation:

$$X(t) = \sum_{i=1}^{m} \mathrm{IMF}_i + p_i. \tag{13}$$

After calculation, the 2–5 in the IMF component diagram needs to be processed. Using the modulus maxima denoising method based on the wavelet transform, the IMF component image after processing is shown in Figure 5.

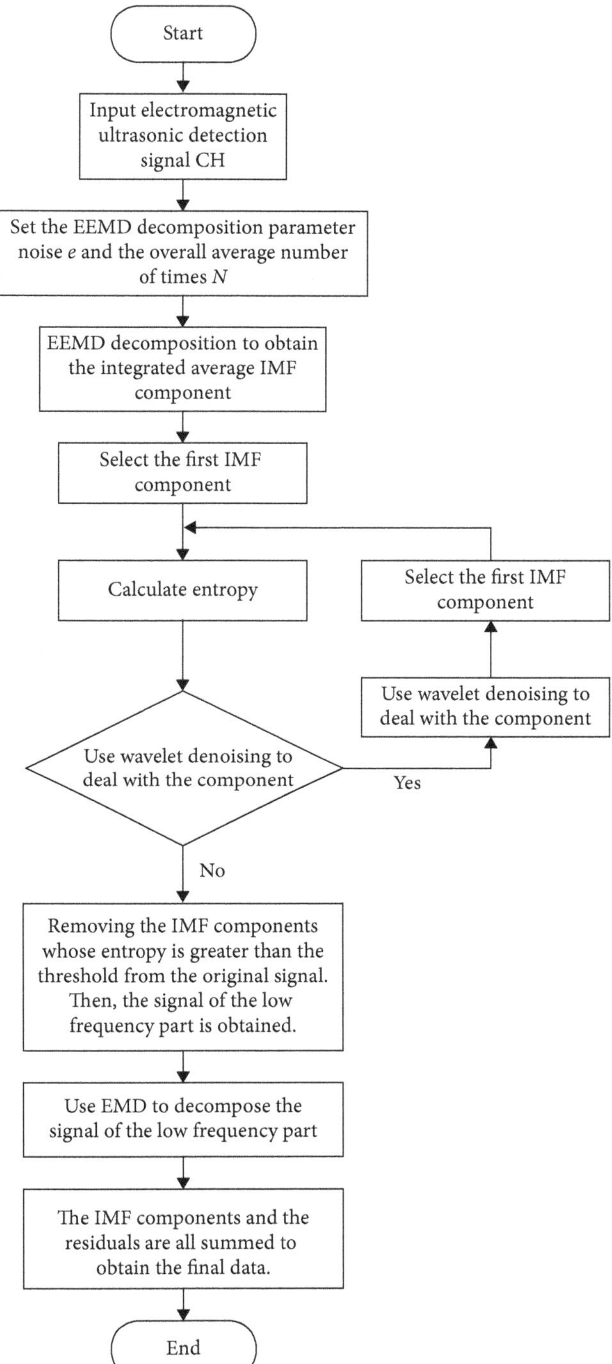

Figure 4: Improved method to handle flowchart.

4.5. Methods Comparison. In order to prove that the improved method is superior to the traditional EEMD method in the processing of electromagnetic ultrasonic detection signals and to verify that it has sufficient stability, the corresponding experiments have been carried out through simulation.

We used the ETG-100 ultrasonic thickness gauge to test three steel plates of the same material as X56 in the laboratory environment. Their length is 50 cm, the width is 30 cm, and the thickness is 12.37 mm, 13.35 mm, and 15.21 mm. Three sets of clean thickness echo signals are obtained.

In the first set of simulation experiments, the original signal is the clear thickness measurement data used in the previous section. The original signal is artificially added with noise and then denoised by the traditional EEMD method and the improved EEMD method, respectively. The comparison chart of denoising effect is shown in Figure 6.

The SNR (signal-noise ratio) of the original signal, the traditional EEMD method, and the improved EEMD method is calculated, which are shown in Table 1.

In the second and third sets of simulation experiments, the original signal uses different clean thickness echo data

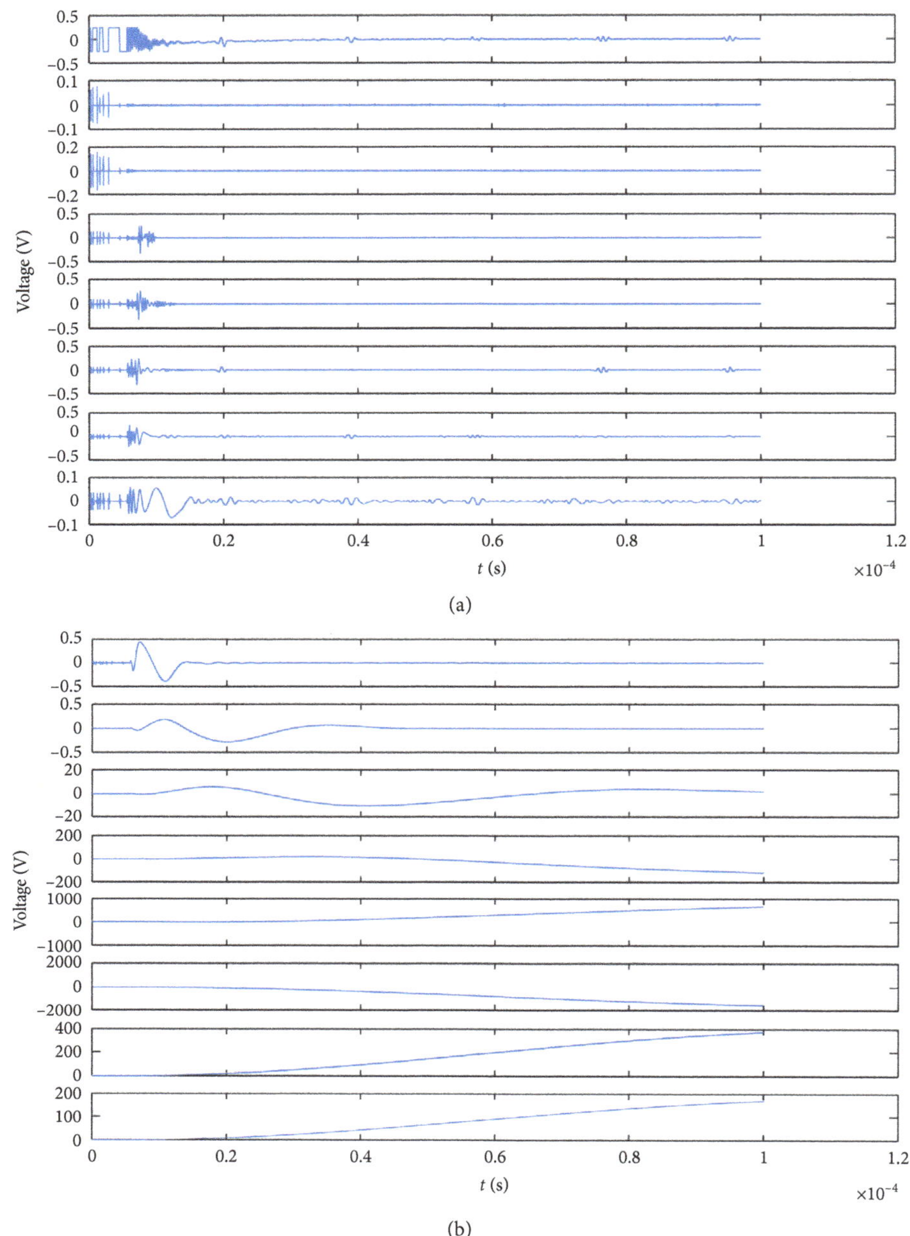

FIGURE 5: IMFs obtained by the improved method. (a) IMF 1–8. (b) IMF 9–16.

which are also artificially added with noise later, and the original signal is processed by the traditional EEMD method and the improved EEMD method, respectively. The SNR of the original signal, the traditional, and the improved EEMD method is presented below.

As can be seen from Tables 1 and 2, the SNR obtained by the improved method is closer to the original SNR. The improved method can not only solve the shortcomings of the EEMD method, but also get closer to the original data, so that the processed signal can better maintain the characteristics of the original signal. Therefore, the improved method is superior to the traditional EEMD method in the processing of electromagnetic ultrasonic detection signals, and it has sufficient stability.

5. Experiment

In order to verify the validity of the improved method, we used the EMAT2000 electromagnetic ultrasonic crack detector on Central Offshore Oil Pipeline Test Platform to test the crack defects at Tanggu, Tianjin. The actual metal spline crack depth of the pipe wall was 0.5 mm. The echo signals obtained from electromagnetic ultrasonic testing are processed by the traditional method and improved method, respectively. Then, the denoising effects of the two methods are compared.

The collected signal data are shown in Figure 7. Obviously, the collected data contain noise, so that it is also necessary to remove singular values and denoising.

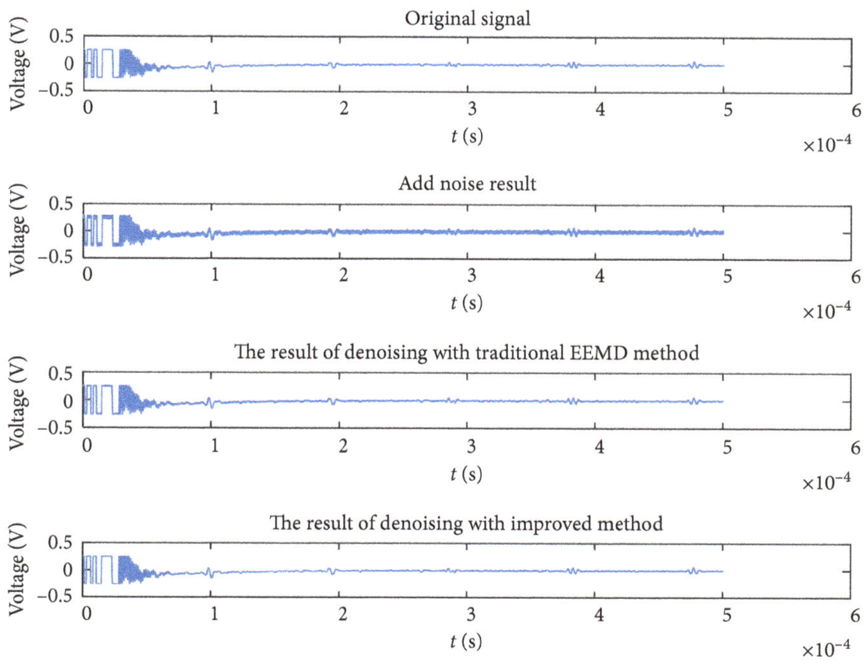

Figure 6: Denoising effect comparison chart.

Table 1: Denoising effect comparison.

Method	The original signal-to-noise ratio	The traditional EEMD method	The improved EEMD method
SNR	11.53	10.12	11.15

Table 2: The other two sets of denoising effects comparison.

Method	The original signal-to-noise ratio	The traditional EEMD method	The improved EEMD method
SNR of the second set of experiments	11.78	11.23	11.59
SNR of the third set of experiments	10.45	9.68	10.23

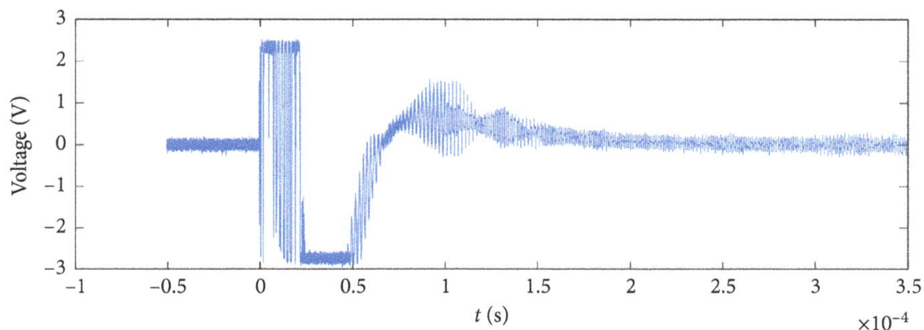

Figure 7: The original electrical signal of crack defects detected.

For the data collected from the crack defect, after eliminating the singular value of the data, the IMF components obtained by the traditional EEMD method are shown in Figure 8, and the IMF components obtained by the improved EEMD method are shown in Figure 9.

Finally, the IMF components and the residual difference are reassembled, and the comparison results of the traditional EEMD method and the improved EEMD method are shown in Figure 10.

As can be seen from Figure 10, the data obtained by the improved EEMD method are cleaner than those obtained by the traditional EEMD method. The valid region is partially enlarged to get Figure 11, from which the resulting echo signal from the EMAT (Electromagnetic Acoustic Transducer)

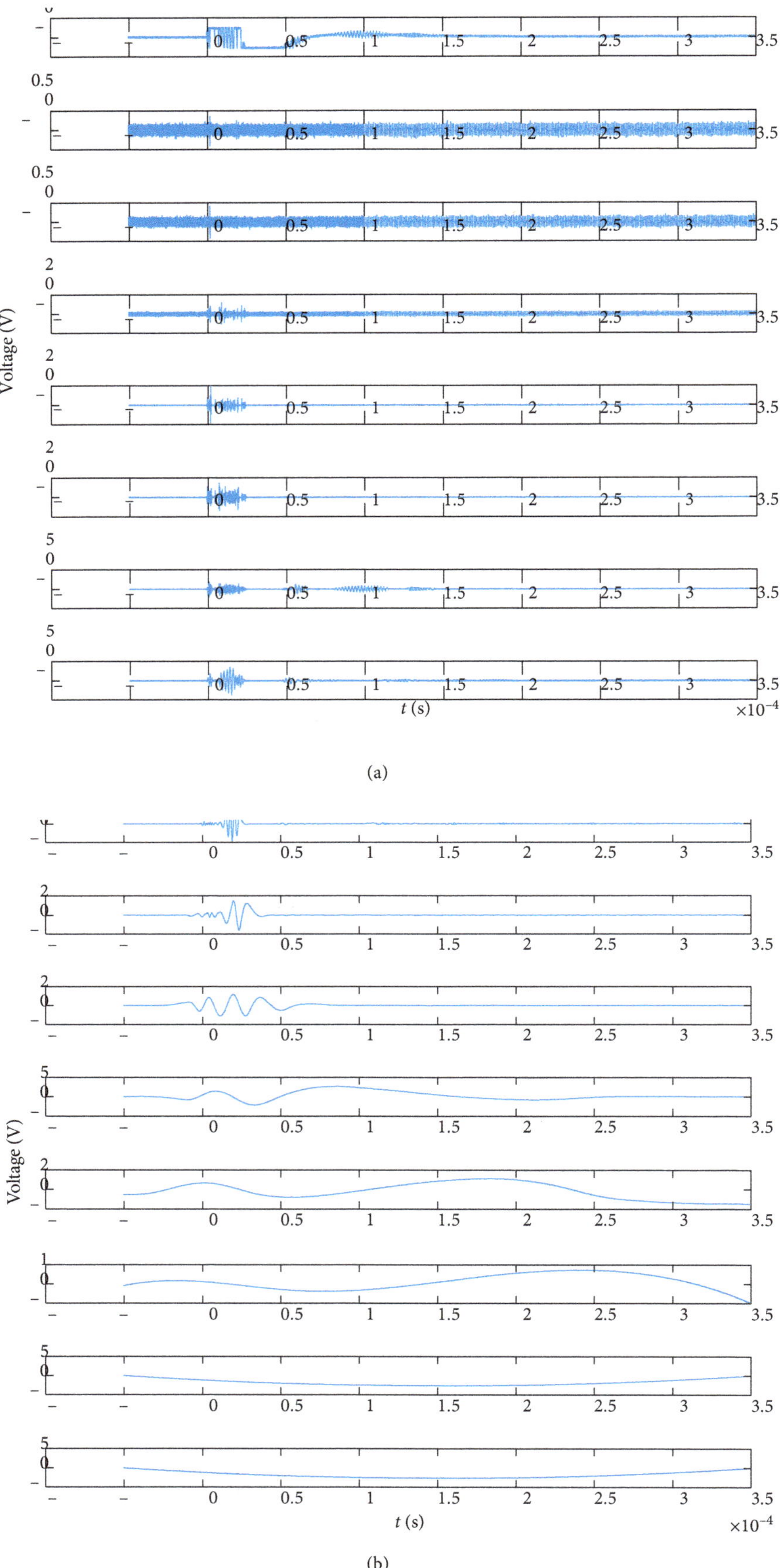

FIGURE 8: The IMF components obtained by the traditional EEMD method. (a) IMF 1–8. (b) IMF 9–16.

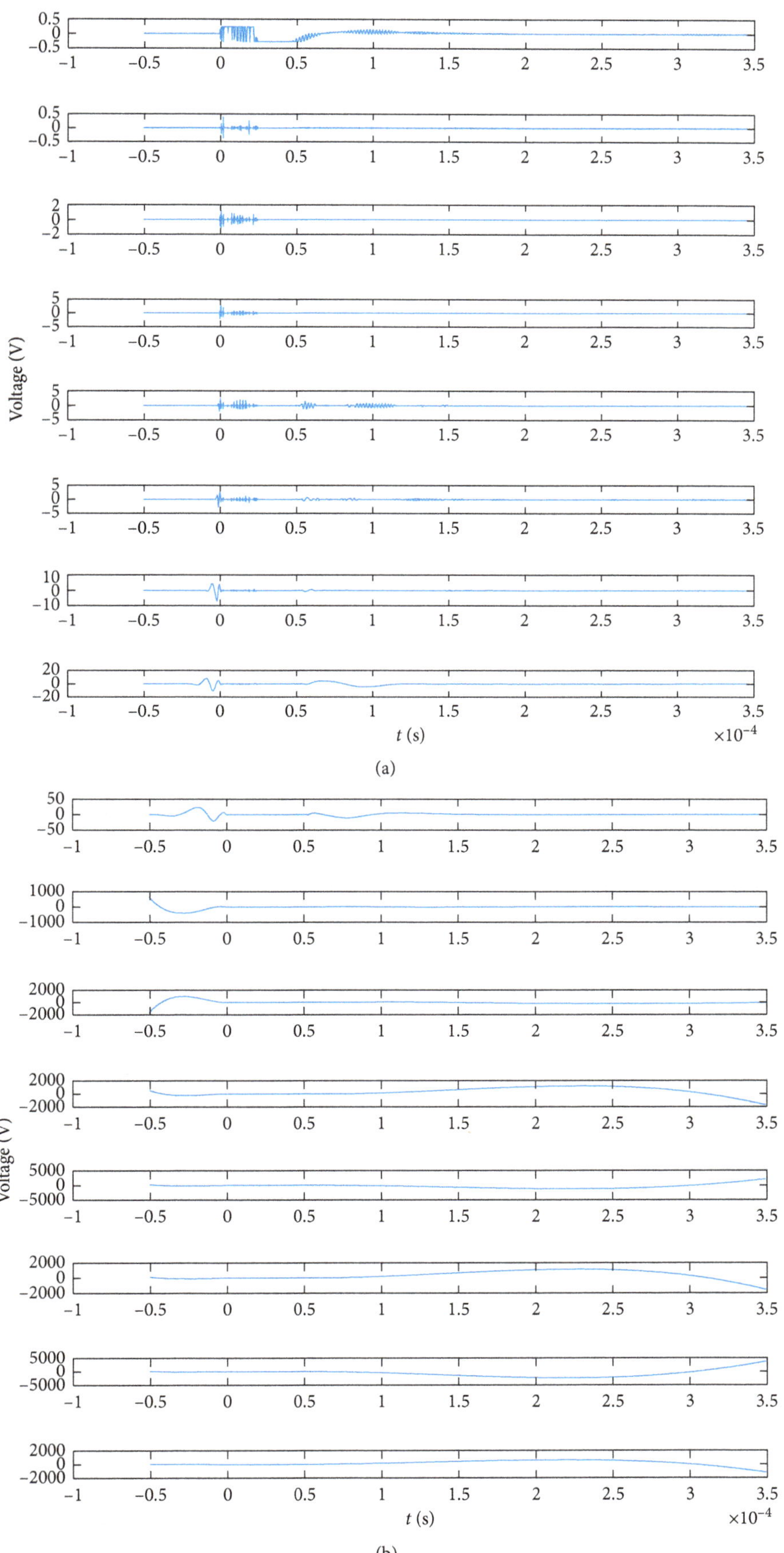

FIGURE 9: The IMF obtained by the improved EEMD method. (a) IMF 1–8. (b) IMF 9–16.

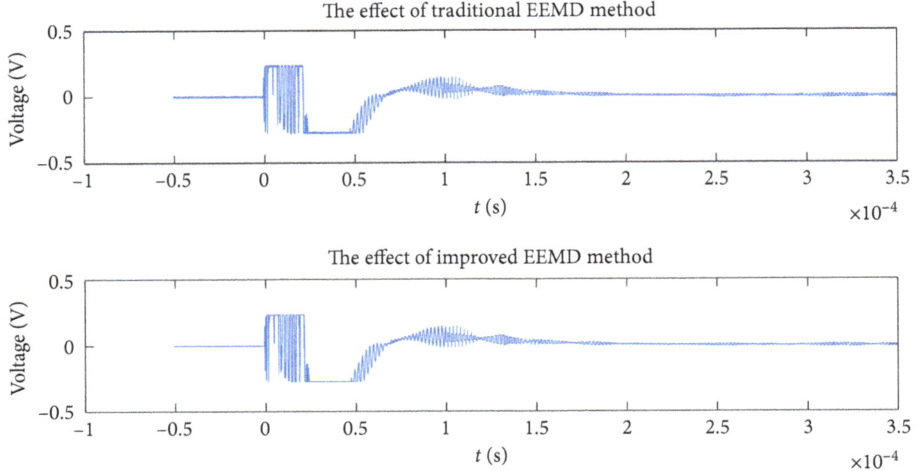

Figure 10: The comparison results of the two methods.

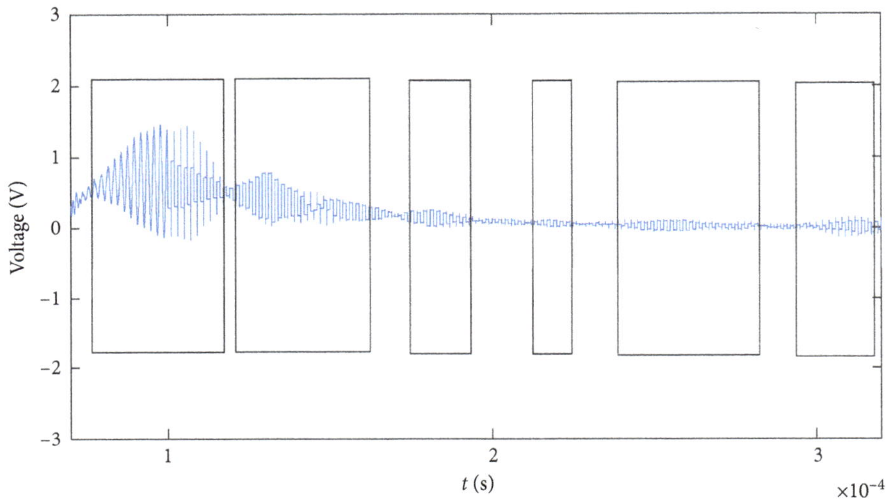

Figure 11: Effective signal denoising location.

receiver can be obtained. The interior of the black box is the received echo signal. Obviously, it can accurately reflect the effective information of the cracked position, and later, it can identify the location information of the defect according to the feature extraction method.

6. Conclusion

In this paper, a new denoising algorithm of electromagnetic ultrasonic testing signal based on the improved EEMD method is used to process the collected data. First of all, singular values in the data are removed. Then, aiming at the way that white noise is added to the EEMD method, the permutation entropy algorithm is used to identify the nature of the components obtained during the decomposition. Furthermore, the components of low-frequency signal are decomposed by EMD directly, while the components of other high-frequency IMF components are decomposed by EEMD. Afterwards, the wavelet transform modulus maximum denoising method is used to deal with the IMF components of the high-frequency part obtained before. Finally, the processed IMF results and residual difference are summed up. In the experiment, crack defect data collected by electromagnetic ultrasonic experiment system were processed by the improved EEMD denoising method. The results show the effectiveness and superiority of the proposed method.

Conflicts of Interest

The authors declare that they have no conflicts of interest.

Acknowledgments

This work was supported by the National Natural Science Foundation of China (61374124, 6147306, and 61627809) and Major Undergraduate Research Project of Northeastern University in 2017 (ZD2017).

References

[1] G. Liang, "A signal de-noising algorithm based on improved wavelet coefficients reconstruction," *Computer Knowledge and Technology*, vol. 12, no. 8, pp. 217–219, 2016.

[2] X. You, L. Du, Y. M. Cheung, and Q. Chen, "A blind watermarking scheme using new nontensor product wavelet filter banks," *IEEE Transactions on Image Processing*, vol. 19, no. 12, pp. 3271–3284, 2010.

[3] X. You, Z. Bao, C. Xing, Y. Cheung, and Y. Tang, "Image denoising using nonseparable wavelet filter banks and 2D-PCA," *Optical Engineering*, vol. 47, no. 10, pp. 107002–107011, 2008.

[4] J. Zhou, B. Xiang, L. Ni, and P. Ai, "Research on wavelet packet threshold denoising of vibration signal based on permutation entropy," *Measurement and Control Technology*, vol. 36, no. 12, pp. 5–9, 2017.

[5] S. Sun, Y. Pang, J. Wang et al., "EEMD harmonic detection method based on the new wavelet threshold denoising pretreatment," *Power System Protection and Control*, vol. 44, no. 2, pp. 42–48, 2016.

[6] N. E. Huang, Z. Shen, S. R. Long et al., "The empirical mode decomposition and the Hilbert spectrum for nonlinear and nonstationary time series analysis," in *Proceedings of the Royal Society A: Mathematical, Physical and Engineering Sciences*, vol. 454, 1971, pp. 903–995, 1998.

[7] Z. Wu and N. E. Huang, "Ensemble empirical mode decomposition: a noise-assisted data analysis method," *Advances in Adaptive Data Analysis*, vol. 1, no. 1, pp. 1–41, 2009.

[8] B. Christoph and P. Bernd, "Permutation entropy: a natural complexity measure for time series," *Physical Review Letters*, vol. 88, no. 17, article 174102, 2002.

[9] Y. Du, J. Wang, and X. Jin, "Defect detection of ultrasonic guided wave pipeline using de-noising method based on wavelet modulus maximum," *Journal of Electronic Measurement and Instrument*, vol. 27, no. 7, pp. 683–688, 2013.

8

Multiorder Fusion Data Privacy-Preserving Scheme for Wireless Sensor Networks

Mingshan Xie,[1,2,3] Yong Bai,[1,2] Mengxing Huang,[1,2] and Zhuhua Hu[1,2]

[1]*College of Information Science & Technology, Hainan University, Haikou 570228, China*
[2]*State Key Laboratory of Marine Resource Utilization in South China Sea, Haikou 570228, China*
[3]*College of Network, Haikou College of Economics, Haikou 571127, China*

Correspondence should be addressed to Yong Bai; bai@hainu.edu.cn

Academic Editor: Xiong Li

Privacy-preserving in wireless sensor networks is one of the key problems to be solved in practical applications. It is of great significance to solve the problem of data privacy protection for large-scale applications of wireless sensor networks. The characteristics of wireless sensor networks make data privacy protection technology face serious challenges. At present, the technology of data privacy protection in wireless sensor networks has become a hot research topic, mainly for data aggregation, data query, and access control of data privacy protection. In this paper, multiorder fusion data privacy-preserving scheme (MOFDAP) is proposed. Random interference code, random decomposition of function library, and cryptographic vector are introduced for our proposed scheme. In multiple stages and multiple aspects, the difficulty of cracking and crack costs are increased. The simulation results demonstrate that, compared with the typical Slice-Mix-AggRegaTe (SMART) algorithm, the algorithm proposed in this paper has a better data privacy-preserving ability when the traffic load is not very heavy.

1. Introduction

Nowadays, the application of Internet of things (IoT) is becoming more diverse. As an important part of the Internet of things, the wireless sensor network (WSN) has been widely used in all aspects of our lives (e.g., military surveillance, patient monitoring, forest monitoring, etc.). A wireless sensor network is a self-organizing network composed of a large number of sensor nodes. It has the characteristics of being resource constrained, distributed, multihop, wireless communication, and so on. WSN sensor nodes are placed in the public, untrusted, and even malicious intrusion environment and exchange the data using wireless communication. These make the data easy to intercept for WSN. WSN faces serious privacy data leakage risk. Especially in medical and military applications, data security requirements are very high. It is necessary to study the privacy protection of WSN.

It is the main task for wireless sensor networks to collect useful information, relying on a large number of nodes scattered in the environment, so that people can analyze and process the information. Data fusion technology is one of the key technologies for WSN. Many of the existing researches have increased the privacy protection function on the basis of data fusion. For instance, in [1], Conti et al. design a private data aggregation protocol that does not leak individual sensed values during the data aggregation process. It enhances the robustness, and the node computing complexity and data transmission are not large.

In [2–5], Wenbo et al. proposed the privacy-preserving algorithms based on data fusion. The SMART algorithm is widely used in these algorithms. In SMART, the data is divided into slices, and then the slices are sent to the randomly selected neighbor nodes; finally, the data fusion is carried out along the data fusion tree. In this paper, the method of decomposing data by random function is proposed. We make some improvements to SMART. We use the random function to decompose the data to increase the difficulty of cracking the data in the data aggregation point.

The purpose of privacy protection is to increase the cost and difficulty of acquisition for eavesdropper. The generation of interference information can greatly increase the cost and difficulty of eavesdropping.

Generally, the energy resources of the sampled sensors are limited. Energy resources are related to the lifetime of the whole WSN, so the energy consumption is usually the key problem of data fusion in WSNs. Now most of the wireless sensor networks take sleep strategies which randomly select a node as a sentinel node, to achieve compressive sampling, but now most of the compressed sampling schemes do not provide privacy protection. In fact, interference information can be generated by nodes randomly selected from the sleep nodes. In this paper, we propose a method to generate interference codes in the multiorder fusion data protection algorithm.

In the traditional password protection work, Girao et al. proposed a privacy-preserving solution for data fusion in [6, 7]. They use homomorphic encryption so that nodes can effectively fuse the data without decrypting data. In order to increase the defeat solution difficulty, this paper puts forward the strategy of password vector protection.

Now the eavesdropper's crack technology is richer. It is difficult to protect the security of wireless sensor networks with only one method. A variety of methods need to be organically combined so that layers of protection are formed. For this purpose, this paper proposes a privacy-preserving mechanism combined with active protection and passive protection and adopts multiorder fusion protection strategy for WSN, to increase the cost and difficulty of interception or eavesdropping in a nonlinear way.

The rest of this paper is organized as follows. Section 2 provides some related work. Section 3 describes the model of privacy-preserving data aggregation in WSN. Section 4 provides our multiorder fusion algorithms for private data aggregation. Section 5 evaluates the proposed schemes for cost and difficulty. We summarize our work and lay out future research direction in Section 6.

2. Related Work

In typical wireless sensor networks, sensor nodes are usually limited in resources and energy. Data aggregation is necessary for wireless sensor networks. Based on the management pattern of cluster structure, in [1], Conti et al. proposed a privacy-preserving algorithm for data fusion. In particular, neither the base station (BS) nor other nodes are able to compromise the privacy of an individual node's sensed value. Bista et al. proposed a new set of data fusion privacy protection solutions. The proposed scheme applies the additive property of complex numbers in [8, 9]. All of them have the advantages of computational complexity and small amount of data transmission.

In [2–5], Wenbo et al. proposed the privacy-preserving data aggregation (PDA), which includes two algorithms, cluster-based private data aggregation (CPDA) and Slice-Mix-AggRegaTe (SMART). But the amount of calculation of CPDA is great, and data traffic of SMART is very large. SMART algorithm in [2] is the most closely related to the algorithm proposed in this paper. The SMART algorithm uses hop-by-hop data fusion mode and node-to-node encryption and decryption mode. It can prevent the invasion of external intruders, in the case of ensuring the accuracy of data fusion. It can guarantee the internal trusted node to obtain privacy data. The SMART algorithm is divided into three steps: slicing, mixing, and aggregating.

In [10–13], some features are added on the basis of SMART. The privacy-preserving algorithm based on fusion is extended. The energy consumption is reduced in document 10. Reference [11] focuses on improving the accuracy of data fusion. Reference [12] adds data integrity verification. The authors add fault tolerance in [13].

In [6, 7], Girao et al. proposed a privacy-preserving solution for data fusion. They adopted homomorphic encryption. This algorithm can make the nodes implement effective fusion of data, without the need of decrypting data. In order to increase the difficulty of crack, this paper puts forward the strategy of using password vector to protect the data.

In [14], the authors proposed a k-anonymization clustering method (k-ACM) that provides a k-anonymity framework with two levels of privacy for WSNs. Zhang et al. proposed a security privacy-preserving data aggregation model, which adopts a mixed data aggregation structure of tree and cluster for the wearable wireless sensors in [15]. In [16], Cao et al. proposed a privacy-preserving and auditing-supporting outsourcing data storage scheme by using encryption and digital watermarking. They used logistic map-based chaotic cryptography algorithm to preserve the privacy of outsourcing data.

Thus, it can be seen that there are many schemes about privacy protection. How to combine a variety of privacy protection strategies is also an urgent need to solve the problem. We propose the multiorder fusion data privacy-preserving (MOFDP) algorithm for wireless sensor networks by extending the privacy protection.

3. System Model

The MOFDP algorithm is based on the cluster structure of wireless sensor networks. It is assumed that there is a cluster head node N_0 which can be a sink node in each cluster structure. It is a high resource node, which can be responsible for data collection and integration. It can also be used as a query node to perform query tasks. In each cluster structure of WSN, other nodes can be the subnodes which are responsible for collecting data. Suppose that the set of subnodes is $\{\text{Node}_1, \text{Node}_2, \ldots, \text{Node}_{\text{NumNode}}\}$, where NumNode is the number of child nodes. The specific structure of the cluster of WSNs is shown in Figure 1.

The MOFDP algorithm achieves multiorder data privacy protection. In Figure 1, "—" denotes order-1 data privacy protection. "- · · -" denotes order-2 data privacy protection. When NumNode = 4, there are four "—" and "- · · -", respectively. "→" represents order-3 data privacy protection.

In the privacy-preserving algorithm of this paper, N_0 node is very important. In this paper, it is assumed that N_0 is very difficult to be captured. In this paper, the wireless sensor network takes the sleeping strategy. The sampling frequency of the whole network is dynamically adjusted by N_0. In each sampling process, all the subnodes are not sampled at the

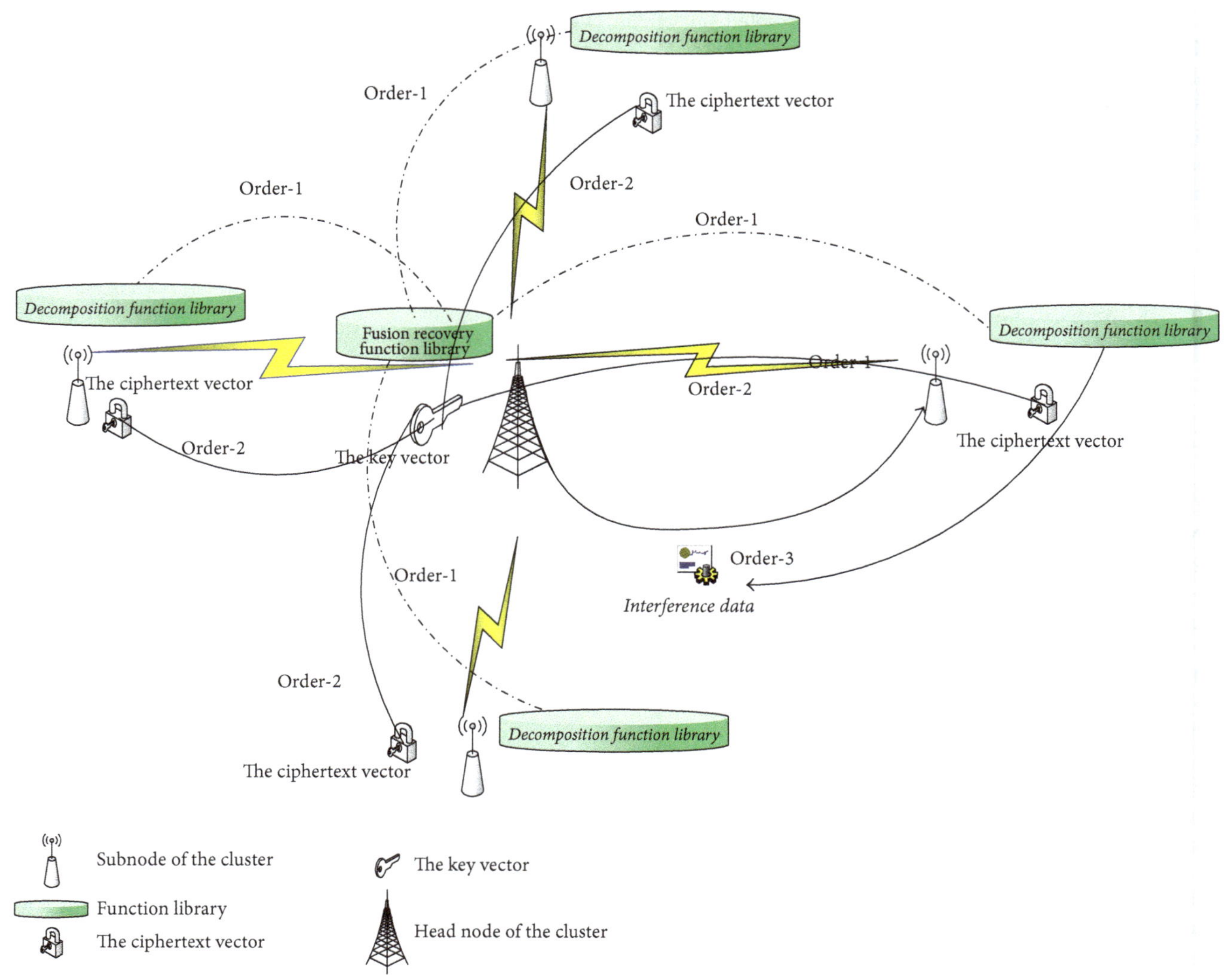

FIGURE 1: MOFDP model based on cluster structure.

same time, but only one random subnode is sampled. Any node that takes the sample is determined by the N_0 node according to the algorithm to generate random values. The fusion recovery function library and the key vector both are in the head node N_0. The detailed algorithm will be described in the rest of this paper.

4. The Protection Mechanisms of Multiorder Fusion Protection

In this paper, we use three order models to protect data privacy. The first step is the use of interference code protection. In the second phase, the data is decomposed by a random function. In the third stage, we use cryptographic vectors to secure data.

4.1. Construction of Decomposition Function Library. In [2], the slice-mixed aggregation (SMART) uses the segmentation and reorganization technology to complete the privacy-preserving data aggregation. The basic idea of SMART is that the sensor node randomly divides the original data into many data slices; the hop-by-hop encryption mechanism is adopted, and the exchanged data slices are randomly selected by the neighbor nodes; it performs the summation operation for all the received data slices and uploads the results to the base station; the base station can obtain all the received data and get the accurate SUM aggregation results.

We do not simply use the fixed method to segment the data in this paper. The decomposition function is randomly extracted from the decomposition function library, and then the collected data are decomposed and sent in separate time periods. It increases the difficulty of cracking data, because the decomposition function is random, and the decomposition function library and the fusion recovery function library must be obtained at the same time in order to recover the signal. Due to the implementation of transmitting data in the interval time, the number of elements in the solution vector of decomposition function is not uniform. It is difficult to determine the number of signals to be intercepted for cracking data; thus, the difficulty of the crack is increased.

Set $D_i(t)$ as the data collected by the ith subnode at t moment. We define the data decomposition function as

$$\begin{aligned}\mathrm{Dec}_q(D_i(t)) &= \mathrm{DD}_i \\ &= \{d_{(1,i)}(t), d_{(2,i)}(t), \ldots, d_{(N,i)}(t)\},\end{aligned} \tag{1}$$

where DD_i is the solution vector. $D_{(j,i)}(t)$ $(j = 1, 2, \ldots, N)$ denotes the elements of DD_i. The number N of components of the solution vector is determined by the decomposition function. The q is the index number of the decomposition function $\mathrm{Dec}(*)$.

The data decomposition function library of subnodes is denoted as

$$\mathrm{Decs} = \{q \in \mathbb{Z} \mid \mathrm{Dec}_q(D_i(t))\}. \tag{2}$$

The data fusion recovery function is interpreted as

$$\begin{aligned}\mathrm{Cec}_p(\mathrm{DD}_i) &= D_i(t) \\ &= \mathrm{Cec}_p(\{d_{(1,i)}(t), d_{(2,i)}(t), \ldots, d_{(N,i)}(t)\}) \\ &= D_i(t),\end{aligned} \tag{3}$$

where p is the index number of the fusion recovery function $\mathrm{Cec}(*)$.

The fusion recovery function library of sink node is defined as

$$\mathrm{Cecs} = \{p \in \mathbb{Z} \mid \mathrm{Cec}_p(\mathrm{DD}_i)\}. \tag{4}$$

A decomposition function corresponds to a fusion recovery function. The fusion recovery function is the inverse operation of the decomposition function. The relationship between them is $\mathrm{Dec}_u(\mathrm{Cec}_u(\mathrm{DD}_i)) = \mathrm{Cec}_u(\mathrm{Dec}_u(D_i(t))) = D_i(t)$, $u \in \mathbb{Z}$. The total number of functions in the decomposition function library corresponds to the total number of functions in the fusion recovery function library.

4.2. End-to-End Encryption Used Key Vectors. In many practical applications, wireless sensor networks must consider privacy protection to ensure that the data collected by each node can only be accessed by the authorized users. Only the authorized nodes can transmit the data to the sink node to ensure the correctness and integrity of the data. In this paper, the idea of encryption is end-to-end encryption.

The sensed data being passed to nonleaf aggregators are revealed for the sake of middle-way aggregation in the hop-by-hop aggregation protocols in [17]. Compared with the hop-by-hop encryption mechanism, the intermediate nodes in the end-to-end encryption mechanism can save the cost of encryption and decryption and reduce the time delay. The end-to-end encryption mechanism is as follows: the sensor node and the sink node share the key encryption, and then the sink node implements the decryption process. The end-to-end encryption mechanism requires data aggregation on encrypted data. Homomorphic encryption can be used to achieve the sum or product operation on the ciphertext, which can effectively support data aggregation in the encrypted data.

The ith sampling node:

$$\begin{aligned}\mathrm{cd}_1 &= \mathrm{Encode}(d_{(1,i)}(t), k_{(i,1)}, M) = d_{(1,i)}(t) + k_{(i,1)} \oplus M; \\ \mathrm{cd}_2 &= \mathrm{Encode}(d_{(2,i)}(t), k_{(i,2)}, M) = d_{(2,i)}(t) + k_{(i,2)} \oplus M; \\ &\vdots \\ \mathrm{cd}_N &= \mathrm{Encode}(d_{(N,i)}(t), k_{(i,N)}, M) = d_{(N,i)}(t) + k_{(i,N)} \oplus M,\end{aligned} \tag{5}$$

where $\oplus$ is denoted as the mode operation and M is the system parameter. $\mathrm{CD}_i = \{\mathrm{cd}_1, \mathrm{cd}_2, \ldots, \mathrm{cd}_N\}$ is ciphertext vector. The key vector $K_i = \{k_{(i,1)}, k_{(i,2)}, \ldots, k_{(i,N)}\}$ is shared by the ith subnode and the head node.

$\mathrm{DD}_i = \{d_{(1,i)}(t), d_{(2,i)}(t), \ldots, d_{(N,i)}(t)\}$ is the plaintext vector of sampling node.. The relationship then follows as

Ciphertext fusion value of head node: $\mathrm{CD}_{1\cdots N} = (\mathrm{Cec}_p(\mathrm{CD}_i)) \oplus M$;

Decryption of head node: $\mathrm{Decode}(\mathrm{CD}_{1\cdots N}, \mathrm{Cec}_p(K_i), M) = \mathrm{CD}_{1\cdots N} - \mathrm{Cec}_p(K_i) \oplus M$.

4.3. Sink Node Randomly Generated Number. Because the sleeping strategy is used in the wireless sensor network in this paper, under the condition that the probability of abnormal event is not high, all nodes are not sampled at the same time, in each sampling process. The sink node randomly generates the really sampling node number by the algorithm and notifies its sampling. This truly sampled node is called a sentinel node.

TSID is denoted as the truly sampling node number. TFID is in terms of the number of decomposition function. Active signal group MES = (TSID, TFID, CONDITION), where CONDITION is the condition to activate node sampling. NumNode is the total number of subnodes. NumFunction is the total number of functions in the decomposition function library. The algorithm of randomly generating number is Algorithm 1.

```
(1) While no error
(2) TSID = Rand(1, NumNode);
(3) TFID = Rand(1, NumFunction);
(4) Send MES to Node(TSID);
(5) Endwhile;
```

ALGORITHM 1: Sink node randomly generated number.

4.4. Generation of Random Interference Data. It is possible to randomly select a node to generate interference data. In this way, the probability that the eavesdropper intercepts the data in the channel and the information of the source is determined by the number of interfering data nodes. Assuming that the number of the real sampling nodes is

```
While no error
t = 0;
RSC = 0;
IF Receive(MES);
    RSC = RSC + 1;
    D_TSID(t) = SAMPLE;
    D_TSID = Dec_TFID(D_TSID(t))
    cd_1 = Encode(d_(1,i)(t), k_(i,1), M);
    cd_2 = Encode(d_(2,i)(t), k_(i,2), M);
    ...
    cd_N = Encode(d_(N,i)(t), k_(i,N), M);
    For tt = 1 to N
        SEND(TSID, TFID, cd_tt) to sink;
        tt = tt + 1;
    Endfor
ELSE IF RSC >= 2||!SAMPLING
        Then NOISYFD;
        RSC = 0;
        SEND FD to sink
        ENDIF
ENDIF
    t = t + 1
Endwhile
```

ALGORITHM 2: Subnode really sample and generate interference data.

$\mathrm{Num}_{\mathrm{datanodes}}$, the number of nodes generating the interference data is $\mathrm{Num}_{\mathrm{datainter}}$, and the probability p of the eavesdropper intercepting the false data is

$$p \approx \frac{\mathrm{Num}_{\mathrm{datainter}}}{\mathrm{Num}_{\mathrm{datanodes}}}. \quad (6)$$

In case there are one node generating the interference data and one real sampling node, the probability p of the eavesdropper intercepting the false data approximately equals 50%. If there are two nodes generating the interference data and one real sampling data, the probability p approximately equals 33.33%.

The selection problem of nodes generating interference data must be considered. If the head node notifies the subnodes to generate random interference codes, this increases the communication cost and instability. The method of this paper is shown in Algorithm 2. In $n-1$ nonsampling nodes, each node's sampling threshold is used to decide whether to transmit random values which can be used as the interference data. Since the real sampling nodes are random, the nodes that generate the interference data by Algorithm 2 are random too.

5. Data Privacy-Preserving Algorithm Based on Multiorder Fusion Protection

In order to ensure that the wireless sensor networks can securely collect data and secure the data collected by wireless sensor networks (user privacy information, especially, is not stolen), MOFDP algorithm has increased the difficulty of cracking from three aspects to achieve the 3-order stereo protection, based on the idea of increasing the difficulty of hacking.

In the aspect of data decomposition and reception, the random selection function in the function library is used to decompose and fuse data. In data encryption, the use of key vector for the end-to-end encryption has increased the difficulty of breaking. In addition, data is protected by sending interference data. Finally, multiorder data privacy protection in wireless sensor networks is achieved.

5.1. The Algorithm of Data Sampling and Sending for Subnode. In the process of sampling, each node is sampled once and then accumulated once. RSC is denoted as a cumulative variable.

The sampling time is t. SAMPLE is denoted as sampling, quantization, and coding to obtain sampled data. $\mathrm{Dec}_{\mathrm{TFID}}(D_{\mathrm{TSID}}(t))$ denotes that the decomposition function is called by TFID from the decomposition function library. In this algorithm the ciphertext is sent out in separate time periods. The symbol tt is accumulator for each time period, which is determined by decomposition function. NOISYFD denotes that false interference data are randomly generated. FD is in terms of false data.

5.2. The Algorithm of Data Fusion Recovering for Head Node. See Algorithm 3.

6. Discussion and Simulation Experiment

6.1. Protection Cost. In this paper, the real sampling node number and the interference node number are randomly generated and distributed uniformly. Thus, the energy balance of wireless sensor networks is guaranteed, and the service lifetime is effectively prolonged.

Most of the computational work of MOFDP is focused on head node. Because the head node is a high resource node, the calculation of energy consumption and communication energy consumption has been adequately supported and guaranteed.

6.2. Crack Cost. Since the aim is to extract the encryption function and the decryption function randomly from the library function in this paper's algorithm, it is necessary to break all the functions to ensure the success rate of crack function. The cost of cracking function is in terms of $\mathrm{cost}_{\mathrm{getfun}}$. The cost of cracking all functions in the library is $\mathrm{NumFuntion} \times \mathrm{cost}_{\mathrm{getfun}}$.

Because of the function decomposition, the cost of collecting data is increased. If the eavesdropper synthesized the data, it is needed that he has to break the communication links between the nodes and spend a certain amount of time gathering the data. The risk that the eavesdropper is found is increased. The cost of collecting one piece of data is $\mathrm{cost}_{\mathrm{getdata}}$. Since MOFDP uses the random function extraction, the eavesdropper has to take the maximum number of function decomposition components in the function library as the

```
(1) While no error
(2)     IF Receive(TSID) == TSID and Receive(TFID) == TFID;
(3)         CD_{1...N} = (Cec_TFID(CD_TSID)) mode M;
(4)         Decode(CD_{1...N}, Cec_TFID(K_i), M);
(5)     Else reject interference data and not receiving
(6)     Endif
(7) End while
```

ALGORITHM 3: Data fusion recovering of head node.

number of times to collect data every time. The cost of collecting data is

$$\mathrm{MAX}\left(N_1, N_2, \ldots, N_{\mathrm{NumFuntion}}\right) \times \mathrm{cost}_{\mathrm{getdata}}. \tag{7}$$

The MOFDP algorithm decomposes the data into many components, which form a solution vector. In order to encrypt data for each component, the key vector is introduced. The number of components of the key vector is equal to the number of decomposition components. The number of key vectors forms a vector: $\{N_1, N_2, \ldots, N_{\mathrm{NumFuntion}}\}$. The cost of breaking a key is set as $\mathrm{cost}_{\mathrm{key}}$. In the process of eavesdropping, the cost of cracking passwords is

$$\mathrm{MAX}\left(N_1, N_2, \ldots, N_{\mathrm{NumFuntion}}\right) \times \mathrm{cost}_{\mathrm{key}} \tag{8}$$

Due to the random interference data, the cost of identifying the authenticity is required. MOFDP uses random nodes to send interference data, so distinguishing the authenticity of the data requires the existence of the characteristics of each interference data of nodes to be analyzed. The cost of setting up one interference data is $\mathrm{cost}_{\mathrm{jarm}}$. The cost of identifying the authenticity is defined as $\mathrm{NumNode} \times \mathrm{cost}_{\mathrm{jarm}}$.

MOFDP preventive measures have three stages of protection. As long as the prevention of any stage cannot break, the right real data cannot be obtained. The total cracking cost of MOFDP is defined as

$$\begin{aligned}\mathrm{COST}_{\mathrm{MOFDP}} &= \mathrm{NumFuntion} \times \mathrm{cost}_{\mathrm{getfun}} \\ &\quad + \mathrm{MAX}\left(N_1, N_2, \ldots, N_{\mathrm{NumFuntion}}\right) \\ &\quad \times \mathrm{cost}_{\mathrm{getdata}} + \mathrm{NumNode} \times \mathrm{cost}_{\mathrm{jarm}} \\ &\quad + \mathrm{MAX}\left(N_1, N_2, \ldots, N_{\mathrm{NumFuntion}}\right) \\ &\quad \times \mathrm{cost}_{\mathrm{key}} \\ &= \mathrm{MAX}\left(N_1, N_2, \ldots, N_{\mathrm{NumFuntion}}\right) \\ &\quad \times \left(\mathrm{cost}_{\mathrm{key}} + \mathrm{cost}_{\mathrm{getdata}}\right) \\ &\quad + \mathrm{NumFuntion} \times \mathrm{cost}_{\mathrm{getfun}} \\ &\quad + \mathrm{NumNode} \times \mathrm{cost}_{\mathrm{jarm}}.\end{aligned} \tag{9}$$

The classic SMART algorithm also uses the idea of data segmentation. It decomposes the data into J slices and then encrypts and transmits $(J-1)$ data. The cracking cost of SMART approximately is defined as

$$\mathrm{COST}_{\mathrm{SMART}} = (J-1) \times \left(\mathrm{cost}_{\mathrm{key}} + \mathrm{cost}_{\mathrm{getdata}}\right). \tag{10}$$

$\mathrm{MAX}(N_1, N_2, \ldots, N_{\mathrm{NumFuntion}})$ and $(J-1)$ are all representatives of the number of decomposed data into components. They can be approximately equal.

Subtract formula (9) and formula (10):

$$\begin{aligned}&\mathrm{COST}_{\mathrm{MOFDP}} - \mathrm{COST}_{\mathrm{SMART}} \\ &\quad = \mathrm{NumFuntion} \times \mathrm{cost}_{\mathrm{getfun}} + \mathrm{NumNode} \\ &\quad \times \mathrm{cost}_{\mathrm{jarm}}.\end{aligned} \tag{11}$$

The actual work found that NumNode is far greater than NumFuntion. But the value of $\mathrm{cost}_{\mathrm{getfun}}$ and $\mathrm{cost}_{\mathrm{jarm}}$ is not small; these two cannot be ignored. It can be seen that the crack cost of MOFDP algorithm is much larger than the cost of the SMART algorithm, mainly because of heavy workload of cracking the function in the library and identifying the true and false data.

6.3. Crack Probability and Simulation Experiment. In the case of random interference data in wireless sensor networks, the probability of obtaining correct data changes with the number of random interference nodes. The true sampled node is also a random node. Set JM to indicate the correct data when sending interference data. The probability is

$$P(\mathrm{JM}) = \frac{1}{1 + N_{\mathrm{jarm}}} \times \frac{1}{\mathrm{NumNode}}, \tag{12}$$

where N_{jarm} denotes the number of joining random interference nodes. In this paper the value of N_{jarm} is set as 1, and NumNode is the total number of subnodes.

In this paper, we use the idea of data fragmentation and synthesis to increase the security of true sampling data. The decomposition function library and the fusion recovery function library are used. The vector of numbers of solution vector components of decomposition function in the function library is $\{N_1, N_2, \ldots, N_q, \ldots, N_{\mathrm{NumFunction}}\}$, where N_q represents the number of components which is the result that the qth decomposition function $\mathrm{Dec}_q(D_i(t))$ decomposes the $D_i(t)$. Since the decomposition function and the fusion recovery function are one-to-one reciprocal operations, decomposition function can be cracked, but also can break the fusion recovery function. The probability that the decomposition function can be cracked is denoted as

$$[\text{Decs}, P] = \begin{bmatrix} \text{Dec}_1 (D_i (t)) & \text{Dec}_2 (D_i (t)) & \cdots & \text{Dec}_q (D_i (t)) & \cdots & \text{Dec}_{\text{NumFunction}} (D_i (t)) \\ P (\text{FunDec}_1) & P (\text{FunDec}_2) & \cdots & P (\text{FunDec}_q) & \cdots & P (\text{FunDec}_{\text{NumFunction}}) \\ \text{pcf}_1 & \text{pcf}_2 & \cdots & \text{pcf}_q & \cdots & \text{pcf}_{\text{NumFunction}} \end{bmatrix}, \quad (13)$$

where Decs(q) is the decomposition function library. NumFunction is the capacity of the function library, or, alternatively, the total number of decomposition functions. The channel is set up between the sampling node and the sink node to transmit the data in separate time. The eavesdropper cannot intercept the correct data at once. They have to intercept the component many times in the channel. The more the components that get decomposed, the smaller the probability of intercepting the correct data. In case the decomposition function is cracked, the probability in which the listener can intercept one component is $P(\text{GetCom}_i \mid \text{FunDec})$, where i is denoted as the ith component. The probability value is determined by the technology and conditions of the eavesdropper. Here we set it as p_{compon}.

$$P (\text{GetCom}_i \mid \text{FunDec}) = p_{\text{compon}}. \quad (14)$$

In this paper, we investigate the probability of capturing each component as equal probability. The probability that the eavesdropper intercepts all components after they break down the decomposition function is $P(\text{GetComs} \mid \text{FunDec})$. There are two steps for our method in this paper: the first step is to extract a function from the decomposition function library. The probability of extracting the function from the decomposition function library obeys uniform distribution. The probability of obtaining the function to implement decomposition is $P(\text{Getchosfun})$; then

$$P (\text{Getchosfun}) = \frac{1}{\text{NumFunction}}. \quad (15)$$

The second step is to decompose the real sampling data with the random function. According to the full probability formula, the probability of information exposure in the procedure of function decomposition protection is denoted as

$$\begin{aligned} P (\text{GetComs}) &= P (\text{Getchosfun}) \times \Bigg(P (\text{FunDec}_1) \\ &\times \prod_{i=1}^{i=N_1} P (\text{GetCom}_i \mid \text{FunDec}_1) + P (\text{FunDec}_2) \\ &\times \prod_{i=1}^{i=N_2} P (\text{GetCom}_i \mid \text{FunDec}_2) + \cdots \\ &+ P (\text{FunDec}_q) \times \prod_{i=1}^{i=N_q} P (\text{GetCom}_i \mid \text{FunDec}_q) \\ &+ \cdots + P (\text{FunDec}_1) \\ &\times \prod_{i=1}^{i=N_{\text{NumFunction}}} P (\text{GetCom}_i \mid \text{FunDec}_{\text{NumFunction}}) \Bigg) \\ &= \frac{1}{\text{NumFunction}} \times \Big(\text{pcf}_1 \times p_{\text{compon}}^{N_1} + \text{pcf}_2 \\ &\times p_{\text{compon}}^{N_2} + \cdots + \text{pcf}_q \times p_{\text{compon}}^{N_q} + \cdots \\ &+ \text{pcf}_{\text{NumFunction}} \times p_{\text{compon}}^{N_{\text{NumFunction}}} \Big). \end{aligned} \quad (16)$$

In this paper, the key vector is promoted in the algorithm. The ith node and the sink node share the key vector $K_i = \{k_{(i,1)}, k_{(i,2)}, \ldots, k_{(i,N)}\}$. The probability of decryption key is

$$\begin{aligned} P (\text{Getkeys}) &= \frac{K_i}{\sum_{j=1}^{j=\text{NumFunction}} N_j} \times \frac{1}{K_i} \\ &= \frac{1}{\sum_{j=1}^{j=\text{NumFunction}} N_j}. \end{aligned} \quad (17)$$

The information exposure probability of our algorithm in this paper is

$$\begin{aligned} P &= P (\text{Getkeys}) \times P (\text{JM}) \times P (\text{GetComs}) \\ &= \frac{1}{\sum_{j=1}^{j=\text{NumFunction}} N_j} \times \frac{1}{1 + N_{\text{jarm}}} \times \frac{1}{\text{NumNode}} \\ &\times \frac{1}{\text{NumFunction}} \times \Big(\text{pcf}_1 \times p_{\text{compon}}^{N_1} + \text{pcf}_2 \\ &\times p_{\text{compon}}^{N_2} + \cdots + \text{pcf}_q \times p_{\text{compon}}^{N_q} + \cdots \\ &+ \text{pcf}_{\text{NumFunction}} \times p_{\text{compon}}^{N_{\text{NumFunction}}} \Big) \\ &= \frac{1}{\sum_{j=1}^{j=\text{NumFunction}} N_j} \times \frac{1}{1 + N_{\text{jarm}}} \times \frac{1}{\text{NumNode}} \\ &\times \frac{1}{\text{NumFunction}} \times \sum_{q=1}^{q=\text{NumFunction}} \text{pcf}_q \times p_{\text{compon}}^{N_q}. \end{aligned} \quad (18)$$

If N_{jarm} is set to be 1, $N_{\text{jarm}} = 1$, formula (18) can be simplified as

$$P = \frac{\sum_{q=1}^{q=\text{NumFunction}} \text{pcf}_q \times p_{\text{compon}}^{N_q}}{2 \times \text{NumFunction} \times \text{NumNode} \times \sum_{j=1}^{j=\text{NumFunction}} N_j}. \quad (19)$$

The information exposure probability and the probability of cracking the function or intercepting the data have positive correlation.

The SMART algorithm and MOFDP algorithm both have the idea of chip integration. At the time of each data aggregation, SMART uses the same function to decompose and merge, while MOFDP uses decomposition and fusion of functions that are randomly extracted from the function library.

In order to further analyze and compare, based on the classical SMART algorithm, the MOFDP algorithm is implemented in this paper. In this section, we set the probability that the link level privacy is broken q to 0.03 from 0.01. We compare the percentage that private data is disclosed in the case of different number of slices J, the number of decomposition functions NumFun, and the number of components of the vector N.

For convenience of comparison, the probability of the crack function pcf is set to a constant. In this simulation experiment, pcf takes 0.01.

In the SMART algorithm, J stands for data slices, but only the $J - 1$ data slices are encrypted. In the MOFDP algorithm, N stands for data fragmentation, and all N data slices are encrypted. NumFun stands for function library capacity. Compare the data exposure rate in the $J - 1$ case of the SMART algorithm with the N of the MOFDP algorithm.

When J takes 4 in SMART, NumFun takes 3, N takes up to 3 in the MOFDP, and the comparison results between them are shown in Figure 2.

When J takes 3 in SMART, NumFun takes 2, N takes up to 2 in the MOFDP, and the comparison results between them are shown in Figure 3.

It is shown that the information exposure rate of MOFDP algorithm is obviously lower than that in the SMART algorithm when the q is greater than 0.022. With the increase of q, the gap between the SMART algorithm and the algorithm of this paper is getting bigger and bigger. With the increase of q, the information exposure rate of SMART algorithm increases greatly, but the information exposure rate of MOFDP algorithm increases slowly. As can be seen from Figure 3, the number of small pieces of this algorithm in this paper is very obvious advantages. The probability of information exposure of MOFDP is very low. The growth rate of SMART algorithm is raising curve, but the curve about MOFDP algorithm is very smooth.

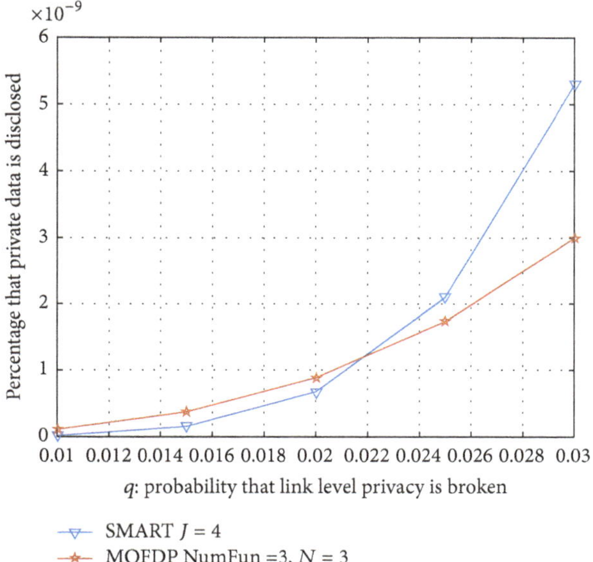

FIGURE 2: Comparison results when $J = 4$, NumFun = 3, $N = 3$.

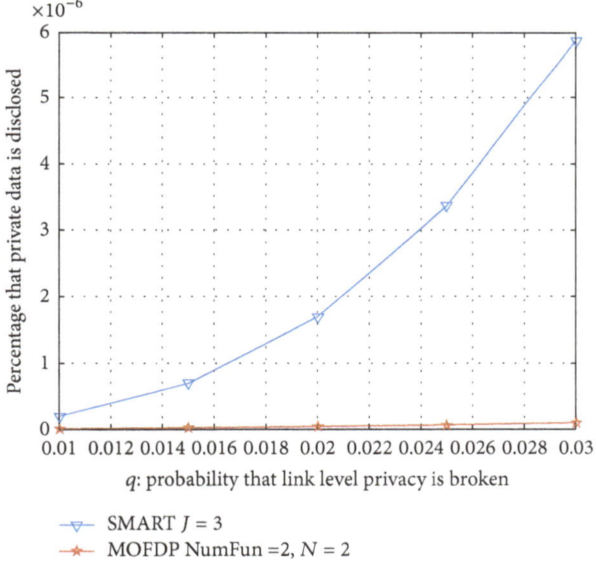

FIGURE 3: Comparison results when $J = 3$, NumFun = 2, $N = 2$.

7. Conclusions

With the rise of the Internet of things, the privacy protection of wireless sensor networks has become more important. In this paper, a multiorder data privacy protection algorithm is proposed based on the idea of SMART algorithm. In this paper, we introduce the interference code protection and adopt the idea of multiorder fusion to implement our proposed scheme. We target the difficulty and cost of cracking to improve effective privacy protection. The simulation results show that the proposed scheme has a better privacy protection function under low traffic.

Conflicts of Interest

The authors declare that there are no conflicts of interest regarding the publication of this paper.

Acknowledgments

This work was financially supported by the Project of Natural Science Foundation of Hainan Province in China (Grant no. 20166232 and Grant no. 617033), the Innovative Research Projects for Graduate Students of Hainan higher education institutions (Grant no. Hyb2017-06), the National Natural Science Foundation of China (Grant no. 61561017 and Grant no. 61462022), and the Open Project of State Key Laboratory of Marine Resource Utilization in South China Sea (Grant no. 2016013B).

References

[1] M. Conti, L. Zhang, S. Roy, R. Di Pietro, S. Jajodia, and L. V. Mancini, "Privacy-preserving robust data aggregation in wireless sensor networks," *Security and Communication Networks*, vol. 2, no. 2, pp. 195–213, 2009.

[2] H. E. Wenbo, X. Liu, H. Nguyen, K. Nahrstedt, and T. Abdelzaher, "PDA: privacy-preserving data aggregation in wireless sensor networks," in *Proceedings of the 26th IEEE International Conference on Computer Communications (INFOCOM '07)*, pp. 2045–2053, Anchorage, AK, USA, May 2007.

[3] W.-B. He, N. Hoang, X. Liu, K. Nahrstedt, and T. Abdelzaher, "iPDA: an integrity-protecting private data aggregation scheme for wireless sensor networks," in *Proceedings of the IEEE Military Communications Conference (MILCOM '08)*, pp. 1–7, San Diego, Calif, USA, November 2008.

[4] W. He, X. Liu, H. Nguyen, and K. Nahrstedt, "A Cluster-Based Protocol to Enforce Integrity and Preserve Privacy in Data Aggregation," in *Proceedings of the 2009 29th IEEE International Conference on Distributed Computing Systems Workshops (ICDCS Workshops)*, pp. 14–19, Montreal, Canada, June 2009.

[5] M. M. Groat, W. Hey, and S. Forrest, "KIPDA: K-indistinguishable privacy-preserving data aggregation in wireless sensor networks," in *Proceedings of the IEEE INFOCOM 2011*, pp. 2024–2032, April 2011.

[6] J. Girao, D. Westhoff, and M. Schneider, "CDA: concealed data aggregation for reverse multicast traffic in wireless sensor networks," in *Proceedings of the IEEE International Conference on Communications (ICC '05)*, vol. 5, pp. 3044–3049, IEEE, May 2005.

[7] C. Castelluccia, E. Mykletun, and G. Tsudik, "Efficient aggregation of encrypted data in wireless sensor networks," in *Proceedings of the 2nd Annual International Conference on Mobile and Ubiquitous Systems: Networking and Services (MobiQuitous '05)*, pp. 109–117, IEEE Computer Society, July 2005.

[8] R. Bista, K.-J. Jo, and J.-W. Chang, "A new approach to secure aggregation of private data in wireless sensor networks," in *Proceedings of the 8th IEEE International Symposium on Dependable, Autonomic and Secure Computing, DASC 2009*, pp. 394–399, December 2009.

[9] R. Bista, K. Hee-Dae, and W. Chang J, "A new private data aggregation scheme for wireless sensor networks," in *Proceedings of the IEEE International Conference on Computer and Information Technology, Cit 2010*, pp. 273–280, Bradford, UK, June 29, 2010.

[10] G. Yang, A.-Q. Wang, Z.-Y. Chen, J. Xu, and H.-Y. Wang, "An energy-saving privacy-preserving data aggregation algorithm," *Chinese Journal of Computers*, vol. 34, no. 5, pp. 792–800, 2011.

[11] L. Sen and Y. Geng, "Research on Precision Aggregation Privacy-preserving Algorithm in Wireless Sensor Networks," *Computer Technology and Development*, no. 9, pp. 139–142, 2013.

[12] S. Lu-sheng and Q. Xiao-lin, "Privacy-preserving data aggregation algorithm with integrity verification," *Computer Science*, vol. 40, no. 11, pp. 197–202, 2013.

[13] W. Tao-chun, L. Yong-long, and Z. Kai-zhong, "Fault-tolerant and privacy-preserving data aggregation algorithm in sensor networks," *Application Research of Computers*, vol. 31, no. 5, pp. 1499–1502, 2014.

[14] H. Bah and A. Lev, "k-anonymity based framework for privacy preserving data collection in wireless sensor networks," *Turkish Journal of Electrical Engineering & Computer Sciences*, vol. 18, no. 2, pp. 241–271, 2010.

[15] C. Zhang, C. Li, and J. Zhang, "A secure privacy-preserving data aggregation model in wearable wireless sensor networks," *Journal of Electrical and Computer Engineering*, vol. 2015, Article ID 104286, 9 pages, 2015.

[16] X. Cao, Z. Fu, and X. Sun, "A privacy-preserving outsourcing data storage scheme with fragile digital watermarking-based data auditing," *Journal of Electrical and Computer Engineering*, vol. 2016, Article ID 3219042, 7 pages, 2016.

[17] E. Mlaih and S. A. Aly, "Secure hop-by-hop aggregation of end-to-end concealed data in wireless sensor networks," in *Proceedings of the 2008 IEEE INFOCOM Workshops*, pp. 1–6, April 2008.

9

Dynamic Search Mechanism with Threat Prediction in a GNSS Receiver

Fang Liu and Meng Liu

School of Information Science and Engineering, Shenyang Ligong University, Shenyang 110159, China

Correspondence should be addressed to Fang Liu; zhqing1019@163.com

Academic Editor: Rajesh Khanna

With the development of GNSS and the application of multimode signals, efficient GNSS receiver research has become very important. However, threat signals in the received signal are inevitable, which will represent important threats to navigation applications and will lead to leakage and fault detection for the receiver. Therefore, the searches with threat prediction in GNSS signals have been regarded as the important problem for GNSS receivers. In view of the limitations and nonadaptability of the current search technologies as well as on the basis of the proportionality peak judgment mechanism, a dynamic search mechanism with threat prediction (DSM-TP) is proposed in this paper, in which we define a series of preset coefficients and a threat index to optimize the decision mechanism. The simulation results demonstrate that the DSM-TP method can predict the threat situation and can adapt to lower SNR environments compared to the traditional method. In addition, the DSM-TP method can avoid the impact of threats, and the detection capability of the new method is better than the detection capability of the traditional method.

1. Introduction

Global navigation satellite systems (GNSS) [1, 2], which include global positioning system (GPS), GALILEO, GLONASS, and COMPASS [3], are currently consolidated and fundamental in satellite communication and mobile communication. GNSS receivers [4] process signals transmitted from satellites to determine user position, velocity, and time. However, to achieve the correct signal reception, it is necessary to determine whether the effective signal exists by using the search detection process. In GNSS, to enhance antijamming and antispoofing capabilities, the longer PRN codes are required. Due to the appearance of long code, the new requirements have been created for the search technology.

Since the clock between transmitter and receiver is different, the local code and received signal may not overlap in time; that is, there is no correlation; thus the search time is uncertain. Additionally, the complexity of the communication channel leads to complexity of the received signal, especially in the conditions of poor environment, human interference, and other threat environments. When there are threat signals in the received signals, the GNSS receiver will be harmed, especially if the signals are counterfeit [5]. Counterfeit signals have a certain purpose: to provide incorrect information to the target receiver. Examples of incorrect information include incorrect navigation messages and incorrect pseudoranges. This can cause the receiver to output incorrect positions or timing information. Thus, counterfeit signals and jamming signals represent an important threat to navigation applications, and the receiver will also output leakage detection, fault detection, and other issues, especially in fields related to the livelihood of a person such as the military, electricity generation, and finance. Therefore, the search and threat prediction of GNSS signals has become an important problem for GNSS receivers.

In recent years, a series of methods have emerged to meet the demand of efficient GNSS signals search. Transformation domain techniques, such as fast Fourier transform- (FFT-) based methods, are commonly used to reduce acquisition times [6, 7]. In addition, aided search schemes through wireless communication networks or other on-board sensors have been proposed to effectively reduce the search space and decrease the search time [8]. Another known method to explore the amplitude-domain freedom is to construct the hypercode by superposing several folded segments of the original code [9].

Then, CCPAZP-FFT (Circular Correlation by Partition and Zero Padding) [10] is proposed, which is based on the storage and zero padding segmentation correlation. This algorithm converts two-dimensional searches of time and frequency to one-dimensional search to improve the acquisition efficiency. However, the algorithm is time-consuming in the uncertain code segment by piecewise searching.

XFAST (Extended Replica Folding Acquisition Search Technique) [11, 12] with average input and local samples (direct average) [13] is proposed. This method reduces the amount of computation for direct acquisition but introduces noise during the superposition process. When an increased number of blocks or samples are folded or summed for averaging, the codes result in single correlations more often. Additionally, when more noise is superimposed and included in the correlation results, detection probability decreases. Increased time for folding or summation also results in increased peak position ambiguity, hence lengthening the mean acquisition time. To overcome these problems, a dual-channel method [14] and dual-folding approach (DF-XFAST) [15] are proposed. To achieve the better SNR, this method suggests folding input samples into blocks before undergoing the XFAST operation, which is essentially equivalent to coherent integration of the XFAST correlation results to achieve additional gain.

In addition, to improve the processing speed, the time-frequency domain combination method [16] and frequency domain parallel method [17] are introduced. In these methods, multichannel parallel processing in the time domain and the frequency domain is performed to improve the search speed. However, in the high-dynamic case, the search efficiency of the methods needs to be improved, and due to the influence of the sampling interval, this series of methods still has the limitation of a high missed detection probability. Although the above methods can improve the search speed, they do not have the ability to suppress and predict complex environments and human threat. Therefore, because of the limitations and nonadaptability of the current technologies, a dynamic search mechanism with threat prediction (DSM-TP) is proposed in this paper.

2. Potential Threat Analysis

The signal not only is affected by noise and multipath of the natural environment but also may be artificially unconscious or have conscious malicious damage. For the GNSS receiver, the synchronous search based on the correlation operation is the core; hence, the effect of the correlation operation is the crucial aspect of synchronous search. Therefore, we analyze potential threat signals from the perspective of the impact of related operations.

Although the communication quality can be improved and the threat can be avoided through the frequency escape or jump of the spread frequency signal, in special cases, there may be threat signals in the communication channel. When there are threat signals in the communication channel, threat signals usually form from the spectrum form of GNSS signals, energy concentration region, and correlation characteristic.

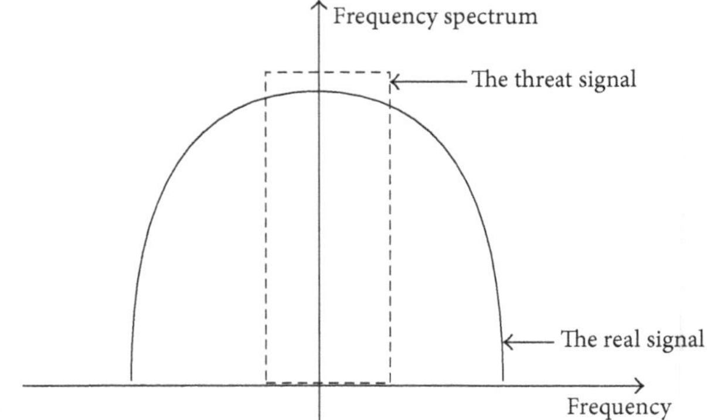

FIGURE 1: Threat signal of partial frequency band.

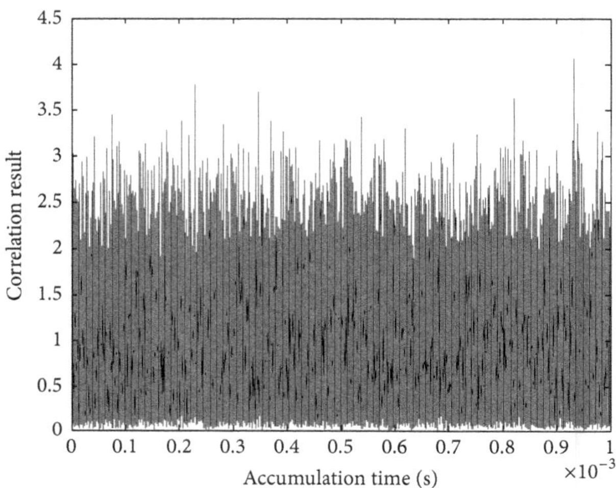

FIGURE 2: Related search influence sample.

The effective energy of the signal may be concentrated in a certain area; therefore, the larger threat is the partial band signal near the center frequency, as shown in Figure 1. When the threat signal frequency covers the center region of the frequency spectrum, signals are most likely to enter the correlator, and when the threat signal energy increases, it may affect correct search of the GNSS signal. A sample of the related search influence is shown in Figure 2. In addition, the influence of the GNSS signal search element is associated with the correlation function of the threat signal. A sample of a threat signal with correlation characteristic is shown in Figure 3. If there is a deviation in the spectrum coverage area, provided that there is partial correlation between the threat signal and the real signals, there is likely a large impact on the search of GNSS signals through the power adjustment. The possible related search effects are shown in Figure 4. If there is no threat analysis and prediction mechanism in the synchronous search, the related decision may fail, which reduces the synchronous receiving efficiency. The threat signal can also be regarded as the real signal, which can cause a significant influence of error application.

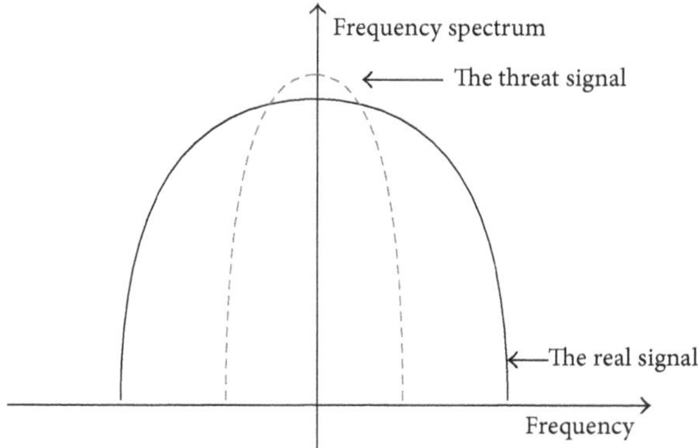

Figure 3: Threat signal with correlation characteristic.

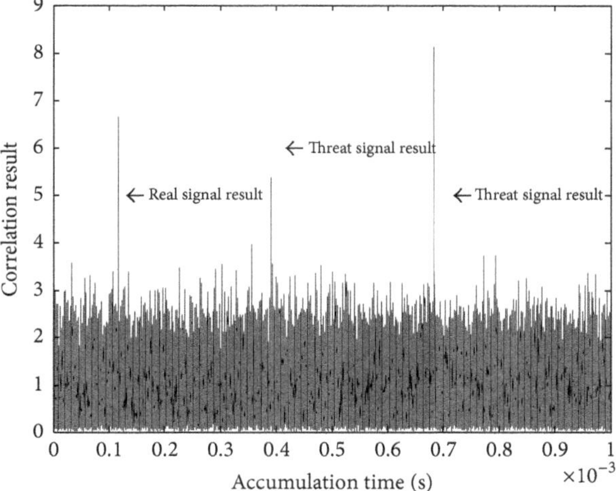

Figure 4: Related search influence sample.

3. Dynamic Search Mechanism with Threat Prediction

For communication signals, the most effective search detection methods use data processing that is based on data segments and FFT. The search decision mechanism is primarily based on relative peak comparison, which is described as traditional method in this paper. Firstly, the received signal is sampled and filtered, and then the local code is generated, extended, and averaged, and another processing is performed to improve the search speed. Then, the frequency estimation error is reduced by multichannel frequency compensation. Furthermore, a multichannel FFT correlation operation is performed between the multiprocessed local and received sequences, and the correlation results are used to calculate the peak value and the average peak value. Finally, a comparison between the maximum peak value and the average peak value is compared with the decision threshold. If the result is successful, the results are output; otherwise, return to the first step, and the signal is rereceived.

Considering the possible drawbacks of the simple absolute energy decision, the decision mechanism uses a proportional peak. Therefore, the proportional peak is defined as the ratio of the maximum peak and average peak. At the same time, the proportional decision mechanism is defined, which compares the proportional peak and the threshold. The proportional peak is expressed as $P = P_{\max}/P_{\text{ave}}$, in which $P_{\max}$ is the maximum peak energy and P_{ave} is the average peak energy. The effects on $P_{\max}$ and P_{ave} are equal, despite a strong or weak mixed signal; thus the proportional decision mechanism is more suitable for acquisition decision in complex environments. The theory of the proportional decision mechanism is shown in Figure 5.

Based on the proportionality peak judgment mechanism, the decision mechanism is further optimized by considering the influence of a complex environment and threat signals. We define a series of preset coefficients and the threat index, so that a dynamic search mechanism with threat prediction (DSM-TP) is established. The overall flow of DSM-TP is shown in Figure 6.

Step 1. The received branch is selected and updated. And the received signal $S(n)$ is defined as

$$S(n) = d(n) C(n) \cos(\omega n + \theta), \tag{1}$$

where $d(n)$ is the navigation data, $C(n)$ is the pseudocode sequence, ω is the carrier frequency, and θ is the carrier phase.

Step 2. The parameters are initialized. Q_i is initialized using the number of samples N_0 of a chip, in which $i \in [1, 2 \cdots N_0]$. The preset coefficient K is set equal to 1, and the residing coefficient T is set equal to 0.

Step 3. The received signal of the selected branch is sequentially accumulated, and filter processing and front-end processing are performed; then the result sequence is denoted as $S_R(n)$.

$$\begin{aligned} S_R(n) &= \text{fil}\left[S(n) \cos(\omega_0 n + \theta_0)\right] \\ &= \text{fil}\left[d(n) C(n) \cos(\omega n + \theta) \cos(\omega_0 n + \theta_0)\right] \\ &\approx d(n) C(n) \cos(\Delta\omega n + \Delta\theta), \end{aligned} \tag{2}$$

where fil[ ] is the low-pass filter processing function, ω_0 is the local carrier frequency, θ_0 is the local carrier phase, $\Delta\omega$ is the result carrier frequency, and $\Delta\theta$ is the result carrier phase.

Step 4. It is determined if the T value is equal to or greater than 5. If the condition is satisfied, Step 19 is executed; otherwise, go back to Step 5.

Step 5. The local pseudocode sequence $C_{Li}(n)$ is generated; then the data result of mean processing and spread processing is denoted as $S_L(n)$.

$$S_L(n) = \sum_{i=1}^{M} C_{Li}(n). \tag{3}$$

Step 6. The correlation operation is performed between $S_R(n)$ and $S_L(n)$, and the result is denoted as $V(n)$.

$$\begin{aligned} V(n) &= S_R(n) * S_L(n) \\ &= \text{abs}\left[\text{IFFT}\left[\text{FFT}(S_R(n)) \cdot \text{FFT}^*(S_L(n))\right]\right]. \end{aligned} \tag{4}$$

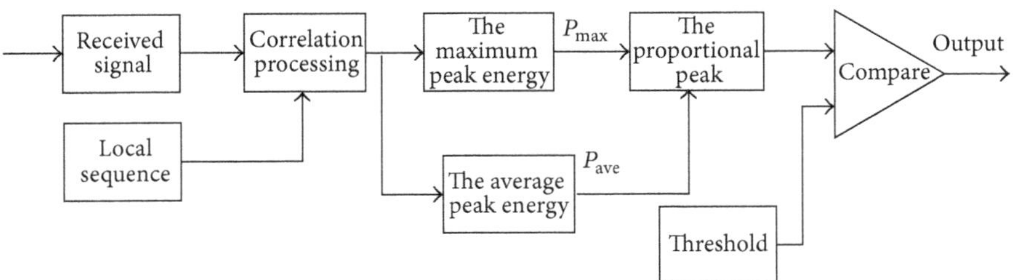

FIGURE 5: The proportional decision mechanism.

Step 7. The peak values are calculated using the correlation results $V(n)$, and the maximum peak is denoted as V_m, and the average peak is denoted as V_v, where L is the length of $V(n)$.

$$V_m = \max[V(n)],$$
$$V_v = \frac{\sum V(n)}{L}. \quad (5)$$

Step 8. The threshold factor is calculated using $V(n)$, V_m, and L.

$$\xi = \frac{\sum V(n)}{\sum V(n) - V_m} \cdot \frac{L-1}{L}. \quad (6)$$

Step 9. The threshold is calculated as $G = \xi \cdot V_v$.

Step 10. The parameter N is defined and set equal to 0; then N is described as

$$N = \begin{cases} N+1, & \text{if } V(n)|_1^L > G \\ N, & \text{other,} \end{cases} \quad (7)$$

where $V(n)|_1^L$ is the traverse function from $V(1)$ to $V(L)$.

Step 11. The removal function $\chi[\ ]$ is defined. It expresses that b position sequence is removed from $V(n)$, which is described as

$$\chi[V(n), b] = \{V(1), V(2) \cdots V(b-1), V(b+1), \ldots, V(L)\}. \quad (8)$$

Then, P_i is defined, whose initial value is equal to 0: $i \in [1, 2 \cdots N]$. And the maximum position function $\max_i[\ ]$ is defined. It expresses the position of the Nth largest peak in $V(n)$, which is described as

$$P_i = \max_i [V(n)] = \max\{\chi[V(n), P_{i-1}]\} = \max\{V(1), V(2) \cdots V(P_{i-1} - 1), V(P_{i-1} + 1), \ldots, V(L)\}. \quad (9)$$

Thus, $P_1, P_2, \ldots, P_N$ can be obtained. An example is used to calculate P_i, which is shown in Figure 7.

Step 12. It is determined if the N value is greater than 1. If the condition is satisfied, Step 13 is executed; otherwise, go to Step 14.

Step 13. It is determined if the N value is equal to 1. If the condition is satisfied and the preset coefficient is adjusted as $K = K + 1$, then Step 16 is executed; otherwise, the new signal is rereceived, and the parameters are adjusted; then go back to Step 1.

Step 14. It is determined if the N value is greater than N_0. If the condition is satisfied, the new signal is rereceived, and the parameters are adjusted, then go back to Step 1; otherwise, go to Step 15.

Step 15. P_i and cache variables Q_i are compared under branching conditions.

$$K = K + 1, \quad \text{if } P_1 == Q_1$$
$$\lambda = \lambda + 1, \quad \text{if } P_{\text{other}} == Q_{\text{other}} \quad (10)$$
$$K = K - 1, \quad \text{if } P_1 == Q_{\text{other}}, \text{ or } P_{\text{other}} == Q_1.$$

Step 16. It is determined if the K value is greater than 1. If the condition is satisfied, this branch signal is continually received, and the residing coefficient is adjusted as $T = T + 1$, then go back to Step 3; otherwise, go to Step 17.

Step 17. Then, Q_i is updated using P_i.

Step 18. It is determined if the K value is greater than 3. If the condition is satisfied, Step 19 is executed; otherwise, this branch signal is continually received, and the residing coefficient is adjusted as $T = T + 1$; then go back to Step 3.

Step 19. The acquisition position parameter P is output and the potential threat index W is analyzed.

$$P = \begin{cases} P = Q_1; & \text{if } K = 3 \\ P = \text{error}; & \text{if } K \neq 3, \end{cases}$$
$$W = \begin{cases} W = 1; & \text{if } \lambda = 0 \\ W = 2; & \text{if } \lambda = 1 \\ W = 3; & \text{if } \lambda > 1. \end{cases} \quad (11)$$

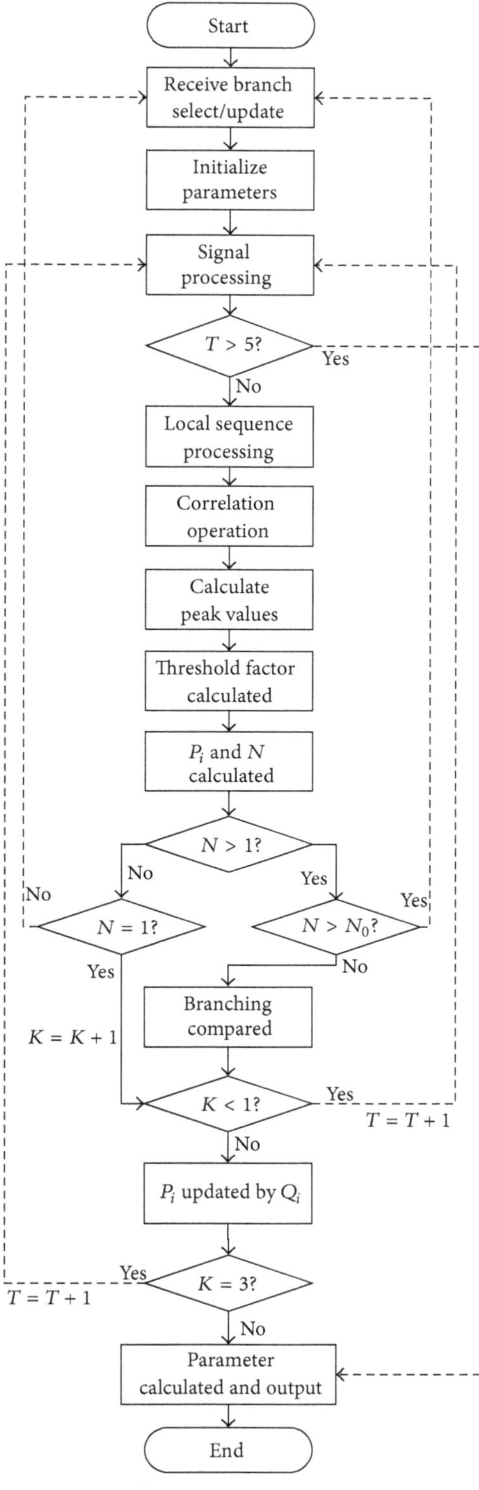

FIGURE 6: Overall flow of DSM-TP.

When W is equal to 1, it indicates that there are no man-made threats in the received signals. When W is equal to 2, it indicates that there may be man-made threats in the received signals, but there are no correlation characteristics between the threat signals and the real signal. When W is equal to 3, it indicates that there are man-made threats in the received signals, and there are correlation characteristics between the threat signals and the real signal.

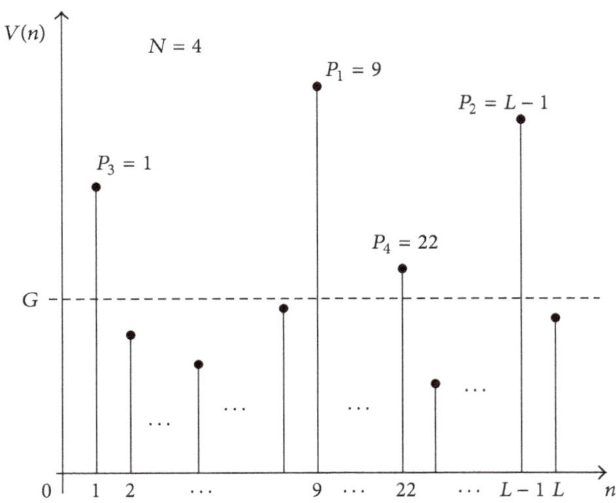

FIGURE 7: The calculated process of P_i.

TABLE 1: Threat prediction results of the two methods.

SNR	Method: DSM-TP method	Method: Traditional method
10 dB	Correctly; $w = 1$, predict nonthreat state	Correctly; No prediction
0 dB	Correctly; $w = 1$, predict nonthreat state	Correctly; No prediction
−14 dB	Correctly; $w = 1$, predict nonthreat state	Incorrectly; No prediction
−17 dB	Alarm and rereceive; $w = 1$, predict nonthreat state	Incorrectly; No prediction

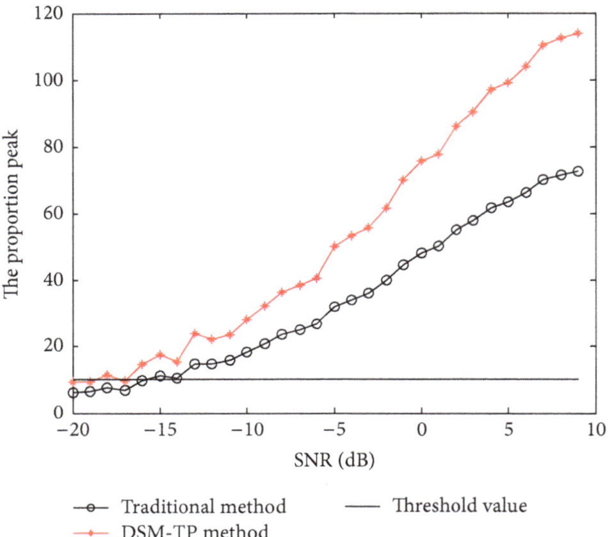

FIGURE 8: Proportion peak for changing SNR.

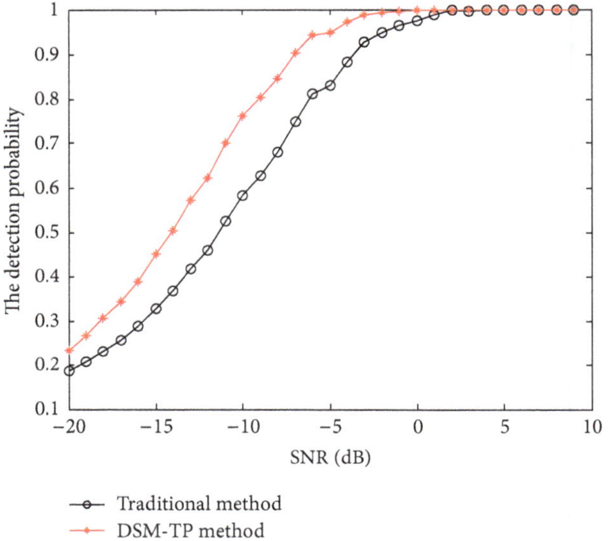

FIGURE 9: Detection probability for changing SNR.

4. Test and Analysis

4.1. Analysis under Safe Conditions. Based on the simulation platform, the DSM-TP method and the traditional method are tested when the receiver is in a safe environment. Input parameters are as follows: BPSK modulation, a signal bandwidth of 10 MHz, and proportional threshold equal to 10. For different values of SNR, the DSM-TP method and the traditional method are tested, and the proportion peak results are shown in Figure 8. For the same SNR value, the correlation result of the new method is superior to the traditional method. For a threshold equal to 10, the effective SNR of the new method is greater than −17 dB, while the conventional method is larger than −14 dB, indicating that SNR adaptability of the new method is better than the traditional method. Further, the detection probability of each method is shown in Figure 9 for different SNR values. For the same SNR condition, the detection probability of the new method is greater than that of the traditional method, indicating that the detection capability of the new method is better than that of the traditional method.

The results of threat prediction ability for the two methods are shown in Table 1. When the SNR is greater than −17 dB, the new method can search correctly and predicts that there are no man-made threats present in the environment, which agrees with the actual input situation. When the SNR is less than −17 dB, the alarm is issued and a research is performed. It also predicts no man-made threats to be present in the environment. However, the traditional method can search correctly when the SNR is greater than −14 dB but cannot predict the threat. When the SNR is less than −14 dB, the traditional method cannot search correctly and cannot predict the threat. This indicates that the new method can adapt to lower SNR environments compared to the traditional method and also predicts the threat situation.

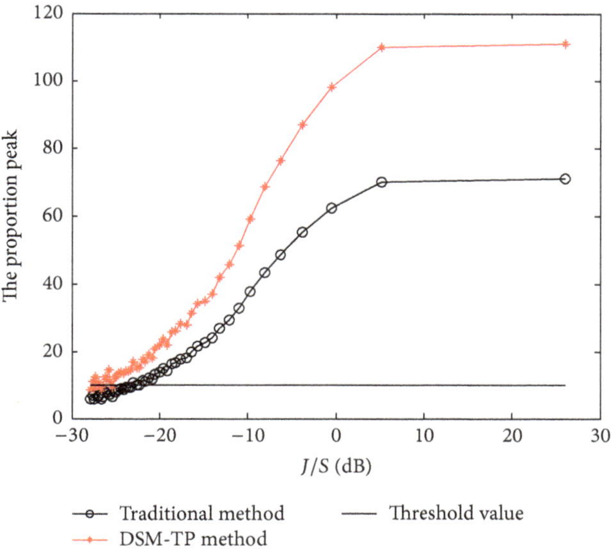

FIGURE 10: Proportion peak for changing S/J.

TABLE 2: Results of threat prediction for the two methods.

J/S	Method: DSM-TP method	Method: Traditional method
25 dB	Correctly; $w = 2$, predict threat state but no correlation	Correctly; No prediction
15 dB	Correctly; $w = 2$, predict threat state but no correlation	Correctly; No prediction
0 dB	Correctly; $w = 2$, predict threat state but no correlation	Correctly; No prediction
−21 dB	Correctly; $w = 2$, predict threat state but no correlation	Incorrectly; No prediction
−28 dB	Alarm and rereceive; $w = 2$, predict threat state but no correlation	Incorrectly; No prediction

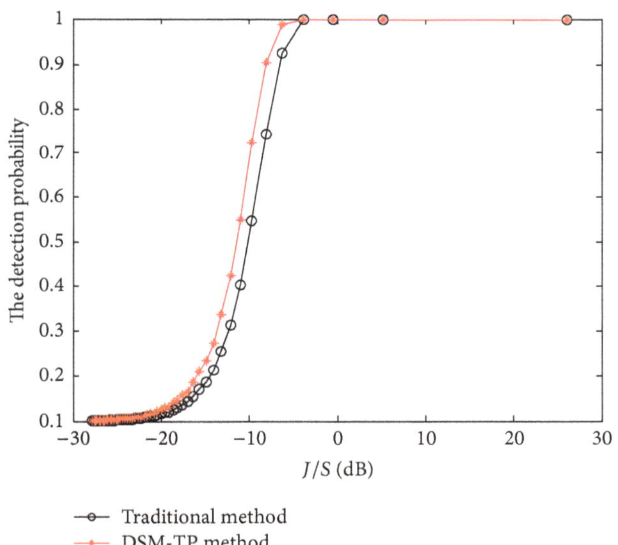

FIGURE 11: Detection probability for changing S/J.

4.2. Analysis under Partial Frequency Band Threat Conditions. The DSM-TP method is tested when the receiver is in a threat environment. Input parameters are as follows: BPSK modulation, signal bandwidth equal to 10 MHz, and proportional threshold equal to 10. Let the form of the threat signals be the uniform spectrum interference, whose frequency is in the center frequency, and the spectrum covers 80% of the main frequency band of the real signal. Let S/J be the ratio of the real signal energy and the threat signal energy, and the larger the S/J value, the less the relative energy of jamming. For different S/J values, the DSM-TP method and the traditional method are tested, and the proportion peak results are shown in Figure 10. When S/J is equal to 0, that is, the energies of the real signal and the threat signal are equal, the correlation value of the new method is greater than that of the traditional method: 30 dB. For a threshold value equal to 10, effective S/J of the new method is larger than −28 dB, while the conventional method is larger than −21 dB, indicating that S/J adaptability of the new method is better than the traditional method. Furthermore, for changes in the S/J value, the detection probability of each method is shown in Figure 11. For the same S/J conditions, the detection probability of the new method is greater than the detection probability of the traditional method, indicating that the detection capability of the new method is better than that of the traditional method.

The results of the threat prediction ability of the two methods are shown in Table 2. When S/J is greater than −28 dB, the new method searches correctly. It can predict that the potential threat index W is equal to 2, which verifies that there are man-made threats in the environment, but there is not a correlation characteristic between the threat signals and the real signal. When the SNR is less than −28 dB, the alarm is issued and a research is performed. It can predict that there are man-made threats in the environment, but there is not a correlation characteristic between the threat signals and the real signal. The results are consistent with the actual input situation. However, the traditional method can search correctly when the SNR is larger than −21 dB, but it cannot predict the threat. When the SNR is less than −21 dB, the traditional method cannot search correctly and cannot predict the threat. These results indicate that the new method can adapt to a threat environment better than the traditional method and can also predict the threat situation.

4.3. Analysis under Correlation Characteristic Threat Conditions. The DSM-TP method is tested when there is correlation characteristic between the threat signals and the real signal, and the pseudocode overlap rate of the threat signal and real signal is set to 30%. For different S/J values, the DSM-TP method and the traditional method are tested, and the

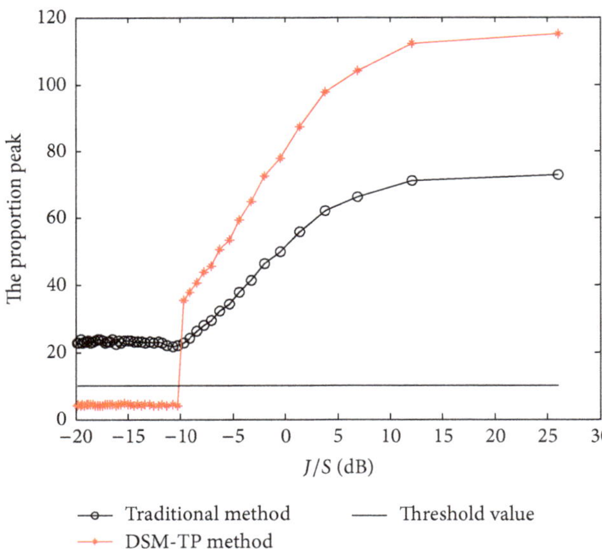

Figure 12: Proportion peak for changing *S/J*.

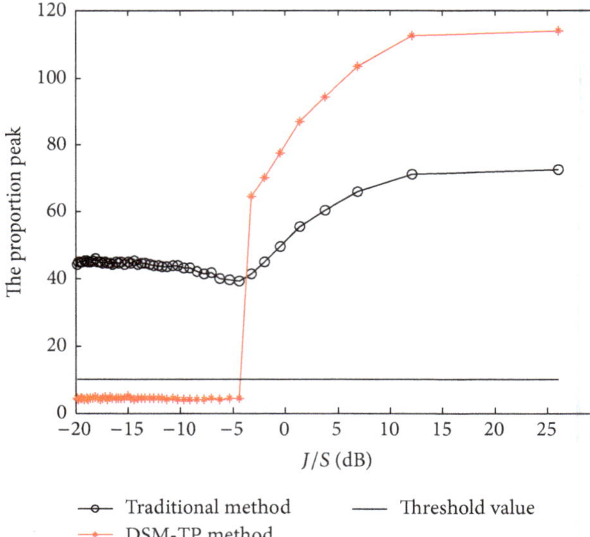

Figure 14: Proportion peak for changing *S/J*.

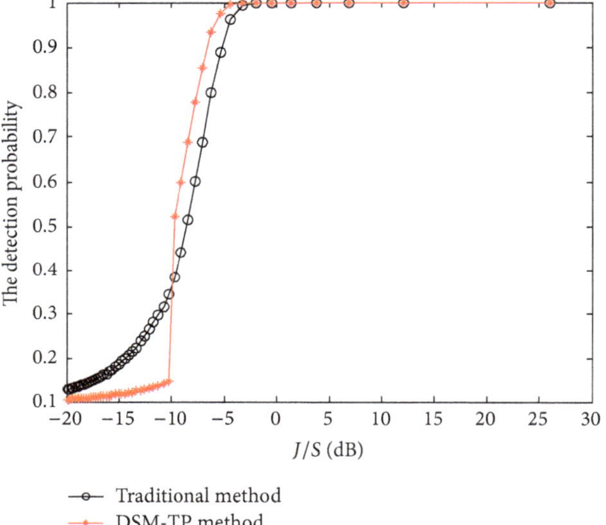

Figure 13: Detection probability for changing *S/J*.

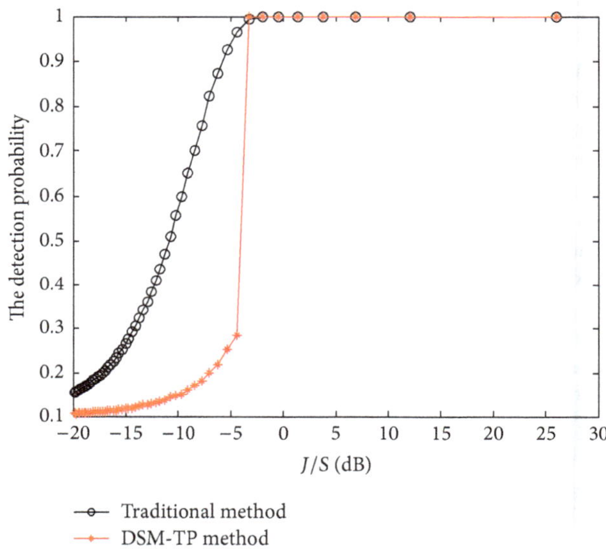

Figure 15: Detection probability for changing *S/J*.

proportion peak results are shown in Figure 12. When *S/J* is equal to 0, the correlation value of the new method is greater than the value of the traditional method, 20 dB. For the same *S/J* conditions, the correlation result of the new method is superior to the traditional method. However, when *S/J* is less than −10 dB, the new method has a very low correlation peak and does not meet the threshold requirement. Thus, the receiver is determined to be in a threat environment, and then the signal is rereceived. However, when *S/J* is less than −10 dB, the correlation peak of the traditional method is larger and meets the threshold requirements. Although the receiver is considered to be secure, the threat signal is received and misinterpreted as a real signal in the traditional method. These results indicate that the new method can search correctly under the relevant threat condition and avoid the false threat. Further, for different values of *S/J*, the detection probability of each method is shown in Figure 13. Under the effective *S/J* condition, the detection probability of the new method is greater than the detection probability of the traditional method, indicating that the detection capability of the new method is better than the traditional method.

Furthermore, the DSM-TP method is tested when the pseudocode overlap rate of threat signal and real signal is set to 50%, with the proportion peak results shown in Figure 14 and the detection probability shown in Figure 15. For a pseudocode overlap rate of the threat signal and real signal equal to 80%, the proportion peak results are shown in Figure 16, and the detection probability is shown in Figure 17. These results indicate that as the pseudocode overlap rate increases, the *S/J* values of the correlation results that do not satisfy the threshold requirement decrease, indicating that as the pseudocode overlap rate increases, the threat of interference increases. Although the impact of this threat is

TABLE 3: Threat prediction results of the two methods.

Overlap rate and J/S		Method: DSM-TP method	Method: Traditional method
30%	20 dB	Correctly; $w = 3$, predict threat state and correlation	Correctly; No prediction
	0 dB	Correctly; $w = 3$, predict threat state and correlation	Correctly; No prediction
	−10 dB	Alarm and rereceive; $w = 3$, predict threat state and correlation	Spoofing; No prediction
50%	20 dB	Correctly; $w = 3$, predict threat state and correlation	Correctly; No prediction
	0 dB	Correctly; $w = 3$, predict threat state and correlation	Correctly; No prediction
	−5 dB	Alarm and rereceive; $w = 3$, predict threat state and correlation	Spoofing; No prediction
80%	20 dB	Correctly; $w = 3$, predict threat state and correlation	Correctly; No prediction
	0 dB	Alarm and rereceive; $w = 3$, predict threat state and correlation	Spoofing; No prediction

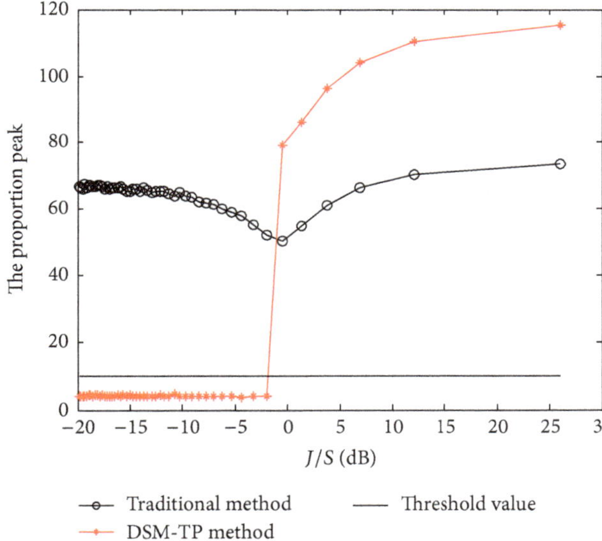

FIGURE 16: Proportion peak for changing S/J.

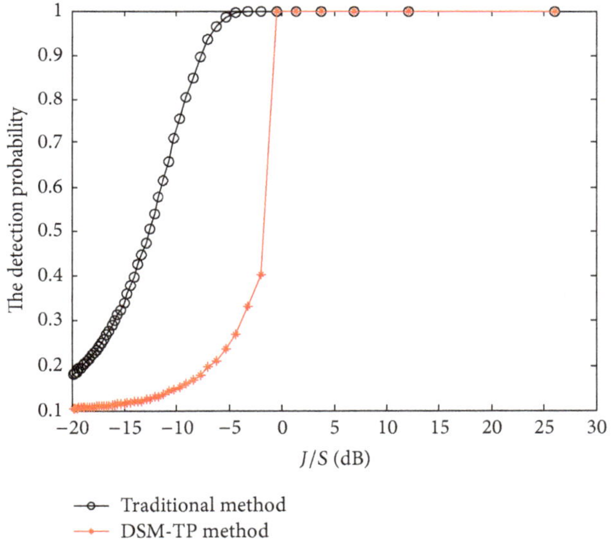

FIGURE 17: Detection probability for changing S/J.

large, the new method can also predict the existence of the threat and can rereceive the signals. However, the traditional method will mistakenly receive the threat signal, and the threat signal is erroneously treated as a real signal. These results indicate that the new method can search correctly under the relevant threat conditions and avoid the false threat, and the detection capability of the new method is better than the traditional method under the effective S/J condition.

The results of threat prediction for the two methods are shown in Table 3 for pseudocode overlap rates of the threat signal and real signal equal to 30%, 50%, and 80%. When S/J is greater than −10 dB, −5 dB, and 0 dB, under different overlap rate conditions, the new method can search correctly. It can predict that the potential threat index W is equal to 3, which verifies that there are man-made threats in the environment and there is a correlation characteristic between the threat signals and the real signal. When S/J is less than −10 dB, −5 dB, and 0 dB, under different overlap rate conditions, the alarm is issued and a research is performed. It can predict that there are man-made threats in the environment and there is a correlation characteristic between the threat signals and the real signal. The results are consistent with the actual input situation. However, the traditional method can search correctly when S/J is greater than −10 dB, −5 dB, and 0 dB, under different overlap rate conditions, but it cannot predict the threat. When S/J is less than −10 dB, −5 dB, and 0 dB, under different overlap rate conditions, the traditional method is affected by the false threat and cannot predict the threat. These results indicate that the new method can adapt to the false threat environment better than the traditional method and can also predict the false threat.

5. Conclusions

In view of the limitations and nonadaptability of current search technology, the proportionality peak judgment mechanism is further optimized considering the influence of complex environment and threat signals. A dynamic search mechanism with threat prediction is proposed, in which we define a series of preset coefficients and a threat index. The simulation results demonstrate that the new method can adapt to lower SNR environments compared to the traditional method in safe conditions, and the detection probability of the new method is higher than the detection probability of the traditional method. Furthermore, these methods are tested in threat conditions, and the results demonstrate that the new method can adapt to the threat environment better than the traditional method and can also predict the threat situation and predict the spoofing threat. In addition, the detection capability of the new method is better than the traditional method in safe conditions and threat conditions.

Conflicts of Interest

The authors declare that they have no conflicts of interest.

Acknowledgments

This work was supported by the National Natural Science Foundation of China (no. 61501309) and the China Postdoctoral Science Foundation (no. 2015M580231).

References

[1] J.-C. Juang and Y.-H. Chen, "Global navigation satellite system signal acquisition using multi-bit codes and a multi-layer search strategy," *IET Radar, Sonar and Navigation*, vol. 4, no. 5, pp. 673–684, 2010.

[2] P. B. S. Harsha and D. V. Ratnam, "Implementation of advanced carrier tracking algorithm using adaptive-extended kalman filter for GNSS receivers," *IEEE Geoscience and Remote Sensing Letters*, vol. 13, no. 9, pp. 1280–1284, 2016.

[3] D. H. Xia, C. H. Liu, Z. J. Wang et al., "Reconstruction progress of the COMPASS-D ECRH system on J-TEXT," *IEEE Transactions on Plasma Science*, vol. 44, no. 9, pp. 1649–1653, 2016.

[4] L. Lestarquit, M. Peyrezabes, J. Darrozes et al., "Reflectometry with an open-source software GNSS receiver: use case with carrier phase altimetry," *IEEE Journal of Selected Topics in Applied Earth Observations and Remote Sensing*, vol. 9, no. 10, pp. 4843–4853, 2016.

[5] J. Li, J. Zhang, S. Chang, and M. Zhou, "Performance evaluation of multimodal detection method for GNSS intermediate spoofing," *IEEE Access*, vol. 4, pp. 9459–9468, 2016.

[6] D. Akopian, "Fast FFT based GPS satellite acquisition methods," *IEE Proceedings—Radar, Sonar and Navigation*, vol. 152, no. 4, pp. 277–286, 2005.

[7] J. B. Y. Tsui, *Fundamentals of Global Positioning System Receivers: A Software Approach*, John Wiley & Son, New York, NY, USA, 2000.

[8] J. B. Bullock, M. Foss, G. J. Geier, and M. King, "Integration of GPS with other sensors and network assistance," in *Understanding GPS Principles and Applications*, Artech House, Boston, Mass, USA, 2006.

[9] A. R. A. Moghaddam, R. Watson, G. Lachapelle, and J. Nielsen, "Exploiting the orthogonality of L2C code delays for a fast acquisition," in *Proceedings of the Institute of Navigation—19th International Technical Meeting of the Satellite Division (ION GNSS '06)*, pp. 1233–1241, Forth Worth, Tex, USA, September 2006.

[10] Y. Wei, *GPS Spread Spectrum Code Synchronization Technology Research*, University of Electronic Science and Technology of China, Chengdu, China, 2007.

[11] Y. Ren, W. Peng, W. Xu, and X. Wang, "The research progress of direct acquisition technology of GPS P(Y)-code," *Global Positioning System*, vol. 2, pp. 2–9, 2003.

[12] C. Yang, J. Vasquez, and J. Chaffee, "Fast direct P(Y)-code acquisition using XFAST," in *Proceedings of the 12th International Technical Meeting of the Satellite Division of the Institute of Navigation (ION GPS '99)*, pp. 317–324, Nashville, Tenn, USA, September 1999.

[13] Q. Zeng, L. Tang, P. Zhang, and L. Pei, "Fast acquisition of L2C CL codes based on combination of hyper codes and averaging correlation," *Journal of Systems Engineering and Electronics*, vol. 27, no. 2, pp. 308–318, 2016.

[14] W. Feng, X. Xing, Q. Zhao, and Z. Wang, "Dual-channel method for fast long PN-code acquisition," *China Communications*, vol. 11, no. 5, pp. 60–70, 2014.

[15] H. Li, X. Cui, M. Lu, and Z. Feng, "Dual-folding based rapid search method for long PN-code acquisition," *IEEE Transactions on Wireless Communications*, vol. 7, no. 12, pp. 5286–5296, 2008.

[16] F. Liu and Y. Feng, "A fast acquisition algorithm overcoming fuzz problems for TDDM spread spectrum signal," *Mathematical Problems in Engineering*, vol. 2014, Article ID 362061, 12 pages, 2014.

[17] F. Liu and Y. Feng, "A long code acquisition algorithm on resolve time-frequency uncertainty problem," *Acta Aeronautica et Astronautica Sinica*, vol. 34, no. 8, pp. 1924–1933, 2013.

10

Performance Analysis of Multiscale Entropy for the Assessment of ECG Signal Quality

Yatao Zhang,[1,2] Shoushui Wei,[1] Yutao Long,[1] and Chengyu Liu[1]

[1]*School of Control Science and Engineering, Shandong University, Jinan 250061, China*
[2]*School of Mechanical, Electrical & Information Engineering, Shandong University, Weihai 264209, China*

Correspondence should be addressed to Shoushui Wei; sswei@sdu.edu.cn

Academic Editor: Mohamad Sawan

This study explored the performance of multiscale entropy (MSE) for the assessment of mobile ECG signal quality, aiming to provide a reasonable application guideline. Firstly, the MSE for the typical noises, that is, high frequency (HF) noise, low frequency (LF) noise, and power-line (PL) noise, was analyzed. The sensitivity of MSE to the signal to noise ratio (SNR) of the synthetic artificial ECG plus different noises was further investigated. The results showed that the MSE values could reflect content level of various noises contained in the ECG signals. For the synthetic ECG plus LF noise, the MSE was sensitive to SNR within higher range of scale factor. However, for the synthetic ECG plus HF noise, the MSE was sensitive to SNR within lower range of scale factor. Thus, a recommended scale factor range within 5 to 10 was given. Finally, the results were verified on the real ECG signals, which were derived from MIT-BIH Arrhythmia Database and Noise Stress Test Database. In all, MSE could effectively assess the noise level on the real ECG signals, and this study provided a valuable reference for applying MSE method to the practical signal quality assessment of mobile ECG.

1. Introduction

ECGs collected via mobile phone are easily contaminated by system noises, body movement, and circumstance disturbing so that the corrupted data could lead to false alarms and misdiagnosis [1]. Thus there is an essential requirement to assess quality of ECG in order to determine whether the ECG can be used for clinical purpose. Recently, ECG quality assessment has been a focus issue [2–8].

ECG waveforms and power spectra are usually in chaos when the ECGs are corrupted badly by noises. So methods currently available employ the metrics mostly based on the characterisation of time or frequency features of the signals. Time based methods aim to identify particular characteristics like RR time interval outliers [4], flat lines, baseline drift, baseline wandering, and steep slopes [5] which can usually compromise the recordings. Frequency based methods can use, for example, the ratio between the low- and the high-frequency content of the signals [6]. A combination of time and frequency methods has also been proposed [7, 8]. The traditional wave form features and power spectrum are relatively simple metrics for assessing quality of ECG; however it is difficult to further improve classification accuracy. So the proper metrics are required to further reflect the content level of noises contained in ECG. Due to ECG being a nonlinear signal, some nonlinear processing methods are reasonable and feasible.

Multiscale entropy (MSE), a nonlinear analysis measure, has been applied to a variety of biomedical signals [9–12]. Costa et al. firstly proposed and employed MSE to biomedical signals analysis [9, 10]. Chung et al. analysed the electroencephalography (EEG) during sleep in patients with Parkinson's disease using MSE [11]. Zhang et al. employed MSE to analyse characterizing different patterns of spontaneous electromyogram (EMG) signals [12] and then got the nonlinear features of these patterns. In these applications, MSE was used for dealing with the signals that were already processed by filtering, correction, and other methods. However ECG quality assessment has to analyse the signals that are without any processing, and the signals are more complex and contain lots of various noises. Besides, the change of the MSE for some common noises contained in ECG and the relationship between the MSE and the content level of noises contained in ECG also need to be analysed in ECG quality assessment. This exemplifies the need to further characterise the factors that affect the MSE and particularly in the context of assessing ECG signal quality.

We clarify that, beyond MSE, other types of dynamic entropy have been used to analyze ECG data. For example, the entropy S defined in a time domain termed natural time differs essentially from MSE since it is defined in an entirely different time domain [13]. By quantifying the S fluctuations and using ratios of "shuffled" and "unshuffled" S fluctuations on fixed time scales or ratios on different time scales complexity measures have been introduced that were found of usefulness not only in the ECG analysis but also to discriminate similar looking electric signals emitted from systems of different dynamics.

The aim of this study was to characterise the MSE in ECG signals and assess the effect of typical ECG noises affecting the ECG, with the help of both artificial and real signals.

2. Materials and Methods

2.1. Databases

2.1.1. Artificial ECG Signals. Four types of artificial signals were considered: ECG representing our actual signal of interest; 0–0.5 Hz low-frequency (LF) noises as representative of baseline wander (BW) and movement artefacts; 50 Hz sinusoid as representative of power-line (PL) noise; 50–180 Hz high-frequency (HF) noises as representative of muscle artefact (MA) and other high-frequency noise. For each of these categories, 50 signals of 50 seconds with a sample rate of 360 Hz were generated and utilised for the analysis.

The artificial ECG signals were generated utilising the open source ECGSYN software described in McSharry et al. [14]. The ECG signals had a heart rate between 50 bpm and 100 bpm which was randomly chosen for each repeat. Each of the low-frequency signals had a frequency which was randomly chosen from the interval 0–0.5 Hz and amplitude which was also randomly chosen. The fifty repeats of PL noise signals were generated with a fixed frequency of 50 Hz sinusoidal signal and random amplitude separately chosen for each repeat. Each of the high-frequency signals had a frequency which was randomly chosen from the interval 50–180 Hz (based on sample rate of 360 Hz) and amplitude which was also randomly chosen.

The study also used three synthetic ECG signals for the analysis of the effect of SNR on the MSE that included the clean artificial ECG plus HF noise, the clean plus LF noise, and the clean plus PL noise. The three types of noisy signals were generated using a SNR in the range from −20 dB to 40 dB in steps of 10 dB. The SNR is defined as

$$\mathrm{SNR} = 10 \times \log_{10}\left(\frac{P_{\text{signal}}}{P_{\text{noise}}}\right), \tag{1}$$

where P_{signal} and P_{noise} denote the power of the clean ECG and the power of the noise, respectively. Figure 1 shows the three types of ECG signals. For each type of signal, SNR was varied from −20 dB to 40 dB, with steps of 10 dB. For each SNR level, 50 repeats were produced and the mean value and standard deviation of each type of signal were calculated. For the effect analysis of signal length, SNR of the ECG was set as a constant of 10 dB. For the effect analysis of SNR, the ECG signal length used the recommendation value from the effect analysis of signal length.

2.1.2. Real ECG Signals. The original real ECG signals were taken from the MIT-BIH Arrhythmia Database made available by the Beth Israel Deaconess Medical Center through the PhysioNet website [15, 16]. Actually, this study had to firstly remove BW from the original real signals because BW contained in the original signals could lead to inaccurate results and then constructed new real ECG signals. This new database contains 48 clean 30-minute 2-channel ambulatory ECG recordings with sample rate of 360 Hz. The MIT-BIH Noise Stress Test Database (NSTDB) [17], also from the PhysioNet website, was also utilised. The NSTDB provides recordings of three typical noise signals which are typically found in ambulatory ECG recordings and include electrode motion artefact (EM), BW, and MA. Because the NSTDB does not include the 50 Hz PL noise, the study also added this type of noise to the real ECG signals for testing. Figure 2 shows the real ECG and the synthetic ECG plus BW, EM, MA, and PL noise.

2.2. MSE Method. The MSE is obtained from a sample entropy analysis of the signals for different time scale factors. Firstly original time series $\{x_1, x_2, \ldots, x_L\}$ (length L) is converted to new multiple time series after a coarse-graining procedure that is performed by averaging the variable data points in nonoverlapping windows by time scale factors. Each element of the new time series y_j^{τ} is computed according to the following equation:

$$y_j^{(\tau)} = \frac{1}{\tau}\sum_{i=(j-1)\tau+1}^{j\tau} x_i, \tag{2}$$

where τ denotes the time scale factor and $1 \le j \le L/\tau$. So the new series $\{y(i) : 1 \le i \le N\}$ is obtained and its length for each scale factor is $N = L/\tau$. For scale 1, the new time series is consistent with the origin time series.

Then the sample entropy of each new time series is computed. For each new time series, a vector $(N - m + 1)$ is constructed as in the following equation:

$$R_i = \{y(i), y(i+1), \ldots, y(i+m-1)\}, \quad i = 1, 2, \ldots, N-m+1, \tag{3}$$

where m indicates the embedding dimension.

The distance $d[R_i, R_j]$ $(i, j = 1, 2, \ldots, N-m+1, i \ne j)$ between the vectors R_i and R_j is computed according to the following equation:

$$d\left[R_i, R_j\right] = \max\left[\left|y(i+k) - y(j+k)\right|\right], \quad k = 0, 1, \ldots, m-1. \tag{4}$$

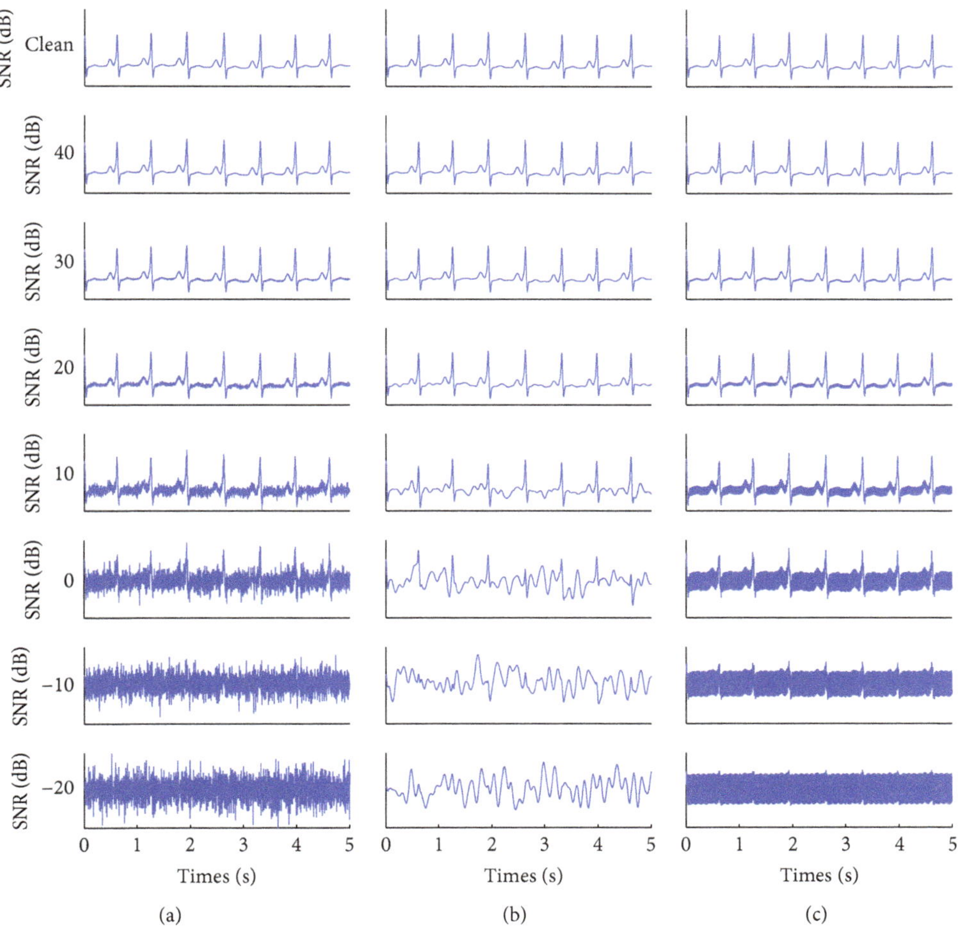

FIGURE 1: The clean artificial ECG and the clean ECG added noises with different SNRs from −20 dB to 40 dB, with a step of 10 dB. (a) The clean ECG and the clean plus HF noise. (b) The clean and the clean plus LF noise. (c) The clean and the clean plus PL noise.

The degree of similarity between R_i^m and the other vector R_j^m within tolerance r can be defined as $B_i^m(r)$, that is, $(N-m)^{-1}$ times the number of the distance $d[R_i, R_j]$ within r, where i range from 1 to $(N-m+1)$ and $i \neq j$ to exclude self-matches, and then the average degree of similarity for all of i is defined as

$$B^m(r) = \sum \frac{B_i^m(r)}{(N-m+1)}, \tag{5}$$

where $i = 1, 2, \ldots, N-m+1$. Similarly, $B^{m+1}(r)$ can also be computed for the embedded dimension of $m+1$. Then, the SampEn is defined as follows:

$$\text{SampEn}(\tau, m, r) = -\ln\left[\frac{B^{\tau,m+1}(r)}{B^{\tau,m}(r)}\right]. \tag{6}$$

Finally, the MSE is the set of SampEn on multiscale and defined as

$$\text{MSE}(s) = \{\tau \mid \text{SampEn}(\tau, m, r)\}. \tag{7}$$

In this study, the values of $m = 2$ and $r = 0.15$ for MSE computation were selected, and the scale factor from 1 to 15 was selected.

2.3. MSE Analysis for the Artificial ECG

2.3.1. The Relationship between MSE and Signal Types. In order to understand how change of MSE of signal reflected signal types, the values of MSE were firstly characterised separately for each of the four typical artificial signals: HF noise, LF noise, the clean ECG, and PL noise, using 50 repeats for each.

2.3.2. Effect of Signal Length on MSE. The possible effect of the length of the recording on the value of its MSE was assessed by calculating the MSE for the clean plus LF, the clean plus PL, and the clean plus HF noise when SNR was 10 dB with a length ranging from 5 s to 85 s, in steps of 5 s, as well as the clean ECG. As before, 50 repeats were considered in each of the four groups. A SNR of 10 dB was used when generating the noisy ECG signals with HF, PL, and LF noise, respectively.

2.3.3. Effect of the SNR on MSE. The possible effect of the SNR of the signals on value of their MSE was assessed by calculating the MSE for the three types of noisy signals

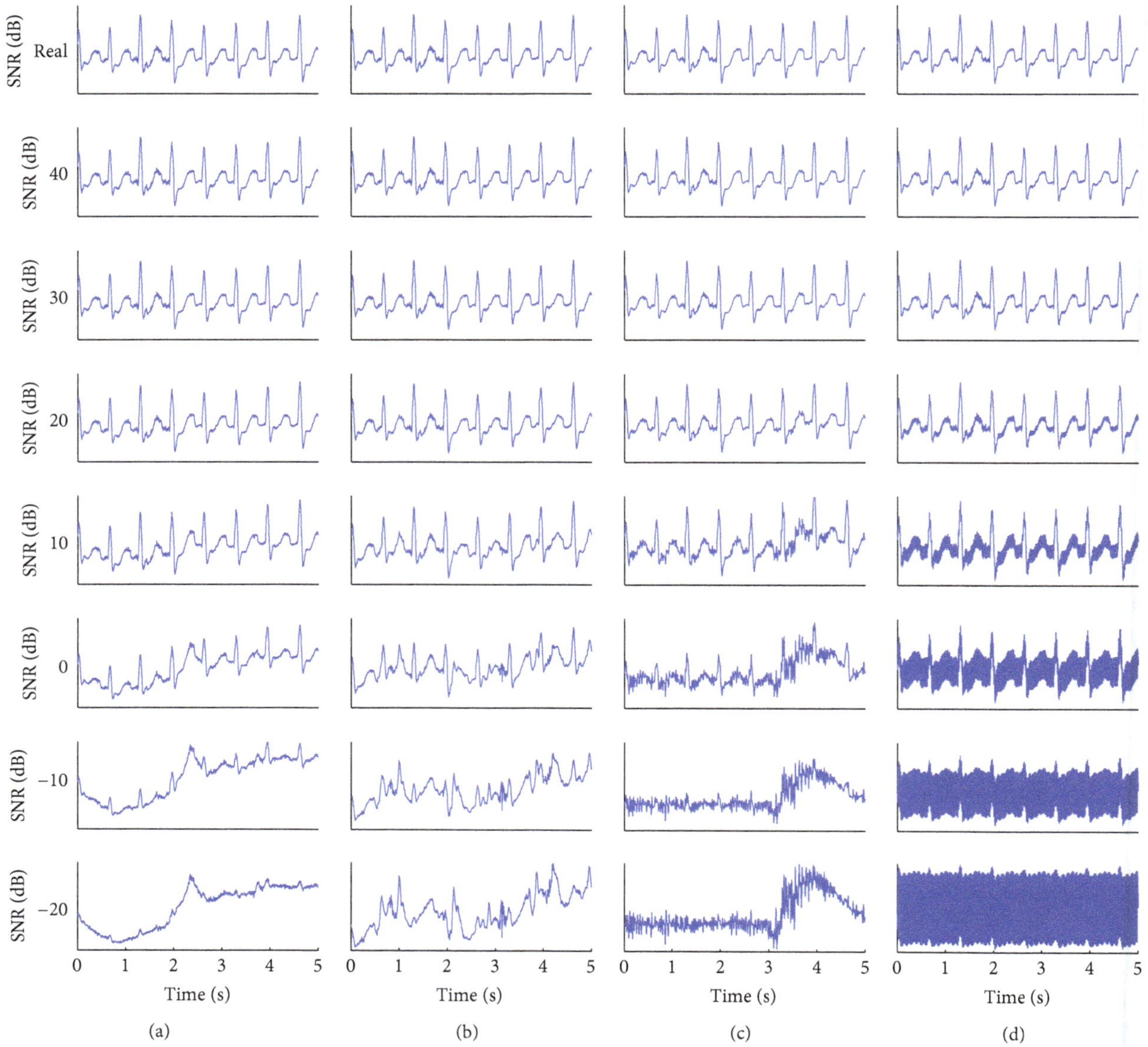

Figure 2: The real ECG and real ECG added noises with different SNRs from −20 dB to 40 dB, with a step of 10 dB. (a) The real ECG and the real plus BW. (b) The real ECG and the real plus EM. (c) The real ECG and the real plus MA. (d) The real ECG and the real plus PL noise.

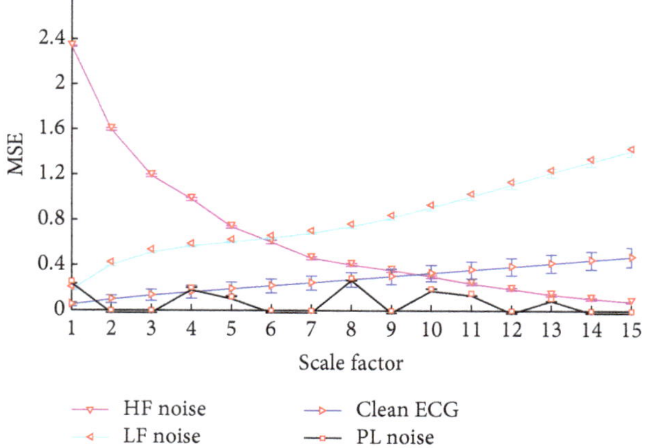

Figure 3: MSE of three typical noises and the clean ECG with scale factor from 1 to 15.

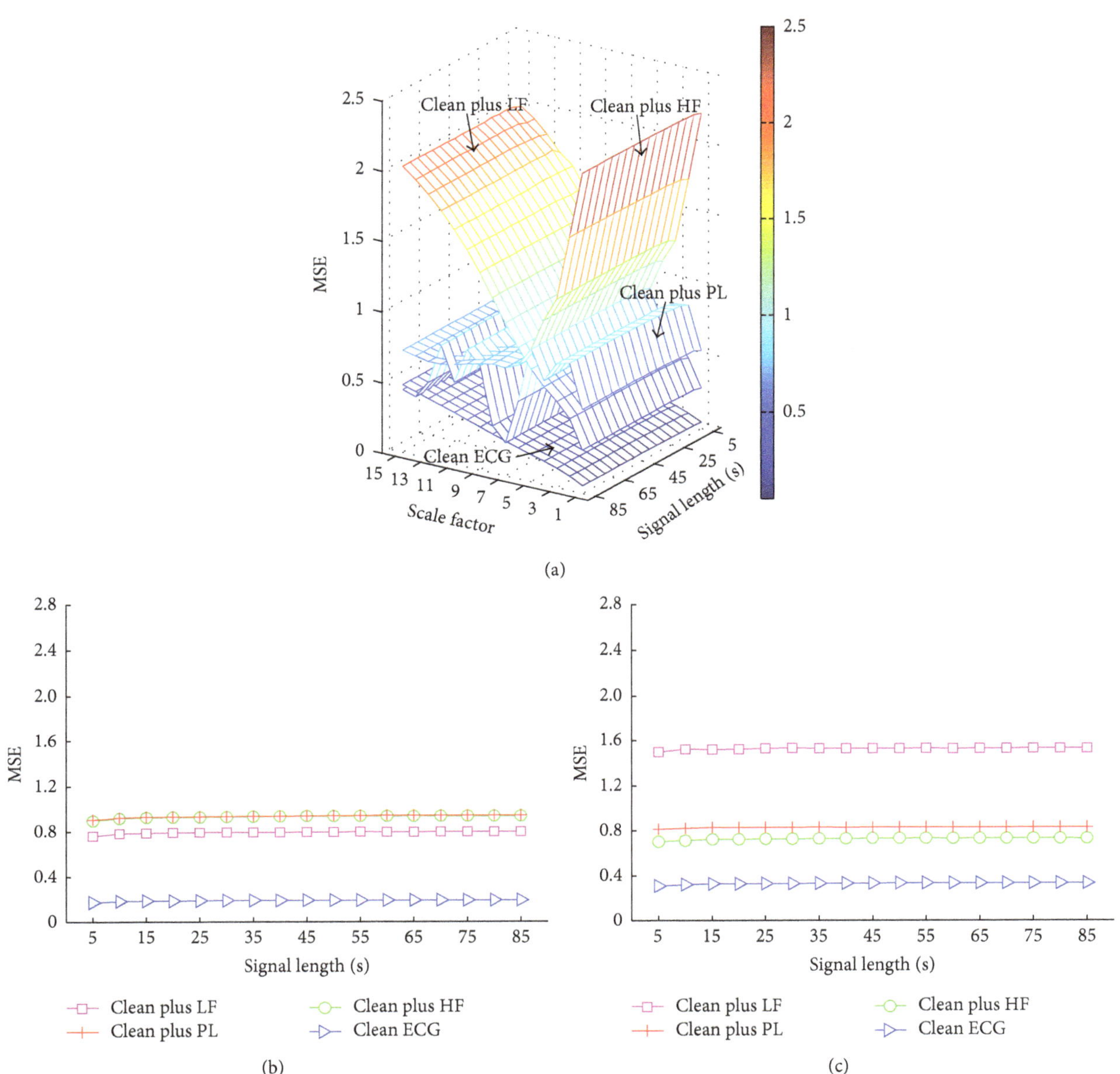

FIGURE 4: The effect of signal length on MSE for the artificial ECG signals. (a) The 3D view of the effect of signal length on all of the four artificial ECGs. (b) Different lengths and their corresponding MSE with a scale factor of 5. (c) Different lengths and their corresponding MSE with a scale factor of 10.

described above, generated using a SNR in the range from −20 dB to 40 dB in steps of 10 dB, as well as the clean ECG.

2.4. MSE Analysis for the Real ECG. We verified the sensitivity of MSE to SNR in the real ECG plus different types of noise (i.e., BW, EM, MA, and PL) using a SNR within the range from 40 dB to −20 dB in steps of 10 dB, as well as the real ECG.

3. Results

Figure 3 shows the mean value and standard deviation of the MSE of four kinds of signals (i.e., HF noise, LF noise, the clean ECG, and PL noise). The MSE value of the HF noise decreased monotonically with scale factor increased. Both of the MSE values of the LF noise and clean ECG increased monotonically with scale factor increased. The MSE of PL interference showed fluctuation when it increased with scale factor increased.

Figure 4(a) shows the effect of signal length on MSE for the artificial ECGs, that is, the clean artificial ECG plus LF, PL, or HF noise, as well as the clean artificial ECG. Figure 4(b) gives examples of the MSE using a scale factor of 5. Similarly, Figure 4(c) gives examples of MSE using a scale factor of 10. The MSE of all types of the ECG had a slight change when the signal length was below 10 s. However, the MSE kept a steady value when signal length was longer than 10 s.

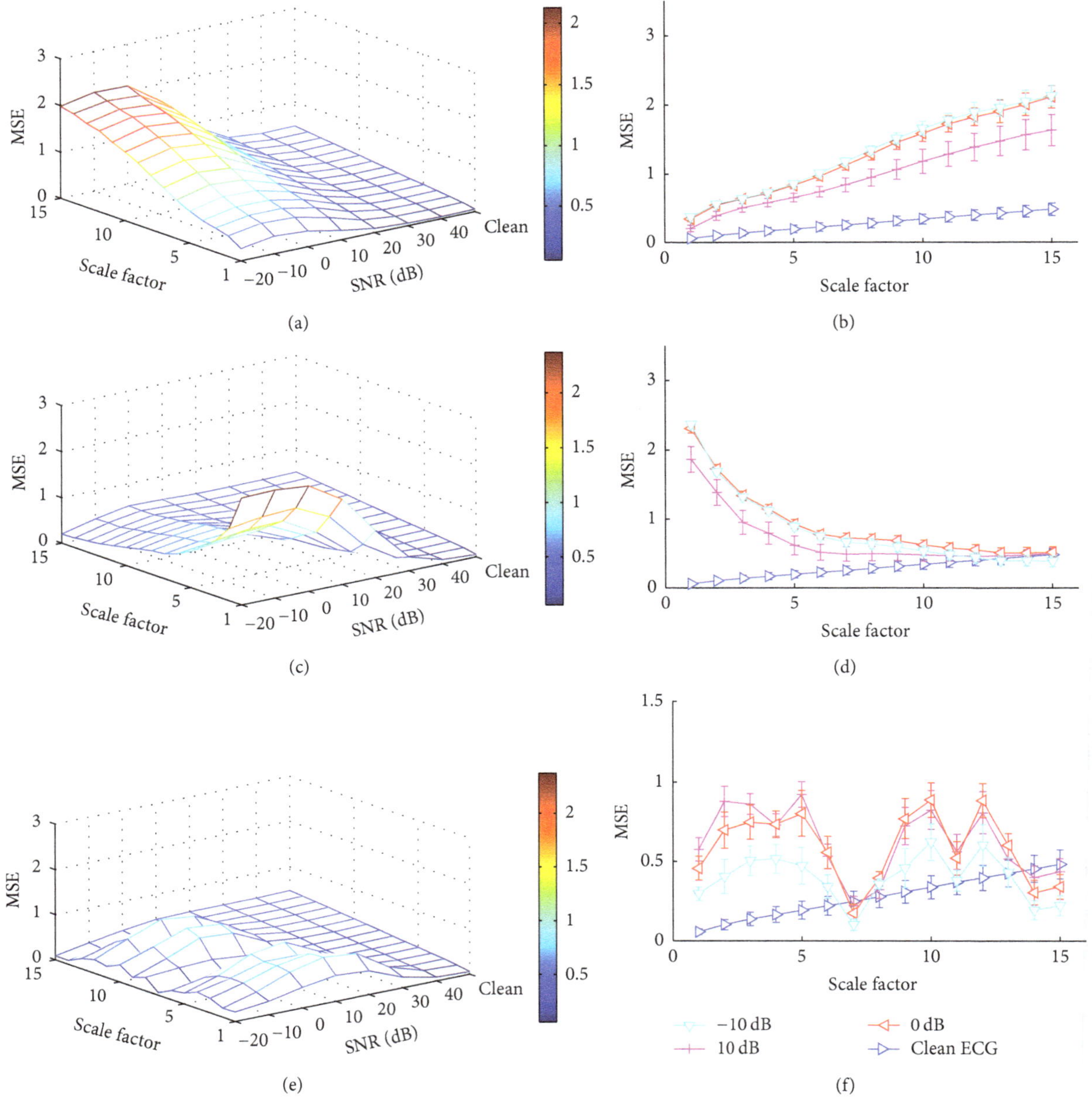

FIGURE 5: The results of SNR effect on the MSE for the artificial signals. (a) The 3D view of the relationship between MSE and SNRs of the clean ECG plus LF noise. (b) Examples of the clean ECG plus LF noise with SNRs of −10 dB, 0 dB, or 10 dB, as well as the clean ECG. (c) The 3D view of the relationship between MSE and SNRs of the clean ECG plus HF. (d) Examples of the clean ECG plus HF with SNRs of −10 dB, 0 dB, or 10 dB, as well as the clean ECG. (e) The 3D view of the relationship between MSE and SNRs of the clean plus PL. (f) Examples of the clean plus PL noise with SNR of −10 dB, 0 dB, or 10 dB, as well as the clean ECG.

Figure 5 shows the relationship between the SNR and the MSE for the artificial ECG signals. For the clean ECG plus LF noise and the clean ECG plus HF noise, the MSE increased with the decrease of the SNR when the SNR was larger than −10 dB and then decreased when the SNR was less than −10 dB. For the clean ECG plus PL noise, the MSE showed fluctuation and increased with the decrease of the SNR when the SNR was larger than 10 dB and then decreased. Figure 5(a) shows that the MSE for the clean ECG plus LF noise increases relatively obviously within the higher range of scale factor, but Figure 5(c) shows that the MSE for the clean ECG plus HF noise significantly increases within the lower range of scale factor.

Figures 5(b), 5(d), and 5(f) show examples of the relationship between the SNR and the MSE for the different artificial ECG signals when the SNR is −10 dB, 0 dB, or 10 dB, as well as the clean ECG.

Figure 6 shows the results of SNR affecting MSE for the real signals. For the real ECG plus BW, the MSE reached the maximum value when the SNR was 10 dB and then decreased

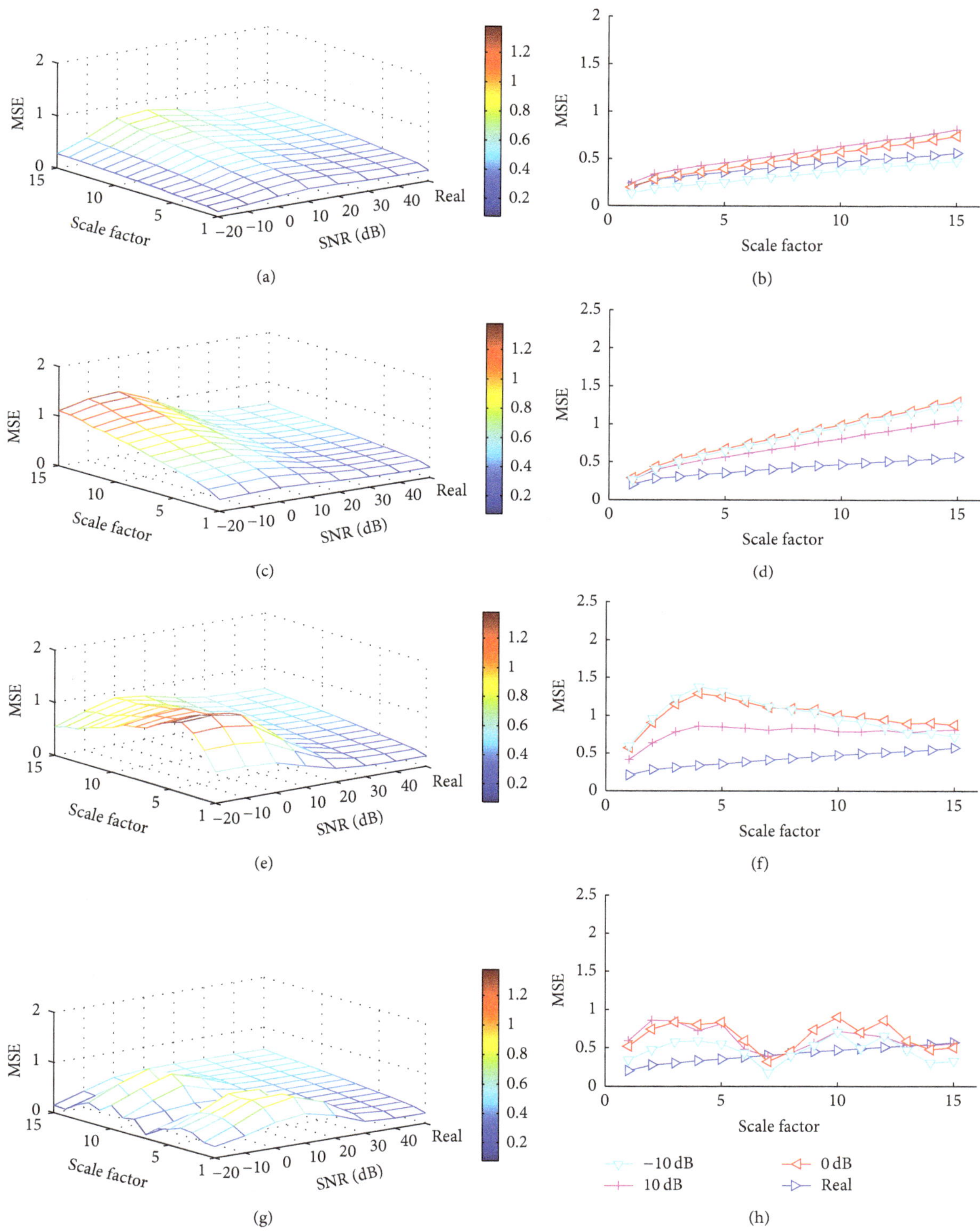

FIGURE 6: The validation results of SNR effect on MSE for the real ECG signals. (a), (c), (e), and (g) are the 3D view of the MSE of the real ECG plus BW, EM, MA, and PL, with SNR from −20 dB to 40 dB with a step of 10 dB, respectively. (b), (d), (f), and (h) are the MSE values of the noisy ECGs with −10 dB, 0 dB, 10 dB, and the real ECG, respectively.

with the decrease of the SNR. However, for the real ECG plus EM or MA, the MSE increased with the decrease of the SNR when the SNR is larger than −10 dB, and then the MSE decreased with the decrease of the SNR. The MSE showed obvious fluctuation and increased with the decrease of the SNR when the SNR was larger than 0 dB and then decreased.

4. Discussions

This study tries to provide interpretation of the MSE measure in the context of quality classification of ECG collected via mobile phone.

The results indicate that the MSE is closely related to complexity of time series, instead of the periodicity and randomicity of time series. Thus, the MSE for LF noise and the clean ECG (quasiperiodic signal) monotonously increase with the increase of scale factor, but the MSE for HF noise monotonously decreases with the increase of scale factor. In addition, the MSE for PL noise (periodic signals) shows fluctuation. The results further indicate that change trends of the MSE for various types of artificial ECG signals are clearly different, and it means that the MSE can be applied to distinguish different noises contained in biomedical signal.

Actually, the MSE for ECG is independent of signal length when the length is longer than 10 s. This point is beneficial for applying MSE to mobile ECG because the length of ECG collected by mobile devices is of relatively short duration. It is noted that the MSE also provides the possibility of real-time analysis since a longer signal means that more time is required to compute its MSE.

The results also show that the MSE is sensitive to SNR; that is, the MSE can reflect the content level of noises contained in the artificial or real ECG signals. For the real ECG signals, BW represents primarily baseline wander, a low-frequency signal usually caused by motion of the subject or the electrodes. So the LF noise represents mostly BW, and EM contains electrode motion artifact, with significant amounts of baseline wander [18, 19]. In this study, we also use the LF noise to approximately simulate EM. The trend of MSE for the real ECG plus BW or EM is consistent with that of the clean artificial ECG plus LF noise except that the corresponding SNR is different when the MSE reaches the maximum value because the real ECG signals always contain considerable LF noise, but the clean artificial ECG signals do not.

MA contains primarily muscle noise, and HF noise represents MA [18, 19]. The trend of MSE for the real ECG plus MA is consistent with that of the clean artificial ECG plus HF noise except that the MSE for the real ECG plus MA noise is relatively larger within high range of scale factor because MA from the NSTDB also contains other noises.

Actually, when content level of noise contained in time series reaches a certain peak value, complexity of time series begins to decrease; that is, structure of time series tends to the steady state. Thus, for all of the synthetic ECG added different types of noise, the MSE firstly increases with the decrease of the SNR and then decreases.

For all of the synthetic ECG signals, the MSE is sensitive to the content level of noise contained in the signals. However the change trend of the MSE for the synthetic ECG added various noises is different. Thus we propose that the MSE should be applied to analyse the signal that is obviously contaminated by a single variety of noise, and the content level of other types of noise is relatively less.

Besides, for the synthetic ECG plus BW or EM, MSE is more sensitive to the change of SNR when scale factor is within higher range. However, for the synthetic ECG plus MA, MSE is more sensitive to SNR within lower range of scale factor. Thus, a proper scale factor has to be chosen in order that MSE is better sensitive to SNR for all types of the synthetic ECGs. This study suggests that a proper scale factor should be selected from 5 to 10.

5. Conclusions

This study verified the performances of MSE for the assessment of ECG signal quality. Our results showed that MSE was sensitive to the noise level within the ECG signals and could, therefore, represent a valuable tool for the assessment of ECG signal quality.

Conflict of Interests

The authors declare that there is no conflict of interests regarding this work.

Acknowledgments

This work was supported by the Young Scientists Fund of the National Natural Science Foundation of China under Grant no. 61201049, the Excellent Young Scientist Awarded Foundation of Shandong Province in China under Grant no. BS2013DX029, and the China Postdoctoral Science Foundation under Grant no. 2013M530323. The authors would like to thank the MIT-BIH Arrhythmia Database for providing the invaluable data used in their research.

References

[1] Q. Li and G. D. Clifford, "Signal quality and data fusion for false alarm reduction in the intensive care unit," *Journal of Electrocardiology*, vol. 45, no. 6, pp. 596–603, 2012.

[2] G. D. Clifford and G. B. Moody, "Signal quality in cardiorespiratory monitoring," *Physiological Measurement*, vol. 33, no. 9, article E01, 2012.

[3] G. B. Moody, *Physionet/Computing in Cardiology Challenge*, 2011, http://physionet.org/challenge/2011.

[4] C. Liu, P. Li, L. N. Zhao, Y. Jing, and L. P. Liu, "Evaluation method for heart failure using RR sequence normalized histogram," in *Proceedings of the 38th Computing in Cardiology*, pp. 305–308, IEEE, Hangzhou, China, September 2011.

[5] P. Langley, L. Y. Di Marco, S. King et al., "An algorithm for assessment of quality of ECGs acquired via mobile telephones," in *Proceedings of the 38th Computing in Cardiology*, pp. 281–284, Hangzhou, China, September 2011.

[6] S. Zaunseder, R. Huhle, and H. Malberg, "Assessing the usability of ECG by ensemble decision trees," in *Proceedings of the 38th Computing in Cardiology (CinC '11)*, pp. 277–280, Hangzhou, China, September 2011.

[7] G. D. Clifford, J. Behar, Q. Li, and I. Rezek, "Signal quality indices and data fusion for determining clinical acceptability of electrocardiograms," *Physiological Measurement*, vol. 33, no. 9, pp. 1419–1433, 2012.

[8] Y. Zhang, C. Liu, S. Wei, C. Wei, and F. Liu, "ECG quality assessment based on a kernel support vector machine and genetic algorithm with a feature matrix," *Journal of Zhejiang University SCIENCE C*, vol. 15, no. 7, pp. 564–573, 2014.

[9] M. Costa, A. L. Goldberger, and C.-K. Peng, "Multiscale entropy analysis of complex physiologic time series," *Physical Review Letters*, vol. 89, no. 6, pp. 0681021–0681024, 2002.

[10] M. Costa, A. L. Goldberger, and C. K. Peng, "Multiscale entropy analysis of biological signals," *Physical Review E: Statistical, Nonlinear, and Soft Matter Physics*, vol. 71, no. 2, Article ID 021906, 18 pages, 2005.

[11] C.-C. Chung, J.-H. Kang, R.-Y. Yuan et al., "Multiscale entropy analysis of electroencephalography during sleep in patients with parkinson disease," *Clinical EEG & Neuroscience*, vol. 44, no. 3, pp. 221–226, 2013.

[12] X. Zhang, X. Chen, P. E. Barkhaus, and P. Zhou, "Multiscale entropy analysis of different spontaneous motor unit discharge patterns," *IEEE Journal of Biomedical and Health Informatics*, vol. 17, no. 2, pp. 470–476, 2013.

[13] P. A. Varotsos, N. V. Sarlis, E. S. Skordas, and M. S. Lazaridou, "Natural entropy fluctuations discriminate similar-looking electric signals emitted from systems of different dynamics," *Physical Review E—Statistical, Nonlinear, and Soft Matter Physics*, vol. 71, no. 1, Article ID 011110, 2005.

[14] P. E. McSharry, G. D. Clifford, L. Tarassenko, and L. A. Smith, "A dynamical model for generating synthetic electrocardiogram signals," *IEEE Transactions on Biomedical Engineering*, vol. 50, no. 3, pp. 289–294, 2003.

[15] G. B. Moody and R. G. Mark, "The impact of the MIT-BIH arrhythmia database," *IEEE Engineering in Medicine and Biology Magazine*, vol. 20, no. 3, pp. 45–50, 2001.

[16] A. L. Goldberger, L. A. Amaral, L. Glass et al., "PhysioBank, PhysioToolkit, and PhysioNet: components of a new research resource for complex physiologic signals," *Circulation*, vol. 101, no. 23, pp. e215–e220, 2000.

[17] G. B. Moody, W. E. Muldrow, and R. G. Mark, "A noise stress test for arrhythmia detectors," in *Proceedings of the 11th Computers in Cardiology Conference*, pp. 381–384, Park City, Utah, USA, 1984.

[18] M. Altuve, O. Casanova, S. Wong, G. Passariello, A. Hernandez, and G. Carrault, "Evaluación de dos Métodos para la Segmentación del Ancho de la Onda T en el ECG," in *Proceedings of the 4th Latin American Congress on Biomedical Engineering Bioengineering Solutions for Latin America Health*, pp. 1254–1258, Margarita Island, Venezuela, 2008.

[19] G. B. Moody and R. G. Mark, "The MIT-BIH Arrhythmia Database on CD-ROM and software for use with it," in *Proceedings of the 17th Computers in Cardiology*, pp. 185–188, Chicago, Ill, USA, September 1990.

11

Compressed Sensing MRI Reconstruction Algorithm based on Contourlet Transform and Alternating Direction Method

Zhenyu Hu, Qiuye Wang, Congcong Ming, Lai Wang, Yuanqing Hu, and Jian Zou

School of Information and Mathematics, Yangtze University, Hubei 434020, China

Correspondence should be addressed to Jian Zou; zoujian@yangtzeu.edu.cn

Academic Editor: Gianluca Setti

Compressed sensing (CS) based methods have recently been used to reconstruct magnetic resonance (MR) images from undersampled measurements, which is known as CS-MRI. In traditional CS-MRI, wavelet transform can hardly capture the information of image curves and edges. In this paper, we present a new CS-MRI reconstruction algorithm based on contourlet transform and alternating direction method (ADM). The MR images are firstly represented by contourlet transform, which can describe the images' curves and edges fully and accurately. Then the MR images are reconstructed by ADM, which is an effective CS reconstruction method. Numerical results validate the superior performance of the proposed algorithm in terms of reconstruction accuracy and computation time.

1. Introduction

CS is a new sampling and compression theory. It utilizes the sparseness of a signal in a particular domain and can reconstruct the signal from significantly fewer samples than Nyquist sampling, which has been the fundamental principle in signal processing for many years [1–3]. Due to the above advantages, CS has received considerable attentions in many areas, one of which is MRI reconstruction [4, 5]. MRI is safer, more frequent, and accurate for clinical diagnosis. However, conventional MRI needs to spend much time scanning body regions, causing the expensive cost and the nonidealized space resolution. In the meantime, the physiological property in the tested body will make the image blurry and distortional. Therefore, under the premise of guaranteeing the image quality, speeding up the MRI compression and reconstruction has been the powerful impetus to promote the development of MRI techniques.

For CS-MRI, there are two key points that need further investigation. The first one is sparse transform. In MRI reconstruction, the MR images themselves are not sparse but have sparse representations in some transform domains. In traditional CS-MRI, wavelet transform is commonly used as a sparse transform [6, 7]. However, as the limitations of direction, wavelet transform can hardly capture the information of image curves and edges fully and accurately. In contrast, curves and edges are mainly features of MR images. Therefore, more effective sparse transform should be considered for CS-MRI. Contourlet transform, also known as Pyramid Directional Filter Bank (PDFB), is put forward to make up for the inadequacy of the wavelet transform [8]. Contourlet transform can describe the image's contour and directional texture information fully and accurately since it realizes any directional decomposition at each scale. Furthermore, contourlet is constructed directly in a discrete domain and has low computing complexity. Thus, contourlet transform can be easily implemented for MR images [9, 10].

The second one is the reconstruction algorithm. In recent years, a number of algorithms have been put forward for the signal reconstruction in CS, for example, interior-point algorithm [11], iterative shrinkage/thresholding algorithm (ISTA) [12], fast iterative shrinkage/threshold algorithm (FISTA) [13], Sparse Reconstruction by Separable Approximation (SpaRSA) [14]. But not all of these algorithms are suitable for CS-MRI since the dimensions of the MR images are huge. Alternating direction method (ADM) is an efficient CS reconstruction algorithm that has a faster convergence speed than some traditional methods [15]. Meanwhile, ADM is able to solve large-scale CS problem since all the iterations of ADM only contain the first-order information of the objective function, which have low computing complexity.

In this paper, we present a new CS-MRI reconstruction algorithm based on contourlet transform and ADM. The proposed algorithm can recover the curves and edges of a MR image more precisely and suit large-scale MRI reconstruction.

The organization of the rest of this paper is as follows: in Section 2, we first introduce CS-MRI model briefly and then present our new algorithm. Numerical results will demonstrate the effectiveness of the proposed algorithm in Section 3. Finally, we conclude the paper in Section 4.

2. ADM for Contourlet-Based CS-MRI

2.1. Contourlet-Based CS-MRI Model. The basic problem of CS is to recover a signal $\mathbf{x}$ from underdetermined linear measurement $\mathbf{y} = \mathbf{\Phi x}$ where $\mathbf{y} \in \mathbf{R}^m$, $\mathbf{x} \in \mathbf{R}^n$, and $\mathbf{\Phi} \in \mathbf{R}^{m\times n}$, $m < n$. This underdetermined linear system has infinite solutions when seen from the aspect of algebra. However, according to the CS theory, under the assumption that $\mathbf{x}$ is sparse, $\mathbf{x}$ can be reconstructed by the following optimization problem:

$$\begin{aligned} &\min_{\mathbf{x}} \quad \|\mathbf{x}\|_0 \\ &\text{s.t.} \quad \mathbf{y} = \mathbf{\Phi x}, \end{aligned} \tag{1}$$

where $\|\mathbf{x}\|_0$ is the ℓ_0 norm which means the nonzero numbers of $\mathbf{x}$.

Problem (1) is difficult to solve since it is NP-hard. It can be relaxed as the following convex problem:

$$\begin{aligned} &\min_{\mathbf{x}} \quad \|\mathbf{x}\|_1 \\ &\text{s.t.} \quad \mathbf{y} = \mathbf{\Phi x}, \end{aligned} \tag{2}$$

where $\|\mathbf{x}\|_1 = \sum_i |x_i|$ is the sum of absolute values of $\mathbf{x}$. Problem (2) is a convex optimization problem and can be solved by many algorithms.

Problems (1) and (2) are under the assumption that $\mathbf{x}$ is sparse. However, in many applications, the signal itself is not sparse but has a sparse representation in some transform domains. For example $\mathbf{x} = \mathbf{\Psi}\boldsymbol{\theta}$, where $\mathbf{x}$ is the original signal which is not sparse and $\boldsymbol{\theta}$ is the sparse coefficient with respect to the sparse transform matrix $\mathbf{\Psi}$. In this case, CS model should be $\mathbf{y} = \mathbf{\Phi}\boldsymbol{\theta} = \mathbf{\Phi\Psi}^*\mathbf{x} = \mathbf{Ax}$, where $\mathbf{\Psi}^*$ denotes the inverse of $\mathbf{\Psi}$. The optimization problem should be changed as follows:

$$\begin{aligned} &\min_{\mathbf{x}} \quad \|\mathbf{\Psi}^*\mathbf{x}\|_1 \\ &\text{s.t.} \quad \mathbf{y} = \mathbf{Ax}. \end{aligned} \tag{3}$$

2.2. ADM for Contourlet-Based CS-MRI. In this subsection, we solve (3) by ADM. We first introduce an auxiliary variable $\mathbf{r}$ and transform (3) into an equivalent problem:

$$\min_{\mathbf{x},\mathbf{r}} \left\{\|\mathbf{\Psi}^*\mathbf{x}\|_1 : \mathbf{Ax} + \mathbf{r} = \mathbf{y},\ \|\mathbf{r}\| \le \delta\right\}. \tag{4}$$

The augmented Lagrangian function of problem (4) is given by

$$\begin{aligned} \min_{\mathbf{x},\mathbf{r}} \Big\{ f(\mathbf{x},\mathbf{r}) &= \|\mathbf{\Psi}^*\mathbf{x}\|_1 - \boldsymbol{\lambda}^{\mathrm{T}}(\mathbf{Ax} + \mathbf{r} - \mathbf{y}) \\ &+ \frac{\beta}{2}\|\mathbf{Ax} + \mathbf{r} - \mathbf{y}\|^2 : \|\mathbf{r}\| \le \delta \Big\}, \end{aligned} \tag{5}$$

where $\boldsymbol{\lambda} \in \mathbf{R}^m$ is a Lagrangian multiplier and $\beta > 0$ is a penalty parameter.

If we fix $\mathbf{x} = \mathbf{x}^{\mathrm{k}}$, $\boldsymbol{\lambda} = \boldsymbol{\lambda}^{\mathrm{k}}$; that is,

$$\begin{aligned} f(\mathbf{x}^{\mathrm{k}},\mathbf{r}) &= \|\mathbf{\Psi}^*\mathbf{x}^{\mathrm{k}}\|_1 - (\boldsymbol{\lambda}^{\mathrm{k}})^{\mathrm{T}}(\mathbf{Ax}^{\mathrm{k}} + \mathbf{r} - \mathbf{y}) \\ &+ \frac{\beta}{2}\|\mathbf{Ax}^{\mathrm{k}} + \mathbf{r} - \mathbf{y}\|^2. \end{aligned} \tag{6}$$

The objective function $f(\mathbf{x}^{\mathrm{k}},\mathbf{r})$ is only connected with $\mathbf{r}$; then (5) is equivalent to

$$\min_{\mathbf{r}} \left\{ f(\mathbf{x}^{\mathrm{k}},\mathbf{r}) : \|\mathbf{r}\| \le \delta \right\}. \tag{7}$$

For problem (7), since $(d/d\mathbf{r}) f(\mathbf{x}^{\mathrm{k}},\mathbf{r}) = -(\boldsymbol{\lambda}^{\mathrm{k}})^{\mathrm{T}} + \beta(\mathbf{Ax}^{\mathrm{k}} + \mathbf{r} - \mathbf{y})^{\mathrm{T}} = 0$, we have $\mathbf{r} = \boldsymbol{\lambda}^{\mathrm{k}}/\beta - (\mathbf{Ax}^{\mathrm{k}} - \mathbf{y})$. The minimization of (7) with respect to $\mathbf{r}$ is shown by

$$\mathbf{r}^{\mathrm{k}+1} = \mathbf{P}_{B_\delta}\left(\frac{\boldsymbol{\lambda}^{\mathrm{k}}}{\beta} - (\mathbf{Ax}^{\mathrm{k}} - \mathbf{y})\right), \tag{8}$$

where $\mathbf{P}_{B_\delta}$ is the projection onto the set $B_\delta : \{\mathbf{r} \in \mathbf{R}^m : \|\mathbf{r}\| \le \delta\}$.

Then we fix $\mathbf{r} = \mathbf{r}^{\mathrm{k}+1}$, $\boldsymbol{\lambda} = \boldsymbol{\lambda}^{\kappa}$; that is,

$$\begin{aligned} f(\mathbf{x},\mathbf{r}^{\mathrm{k}+1}) &= \|\mathbf{\Psi}^*\mathbf{x}\|_1 - (\boldsymbol{\lambda}^{\mathrm{k}})^{\mathrm{T}}(\mathbf{Ax} + \mathbf{r}^{\mathrm{k}+1} - \mathbf{y}) \\ &+ \frac{\beta}{2}\|\mathbf{Ax} + \mathbf{r}^{\mathrm{k}+1} - \mathbf{y}\|^2. \end{aligned} \tag{9}$$

The objective function $f(\mathbf{x},\mathbf{r}^{\mathrm{k}+1})$ is only relevant to $\mathbf{x}$; then problem (5) is equivalent to

$$\min_{\mathbf{x}} f(\mathbf{x},\mathbf{r}^{\mathrm{k}+1}). \tag{10}$$

Equation (10) is equivalent approximately to

$$\min_{\mathbf{x}} \|\mathbf{\Psi}^*\mathbf{x}\|_1 + \frac{\beta}{2}\left\|\mathbf{Ax} + \mathbf{r}^{\mathrm{k}+1} - \mathbf{y} - \frac{\boldsymbol{\lambda}^{\mathrm{k}}}{\beta}\right\|^2. \tag{11}$$

Require: $\mathbf{A}, \mathbf{y}, \mathbf{r}^0, \mathbf{x}^0, \boldsymbol{\lambda}^0, \beta > 0, \gamma > 0, \Gamma > 0.$
Ensure: $\mathbf{x}$
while "stopping criterion is not met" **do**
$\mathbf{r}^{k+1} = \mathbf{P}_{B_\delta}\left(\frac{\boldsymbol{\lambda}^k}{\beta} - (\mathbf{A}\mathbf{x}^k - \mathbf{y})\right);$
$\mathbf{x}^{k+1} = \text{Shrink}\left(\mathbf{x}^k - \Gamma\mathbf{g}^k, \frac{\Gamma}{\beta}\boldsymbol{\Psi}^*\right);$
$\boldsymbol{\lambda}^{k+1} = \boldsymbol{\lambda}^k - \gamma\beta\left(\mathbf{A}\mathbf{x}^{k+1} + \mathbf{r}^{k+1} - \mathbf{y}\right);$
end while

ALGORITHM 1

Let $h(\mathbf{x}) = (\beta/2)\|\mathbf{A}\mathbf{x} + \mathbf{r}^{k+1} - \mathbf{y} - \boldsymbol{\lambda}^k/\beta\|^2$; then

$$\begin{aligned} h(\mathbf{x}) &= \frac{\beta}{2}\left[\left\|\mathbf{A}\mathbf{x}^k + \mathbf{r}^{k+1} - \mathbf{y} - \frac{\boldsymbol{\lambda}^k}{\beta}\right\|^2 \right. \\ &\quad + \left(2\mathbf{A}^T\left(\mathbf{A}\mathbf{x}^k + \mathbf{r}^{k+1} - \mathbf{y} - \frac{\boldsymbol{\lambda}^k}{\beta}\right)\right)^T \\ &\quad \left. + \frac{1}{2}\left(\mathbf{x} - \mathbf{x}^k\right)^T 2\mathbf{A}^T\mathbf{A}\left(\mathbf{x} - \mathbf{x}^k\right)\right] \\ &= \beta\left[\left(\mathbf{g}^k\right)^T\left(\mathbf{x} - \mathbf{x}^k\right) + \frac{1}{2}\left\|\mathbf{A}\left(\mathbf{x} - \mathbf{x}^k\right)\right\|^2\right] \\ &= \beta\left[\left(\mathbf{g}^k\right)^T\left(\mathbf{x} - \mathbf{x}^k\right) + \frac{1}{2\Gamma}\left\|\mathbf{x} - \mathbf{x}^k\right\|^2\right], \end{aligned} \tag{12}$$

where $\mathbf{g}^k \triangleq \mathbf{A}^T(\mathbf{A}\mathbf{x}^k + \mathbf{r}^{k+1} - \mathbf{y} - \boldsymbol{\lambda}^k/\beta)$ is the gradient of quadratic items about (12), $\Gamma > 0$ is a parameter, and the first equation of (12) is based on the Taylor expansion. Then (11) is equivalent to

$$\min_{\mathbf{x}}\left\{\left\|\boldsymbol{\Psi}^*\mathbf{x}\right\|_1 + \beta\left(\left(\mathbf{g}^k\right)^T\left(\mathbf{x} - \mathbf{x}^k\right) + \frac{1}{2\Gamma}\left\|\mathbf{x} - \mathbf{x}^k\right\|^2\right)\right\}. \tag{13}$$

Problem (13) is equivalent approximately to

$$\min_{\mathbf{x}}\left\{\boldsymbol{\Psi}^*\left(\|\mathbf{x}\|_1 + \frac{\beta}{2\Gamma\boldsymbol{\Psi}^*}\left\|\mathbf{x} - \left(\mathbf{x}^k - \Gamma\mathbf{g}^k\right)\right\|\right)\right\}, \tag{14}$$

which has a close form solution by shrinkage (or soft thresholding) formula

$$\mathbf{x}^{k+1} = \text{Shrink}\left(\mathbf{x}^k - \Gamma\mathbf{g}^k, \frac{\Gamma}{\beta}\boldsymbol{\Psi}^*\right). \tag{15}$$

Finally, we update the multiplier $\boldsymbol{\lambda}$ through

$$\boldsymbol{\lambda}^{k+1} = \boldsymbol{\lambda}^k - \gamma\beta\left(\mathbf{A}\mathbf{x}^{k+1} + \mathbf{r}^{k+1} - \mathbf{y}\right), \tag{16}$$

where $\gamma > 0$ is a constant.

Now, we present alternating direction method for problem (3) as Algorithm 1.

TABLE 1: Comparisons of different algorithms.

	PSNR (dB)	CPU time (s)
Algorithm 1 with Contourlet	48.28	4.52
SparseMRI	36.71	16.17
ICOTA	45.79	15.46
FICOTA	46.93	7.79
SpaRSA	38.46	14.52
Algorithm 1 with wavelet	42.27	4.21

3. Numerical Experiments

In this section, we present numerical results to illustrate the performance of the proposed algorithm for MRI reconstruction. All experiments are made by using MATLAB 7.8.0 on the PC with Intel Core 3.4 GHz and 4 G memory.

We compare the proposed Algorithm 1 with the state-of-the-art MRI reconstruction algorithms: SparseMRI [5], ICOTA [9], FICOTA [10], and SpaRSA [14]. SparseMRI is based on conjugate gradient (CG), ICOTA is based on contourlet transform and ISTA, FICOTA is based on contourlet transform and FISTA, and SpaRSA is an effective algorithm to solve the CS problem. We also compare the reconstruction performance of contourlet and wavelet. We quantify the reconstruction performance by peak signal to noise ratio (PSNR) and CPU time. PSNR is defined as

$$\text{PSNR} = 10\log_{10}\left(\frac{255^2}{\text{MSE}}\right), \tag{17}$$

where $\text{MSE} = (1/mn)\sum_{i=0}^{m-1}\sum_{j=0}^{n-1}(I_{\text{ori}}(i,j) - I_{\text{rec}}(i,j))^2$, I_{ori} and I_{rec} are the original image and reconstructed image, respectively, and m, n are the size of the images.

In the following experiments, we choose the measurement matrix $\boldsymbol{\Phi}$ as a partial Fourier transform matrix with m rows and n columns and define the sampling ratio as m/n. The MR scanning time is less if the sampling ratio is lower. We use the variable density sampling strategy just as the previous works in papers [4, 5, 9, 10], which randomly choose more Fourier coefficients from low frequencies and less coefficients from high frequencies.

In the first experiment, we use the MR image as Figure 1(a) shows, and the variable density sampling pattern as Figure 1(b) shows. For wavelet transform, we use Daubechies wavelet with 4 vanishing moments, and contourlet transform with decomposition [5,4,4,3], just the same as paper [9].

Figure 1 shows the reconstructed images using different algorithms. Table 1 summarizes the comparisons of different algorithms. From Table 1 we can see that Algorithm 1 with contourlet transform outperforms Algorithm 1 with wavelet transform in terms of PSNR, although its running time is slightly slower, and Algorithm 1 outperforms other algorithms in terms of both PSNR and CPU time.

In the second experiment, we illustrate the reconstruction performance of each algorithm as the sampling rate varies from 0.1 to 0.9.

Figure 2 shows the variations of PSNR and CPU time of CS-MRI reconstruction versus sampling rates for different

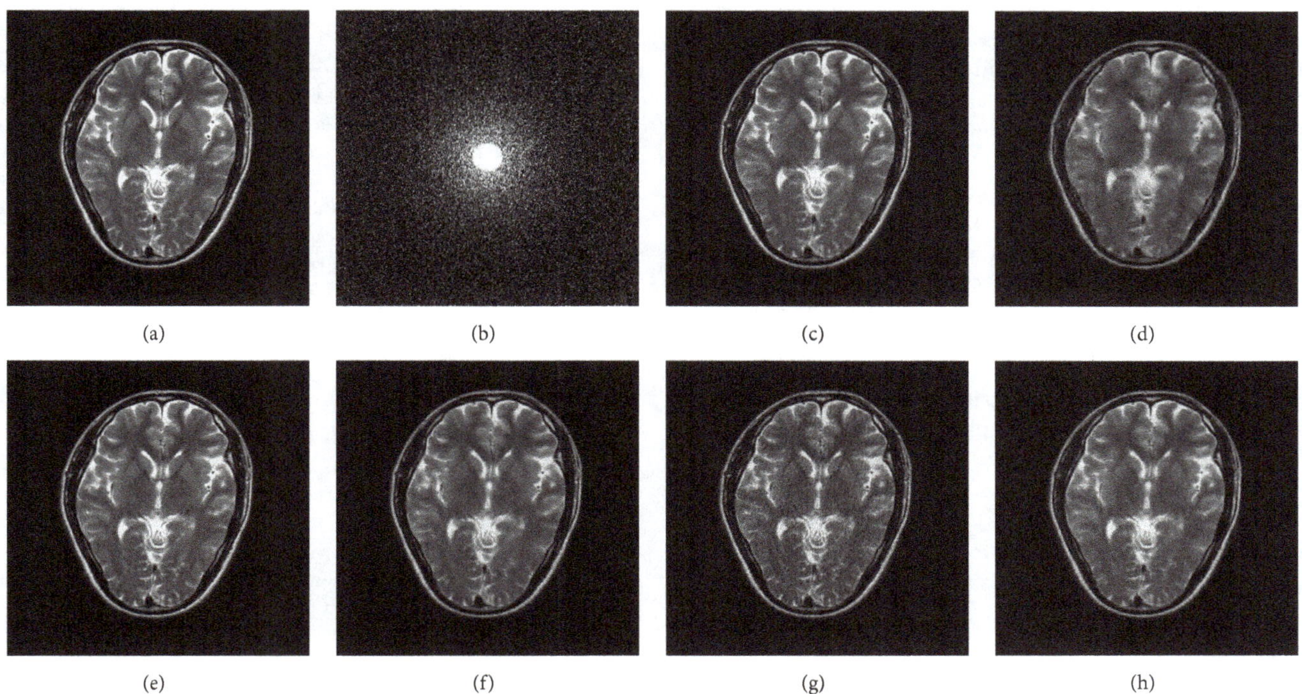

FIGURE 1: Comparison of CS-MRI reconstruction results obtained from different algorithms. (a) Original MR image. (b) Variable density sampling pattern with sampling rate 0.2. ((c)–(h)) Reconstructed result from Algorithm 1 with contourlet transform, SparseMRI, ICOTA, FICOTA, SpaRSA, and Algorithm 1 with wavelet transform.

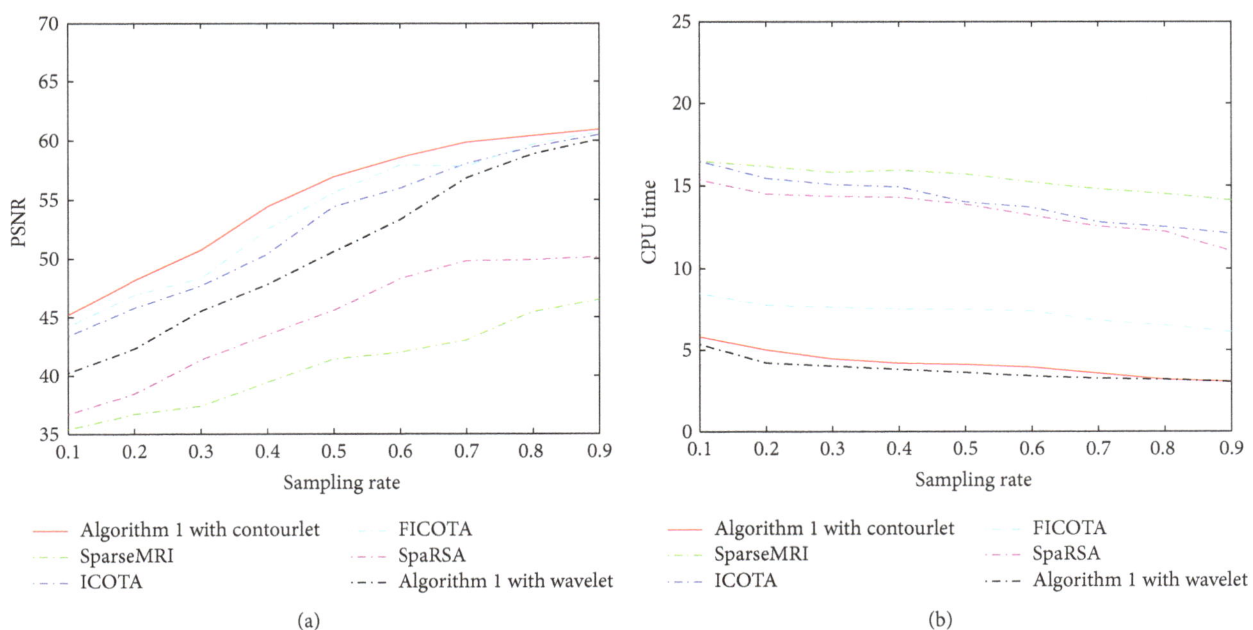

FIGURE 2: Comparison of CS-MRI reconstruction results obtained from different algorithms with different sampling rates. The results are average of 50 runs. (a) PSNR versus sampling rate. (b) CPU time versus sampling rate.

algorithms. Figure 2(a) shows that PSNR of Algorithm 1 with contourlet is better than other algorithms with sampling growing. Furthermore, Figure 2(a) also shows that the advantage of contourlet becomes less obvious with the increase of sampling rate. Figure 2(b) indicates that Algorithm 1 with contourlet is slightly slower than Algorithm 1 with wavelet but is much faster than other algorithms.

4. Conclusion

In this paper, we propose a novel algorithm based on contourlet transform and the classic alternating direction method to solve CS-MRI problem. The proposed algorithm has low computational complexity and is suitable for large scale problem. Our numerical results show that the proposed algorithm compares favorably with these algorithms referred to in terms of PSNR and CPU time.

Conflict of Interests

The authors declare that there is no conflict of interests regarding the publication of this paper.

Acknowledgments

This work is supported by Undergraduate Training Program of Yangtze University for Innovation and Entrepreneurship (104892014034) and Research Funding of Yangtze University for Basic Subject (2013cjp21).

References

[1] D. L. Donoho, "Compressed sensing," *IEEE Transactions on Information Theory*, vol. 52, no. 4, pp. 1289–1306, 2006.

[2] Y. C. Eldar and G. Kutyniok, *Compressed Sensing: Theory and Applications*, Cambridge University Press, 2012.

[3] E. J. Candes and M. B. Wakin, "An introduction to compressive sampling," *IEEE Signal Processing Magazine*, vol. 25, no. 2, pp. 21–30, 2008.

[4] M. Lustig, D. L. Donoho, J. M. Santos, and J. M. Pauly, "Compressed sensing MRI: a look at how CS can improve on current imaging techniques," *IEEE Signal Processing Magazine*, vol. 25, no. 2, pp. 72–82, 2008.

[5] M. Lustig, D. Donoho, and J. M. Pauly, "Sparse MRI: the application of compressed sensing for rapid MR imaging," *Magnetic Resonance in Medicine*, vol. 58, no. 6, pp. 1182–1195, 2007.

[6] D. Zhao, H. Du, Y. Han, and W. Mei, "Compressed sensing MR image reconstruction exploiting TGV and wavelet sparsity," *Computational and Mathematical Methods in Medicine*, vol. 2014, Article ID 958671, 11 pages, 2014.

[7] C. Deng, S. Wang, W. Tian, Z. Wu, and S. Hu, "Approximate sparsity and nonlocal total variation based compressive MR image reconstruction," *Mathematical Problems in Engineering*, vol. 2014, Article ID 137616, 13 pages, 2014.

[8] M. N. Do and M. Vetterli, "The contourlet transform: an efficient directional multiresolution image representation," *IEEE Transactions on Image Processing*, vol. 14, no. 12, pp. 2091–2106, 2005.

[9] X. Qu, W. Zhang, D. Guo, C. Cai, S. Cai, and Z. Chen, "Iterative thresholding compressed sensing MRI based on contourlet transform," *Inverse Problems in Science and Engineering*, vol. 18, no. 6, pp. 737–758, 2010.

[10] W. Hao, J. Li, X. Qu, and Z. Dong, "Fast iterative contourlet thresholding for compressed sensing MRI," *Electronics Letters*, vol. 49, no. 19, pp. 1206–1208, 2013.

[11] S.-J. Kim, K. Koh, M. Lustig, S. Boyd, and D. Gorinevsky, "An interior-point method for large-scale l_1-regularized least square," *IEEE Journal on Selected Topics in Signal Processing*, vol. 1, no. 4, pp. 606–617, 2007.

[12] I. Daubechies, M. Defrise, and C. De Mol, "An iterative thresholding algorithm for linear inverse problems with a sparsity constraint," *Communications on Pure and Applied Mathematics*, vol. 57, no. 11, pp. 1413–1457, 2004.

[13] A. Beck and M. Teboulle, "A fast iterative shrinkage-thresholding algorithm for linear inverse problems," *SIAM Journal on Imaging Sciences*, vol. 2, no. 1, pp. 183–202, 2009.

[14] S. J. Wright, R. D. Nowak, and M. A. Figueiredo, "Sparse reconstruction by separable approximation," *IEEE Transactions on Signal Processing*, vol. 57, no. 7, pp. 2479–2493, 2009.

[15] J. Yang and Y. Zhang, "Alternating direction algorithms for l_1-problems in compressive sensing," *SIAM Journal on Scientific Computing*, vol. 33, no. 1, pp. 250–278, 2011.

12

Comparing Digital Phase-Locked Loop and Kalman Filter for Clock Tracking in Ultrawideband Location System

Qian Gao,[1,2] **Chong Shen**,[1,2] **and Kun Zhang**[1,2,3]

[1]*State Key Laboratory of Marine Resources Utilization in South China Sea, Hainan University, Haikou, Hainan 570228, China*
[2]*College of Information Science and Technology, Hainan University, Haikou, Hainan 570228, China*
[3]*College of Ocean Information Engineering, Hainan Tropical Ocean University, Sanya, Hainan 572022, China*

Correspondence should be addressed to Chong Shen; sc_hainu@163.com

Academic Editor: Jose R. C. Piqueira

For timing and synchronization system, digital phase-locked loop (DPLL) and Kalman filter all have been widely used as the clock tracking and clock correction schemes for the similar structure and properties. This paper compares the two schemes used for ultrawideband (UWB) location system. The improved Kalman filter is more immune to interference.

1. Introduction

Impulse radio ultrawideband (IR-UWB) [1] is considered to be promising for indoor location. To estimate the tags location using time difference of arrival- (TDOA-) based localization, the anchors' local clocks are required to be fully synchronized with each other [2], but the anchors' clocks are varied with the running time and temperature drift [3]. The anchors must be synchronized periodically [4]. The location system is as Figure 1 shows. There are four anchors: anchor 1 is selected as the reference anchor and the other three anchors are the passive anchors. The reference anchor sends the clock synchronization packets to its passive anchor, which are represented by the orange lines in Figure 1. The clock synchronization algorithm (Algorithm 1) that we used is one-way message dissemination [5]. The clock variance between the passive anchor's local clock and its reference anchor is tracked. The data's arrival time from tag to anchors is corrected for the same time base between the reference and passive anchors. Then the TDOA algorithm effectively gets the tag's location. So how to track the clock variance between reference-passive anchors is important for UWB location.

Traditionally, we model the clock time as a continuous function of clock skew (frequency difference) γ and the clock offset (phase difference) θ [6].

$$\begin{aligned} C_m(t) &= t, \\ C_s(t) &= \gamma \cdot t + \theta, \end{aligned} \tag{1}$$

where $C_m(t)$ denotes a reference clock of the sending anchor and $C_s(t)$ denotes the local clock of the receiving anchor. In digital clocks, time is recorded by counting the number of periods of a repeating clock signal. At each rising clock edge of the periodic signal, an integer time counter is incremented.

The main problem of network synchronization is to resolve the observed time in (1). The algorithms considered here use a one-way message dissemination approach at the level of discrete clock ticks.

Suppose that the anchors all have the features of transmitting and receiving the clock check packets (CCP) with the time stamps, and the initial master anchor transmits a CCP with period T, as shown in Figure 2. In the jth round of broadcast message, reference anchor broadcasts a synchronization message CCP at $T_{1,j}$ and the passive anchor records its time $T_{2,j}$ at the reception of that message. Δ_j denotes the interval between receiving a signal and the following initial local clock tick caused by the clock offset. According to [5], the timing model of the jth broadcast message is given by

$$T_{2,j} \approx \gamma \cdot T_{1,j} + \theta + \psi_j, \tag{2}$$

where ψ_j is the random variable delay in the transmission.

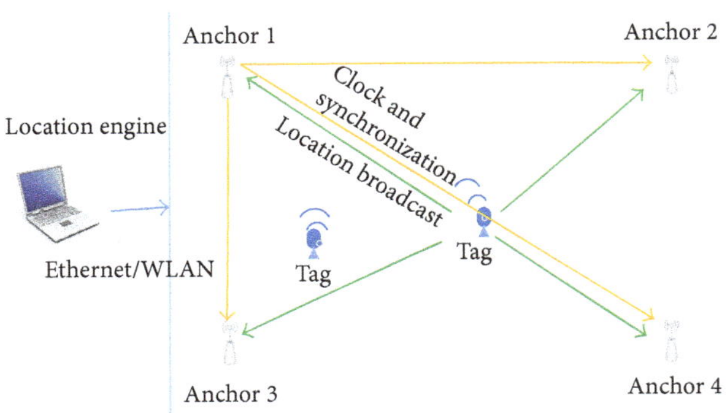

FIGURE 1: IR-UWB location system diagram.

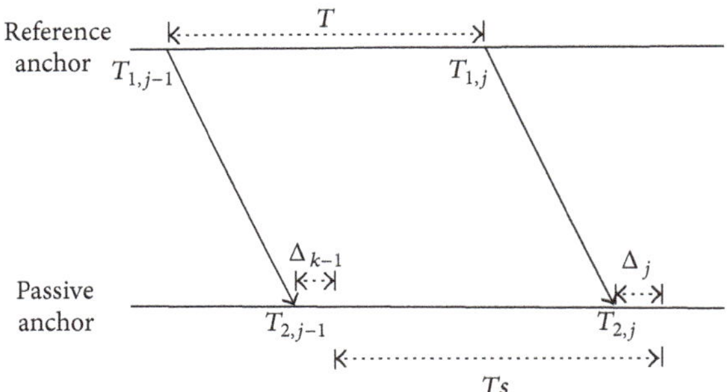

FIGURE 2: Space-time of reference-passive anchors.

The clock tracking is implemented with the main "process" function taking two inputs: (1) The slave anchor CCP receiving time with its time base. (2) The master anchor CCP transmitting time with its time base and the CCP time of flight (TOF).

According to Figure 3, the clock tracking process uses CCP receiving time and CCP transmitting time and the best estimated time between the master unit and the slave unit. If given the master and slave anchors' (X, Y, Z) coordinates, the CCP TOF will be obtained by dividing the distance by the speed. At last, clock tracking process draws the real relative clock offset and the best estimated relative clock offset between master and slave units. Digital phase-locked loop (DPLL) and Kalman filter both have been widely used as the clock tracking and clock correction schemes for the similar structure and properties. This paper compares the two schemes used for UWB location system.

2. Digital Phase-Locked Loop

Digital phase-locked loop (DPLL) is a digital closed-loop automatic control system that can follow the frequency and phase of the input signals [7, 8]. For UWB location system, we consider a second-order DPLL based on ZC-DPLL, as Figure 4 shows.

Assume that $s(t)$ is the input signal, $n(t)$ is zero mean additive white Gaussian noise, and T_0 is the input signal clock period without correction. The input signal $s(t)$ with $n(t)$ is sampled at t_k by digital clock to output the loop phase

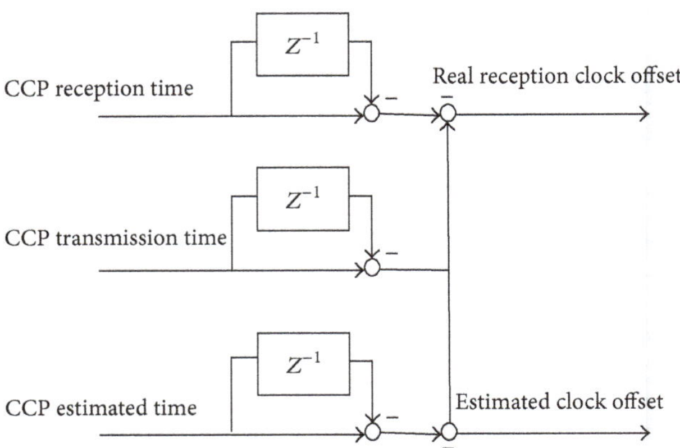

FIGURE 3: The relative clock variation by clock tracking.

```
P = A * P * trans(A) + Q;
J = 1/(R + H * P * trans(H));
OM = measuredError * J * measuredError;
if (OM > threshold) && (counter > 20),
outlier = 1;
measuredError = 0.0;
        outlier_counter = outlier_counter + 1;
if outlier_counter > 8,
          outlier_counter = 0;
counter = 0;
end
x0 = x_0 + dt;
EstimatedTime(i) = x0;
else
K = P * trans(H) * inv(H * P * trans(H) + R);
x = x + K * measuredError;
```

ALGORITHM 1

error z_k. Ignore the impact of the quantizer; the sequence $\{z_k\}$ directly comes into the digital filter; by smoothing, the digital filter outputs a more reliable correcting sequence $\{y_k\}$ to digital clock: $y_k = D(z)z_k$. The second-order z operator function is

$$D(z) = G_1 + G_2\left(1 - z^{-1}\right)^{-1}, \quad (3)$$

where G_1 and G_2 are the loop gain factors. Assume that the loop gains of second-order DPLL are K_{0f} and K_{1f}, respectively.

$$y_k = K_{0f}z_k + K_{1f}\sum_{i=0}^{k} z_i. \quad (4)$$

In DPLL, correction signal y_k is used to control the next period: $T_{k+1} = T_0 - y_k$. Adjust T_k until the loop into the locked state T_k is the sample interval: $T_k = t_k - t_{k-1}$, $k = 1, 2, \ldots$.

The sampling time t_k is deduced:

$$\hat{t}_{k+1} = \hat{t}_k + T_0 + K_{0f}z_k + K_{1f}\sum_{i=0}^{k} z_i. \quad (5)$$

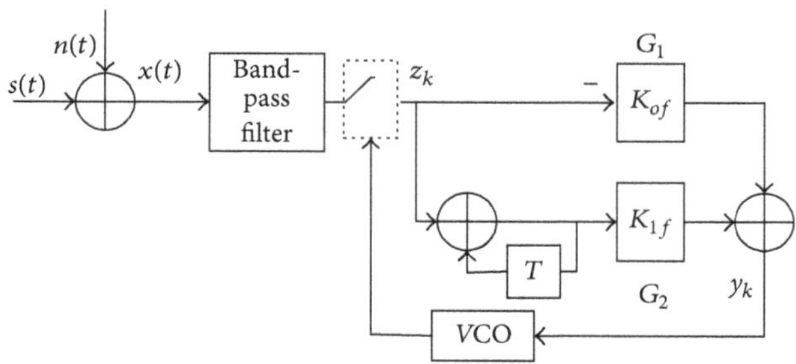

FIGURE 4: DPLL structure block diagram.

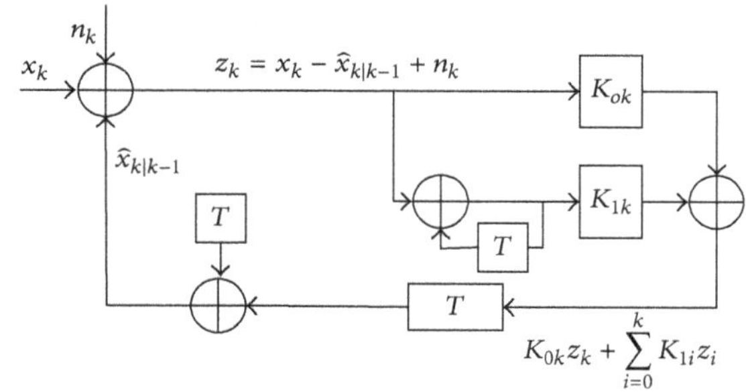

FIGURE 5: Kalman filter structure block diagram.

3. Kalman Filter

Kalman filter is the solution by the minimum mean square error (MMSE) of the optimal linear filtering [9, 10]. It estimates the current signal value according to the previous estimation and a recent observation data. In the concrete implementation process, the $(k+1)$th period clock skew and clock drift of the master-slave clock is estimated according to the kth sync cycle information. T is the clock synchronization period; $U_{\theta,k}$ and $U_{\gamma,k}$ are the correction of the clock skew and clock offset at the kth clock period, respectively. θ_k and γ_k are the kT clock skew and clock offset, respectively. At the moment of $(k+1)T$, the clock relations between the adjacent clock periods are

$$\theta_{k+1} = \theta_k - U_{\theta,k} + (\gamma_k - U_{\gamma,k})T + \omega_{\theta,k}$$
$$\gamma_{k+1} = \gamma_k - U_{\gamma,k} + \omega_{\gamma}, \quad (6)$$

where $\omega_{\theta,k}$ is the clock skew variance and $\omega_{\gamma,k}$ is the clock offset variance. Assume that $\omega_k = [\omega_{\theta,k}\ \omega_{\gamma,k}]^T$; its additive covariance matrix is Q. We define the vector and matrix as follows:

$$x_k = [\theta_k; \gamma_k]^T,$$
$$u_k = [U_{\theta,k}; U_{\gamma,k}]^T. \quad (7)$$

Kalman filter equations by iteration are as follows.

(1) Estimation

$$\widehat{x}_{k+1|k} = Ax_k + Bu_k, \quad (8)$$

where $A = \begin{bmatrix} 1 & T \\ 0 & 1 \end{bmatrix}$, $B = \begin{bmatrix} -1 & -T \\ 0 & -1 \end{bmatrix}$, x_k is the state to be estimated, and u_k is the input control vector.

(2) MMSE Matrix of the Estimation

$$P_{k+1|k} = AP_kA^T + Q, \quad (9)$$

where P_k is the MMSE matrix of the estimated x_k.

(3) Kalman Filter Gain Matrix

$$K_{k+1} = P_{k+1|k}(H_{k+1})^T \left(R_{k+1} + H_{k+1}P_{k+1|k}(H_{k+1})^T\right)^{-1}, \quad (10)$$

where R_{k+1} is the covariance matrix of the observation noise and the measurement matrix H_{k+1} is a unit matrix.

(4) Correction

$$\widehat{x}_{k+1} = \widehat{x}_{k+1|k} + K_{k+1}(z_{k+1} - H_{k+1}\widehat{x}_{k+1|k}). \quad (11)$$

(5) MMSE Matrix

$$P_{k+1} = (1 - K_{k+1})P_{k+1|k}. \quad (12)$$

After Kalman filtering, the correction is $\widehat{x}_{k+1} = [\widehat{\theta}_{k+1}; \widehat{\gamma}_{k+1}]^T$ at the $(k+1)$th clock period. $u_{k+1} = \widehat{x}_{k+1}$ is set to make up for the clock skew and clock offset. So the slave anchor's clock base will make up to the same clock base when the tag's data arrives.

Relative to DPLL, we define $X_k = [x_k; x_k]$ and define the Kalman gain vector as $K_k = [K_{0k}; K_{1k}]$. $P_0 = [T_0^2/12\ 0; 0\ T_0^2 f_{\Delta k}^2]$, where $T_0^2/12$ is the variance of the initial phase X_0; $E[(X_0)^2] = T_0^2 f_{\Delta k}^2$; $f_{\Delta k}^2 = E[(f_\Delta/f_0)^2]$ is the normalized variance values of f_Δ with zero average distribution. The error is $z_k = x_k - \widehat{x}_{k|k-1} + n_k = y_k - H\widehat{X}_{k|k-1}$.

According to (4)–(7), as $k = 0, 1, 2, \ldots$, we will get

$$\widehat{x}_{k+1} = \widehat{x}_k + K_{0k}z_k \sum_{i=0}^{k} K_{1i}z_i. \quad (13)$$

Comparing (5) and (13), the recursive types are very similar. The Kalman filter structure is depicted in Figure 5.

4. Comparison and Analysis

According to the above descriptions about Kalman filter and DPLL, we compare the two schemes for UWB indoor location. In DPLL, correction sequences as the output of the signal through digital filtering control the digital clock period until the loop is locked. Kalman filter also abstracts the needed signal through the feedback loop, which uses the former data to estimate the current data. The two schemes use the error z_k through gain factor K to find the optimal estimation. By comparing (5) and (13), we just need to adjust gain factor K so as to get the similar results.

We define the passive anchor clock variance error between the real clock variance and the optimal estimated clock variance as $e_k = x_k - \hat{x}_{k|k-1}$. Using the reference-passive anchors data backhaul sending time, data receiving time, and the optimal estimated time with the data flight time, the location engine in the server will calculate the relative clock offset variance.

The paper uses Matlab for simulation. Assume that the anchors' coordinates are anchor1 (1.1, 1.17, 1.93) and anchor2 (11.3, 1.17, 1.21). The TOF of the reference anchor1 to the passive anchor2 is $\text{TOF}_{1,2} = 0.000000034123193$ s. For second-order DPLL, variable loop gains with lower bounds $K_{0f} = 0.2$, $K_{1f} = 0.05$, $E_b/N_0 = 10.6$ dB, $\sigma_n^2/T_0^2 = \hat{\sigma}_n^2/T_0^2 = 0.001$, and $f_\Delta/f_0 = 0.1$. The clock synchronization period is 150 ms. The measurement noise variance is $3e-12$; the process noise variance is $5e-12$. Alternatively, a more complex multistage in-lock and out-of-lock detection algorithm may be employed, which trades off acquisition, tracking, and false lock performance according to the system requirements. Such tradeoff issues are beyond the scope of this paper [11, 12].

As Figure 6 shows, the green line is the clock variance difference by Kalman filter and the black line is that by DPLL. They all tend to be stable over time, but the properties of Kalman filter are significantly better than those of DPLL. Kalman filter requires shorter capture time and smaller error.

By the theory of hypothesis, in order to give sufficient information for the DPLL to stay locked for continued real-time location system operation with good performance (including coping with a certain packet error/loss rate), we need to send the clock synchronization message more frequently than for Kalman filter. It reduces the air-occupancy needed for clock synchronization messages, which allows more air-time for receiving blink messages. This essentially increases the system tag capacity, especially in the lower data rate and longer preamble modes.

5. Improved Algorithms on Kalman Filter

As stated above, Kalman filter is better for clock synchronization indoor UWB location system. Its calculation is based on such an assumption: all measurements are composed of the real signal and additive Gaussian noise. If these assumptions are correct, Kalman filter will effectively get signal from the measurements containing noise. But if the reference anchor's clock check packets collide with tag's data packets with TOA or some other mistake challenges in [13], the assumptions are incorrect. Kalman filter will treat the collision or mistake as credible clock variance data, and it calculates by these data. And Kalman filter itself is a kind of low-pass filter; its response and correcting speed are slower. Therefore, the errors generated by the collision will for a long time seriously degrade the performance of the clock synchronization algorithm.

This paper proposes a method of monitoring and avoiding the wrong of collisions. Kalman filter gain is as (10) shows, defining an information matrix J as

$$J = \left(R_{k+1} + H_{k+1}P_{k+1|k}\left(H_{k+1}\right)^T\right)^{-1}. \quad (14)$$

J is used to represent the difference between estimated clock error and actual clock error. This information will be used to prompt how well the current input fits the current state of filter.

$$\text{OM}_{k+1} = \left(\hat{x}_{k+1} - x_{k+1}\right) * J * \left(\hat{x}_{k+1} - x_{k+1}\right). \quad (15)$$

If the OM (outlier metric) rises above a preset threshold which is an empirical value, the current input is untrusted. The improved Kalman filter does not update current state but discards this data directly to avoid error packet having a big impact for filter output.

In Figures 7 and 8, the blue lines are the real clock offsets and the red lines are the estimated clock offset by Kalman filter. We set a big data mistake at 150 s; the estimated clock offset is unable to keep pace with the real clock offset and up and down shocks with Kalman filter in Figure 7. With the

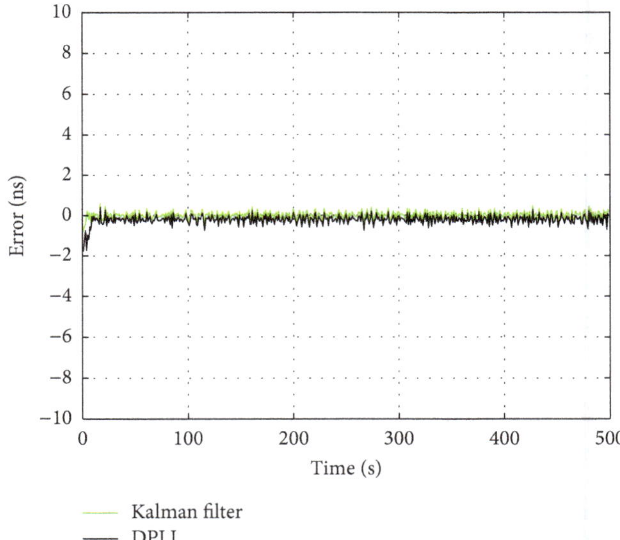

Figure 6: The difference between estimated time and real time.

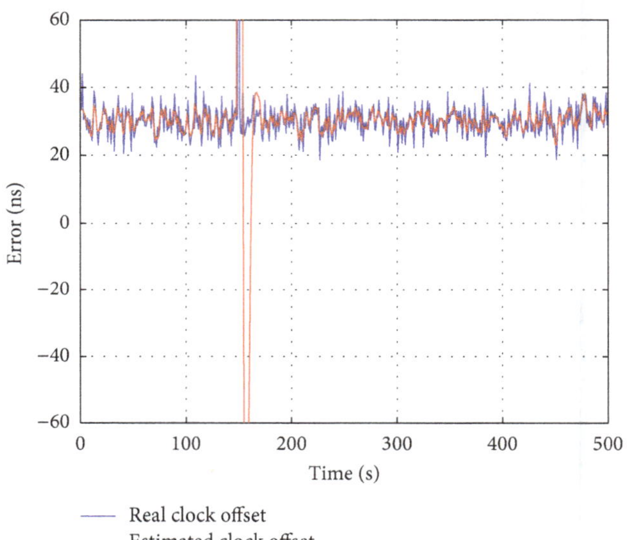

Figure 7: The relative clock offset with big disturbance by Kalman filter.

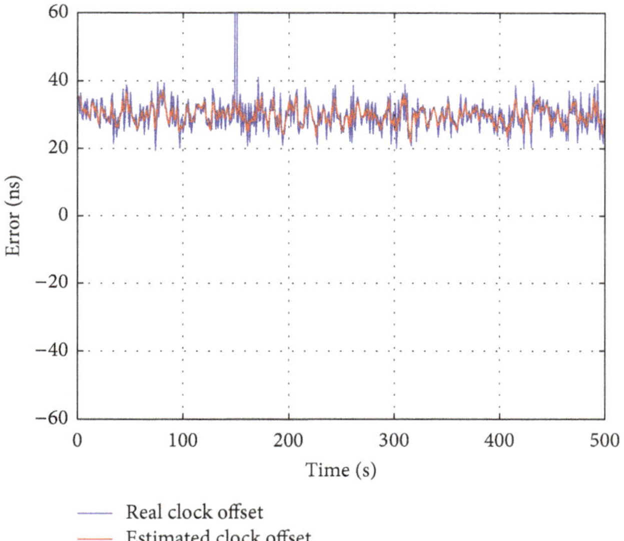

FIGURE 8: The relative clock offset with big disturbance by improved Kalman filter.

resolution of the trustless input, the estimated clock offset is smooth in Figure 8. By comparing Figures 7 and 8, it is clearly seen that the improved Kalman filter enhances the capacity of resisting disturbance.

6. Conclusion

We have compared DPLL and Kalman filter for UWB indoor location network clock synchronization, and the analysis results show that Kalman filter copes better with clock errors and has better lock performance. And the improved Kalman filter is more immune to interference as the simulation results show.

Conflicts of Interest

The authors declare that they have no conflicts of interest.

Acknowledgments

This work was supported in part by Major Research and Development Plan of Hainan Province (ZDYF2016002), the National Natural Science Foundation of China (61461017), Hainan Province Natural Science Foundation of Innovation Team Project (2017CXTD004), and Innovative Research Project of Postgraduates in Hainan Province (Hyb2017-04).

References

[1] "Standard IEEE 802.15.4-2011. Part 15.4: Low-rate wireless personal area networks (LR-WPANs)," September 2011.

[2] S. Gezici, "A survey on wireless position estimation," *Wireless Personal Communications*, vol. 44, no. 3, pp. 263–282, 2008.

[3] D. Dardari, A. Conti, U. Ferner, A. Giorgetti, and M. Z. Win, "Ranging with ultrawide bandwidth signals in multipath environments," *Proceedings of the IEEE*, vol. 97, no. 2, pp. 404–425, 2009.

[4] D. Zachariah, S. Dwivedi, P. Handel, and P. Stoica, "Scalable and Passive Wireless Network Clock Synchronization in LOS Environments," *IEEE Transactions on Wireless Communications*, vol. 16, no. 6, pp. 3536–3546, 2017.

[5] Y. Wu, Q. Chaudhari, and E. Serpedin, "Clock synchronization of wireless sensor networks," *IEEE Signal Processing Magazine*, vol. 28, no. 1, pp. 124–138, 2011.

[6] T.-D. Tran, J. Oliveira, J. Sá Silva et al., "A scalable localization system for critical controlled wireless sensor networks," in *Proceedings of the 2014 6th International Congress on Ultra Modern Telecommunications and Control Systems and Workshops, ICUMT 2014*, pp. 302–309, Russia, October 2014.

[7] C.-H. Shan, Z.-Z. Chen, and J.-X. Jiang, "An all digital phase-locked loop system with high performance on wideband frequency tracking," in *Proceedings of the 2009 9th International Conference on Hybrid Intelligent Systems, HIS 2009*, pp. 460–463, China, August 2009.

[8] S. Bhattacharyya, R. N. Ahmed, B. B. Purkayastha, and K. Bhattacharyya, "Zero crossing DPLL based phase recovery system and its application in wireless communication," in *Proceedings of the IEEE International Conference on Computer, Communication and Control, IC4 2015*, India, September 2015.

[9] C. McElroy, D. Neirynck, and M. McLaughlin, "Comparison of wireless clock synchronization algorithms for indoor location systems," in *Proceedings of the 2014 IEEE International Conference on Communications Workshops, ICC 2014*, pp. 157–162, Australia, June 2014.

[10] S. Y. Chen, "Kalman filter for robot vision: a survey," *IEEE Transactions on Industrial Electronics*, vol. 59, no. 11, pp. 4409–4420, 2012.

[11] G. A. Leonov, N. V. Kuznetsov, M. V. Yuldashev, and R. V. Yuldashev, "Hold-in, pull-in, and lock-in ranges of PLL circuits: Rigorous mathematical definitions and limitations of classical theory," *IEEE Transactions on Circuits and Systems I: Regular Papers*, vol. 62, no. 10, pp. 2454–2464, 2015.

[12] F. Ren, C. Lin, and F. Liu, "Self-correcting time synchronization using reference broadcast in wireless sensor network," *IEEE Wireless Communications Magazine*, vol. 15, no. 4, pp. 79–85, 2008.

[13] H. Soganci, S. Gezici, and H. V. Poor, "Accurate positioning in ultra-wideband systems," *IEEE Wireless Communications Magazine*, vol. 18, no. 2, pp. 19–27, 2011.

13

Design of Wireless Automatic Synchronization for the Low-Frequency Coded Ground Penetrating Radar

Zhenghuan Xia,[1,2] Qunying Zhang,[1] Shengbo Ye,[1] Zhiwu Xu,[1] Jie Chen,[1] Guangyou Fang,[1] and Hejun Yin[3]

[1] *Key Laboratory of Electromagnetic Radiation and Sensing Technology, Chinese Academy of Sciences, Beijing 100190, China*
[2] *University of Chinese Academy of Sciences, Beijing 100049, China*
[3] *Chinese Academy of Sciences, Beijing 100864, China*

Correspondence should be addressed to Zhenghuan Xia; maxwell_xia@126.com

Academic Editor: Ahmed El Wakil

Low-frequency coded ground penetrating radar (GPR) with a pair of wire dipole antennas has some advantages for deep detection. Due to the large distance between the two antennas, the synchronization design is a major challenge of implementing the GPR system. This paper proposes a simple and stable wireless automatic synchronization method based on our developed GPR system, which does not need any synchronization chips or modules and reduces the cost of the hardware system. The transmitter omits the synchronization preamble and pseudorandom binary sequence (PRBS) at an appropriate time interval, while receiver automatically estimates the synchronization time and receives the returned signal from the underground targets. All the processes are performed in a single FPGA. The performance of the proposed synchronization method is validated with experiment.

1. Introduction

Low-frequency ground penetrating radar (GPR) is an advanced geophysical technique that is rapidly and widely applied in deep detection [1, 2]. In order to achieve high signal-to-noise ratio (SNR), pseudorandom coded signals had been applied in some detection applications, such as through-the-wall tracking and life detection [3–6]. Meanwhile, pseudorandom coded signals had also been used in GPR to obtain the deeper detection [7, 8]. In practical applications, wire dipole antennas are more suitable for low-frequency GPR system with large transmit power. However, the synchronization between transmitter and receiver is a major challenge.

Synchronization with long cables introduces severe interference to the returned signals, which could mask the returned signals from the underground targets. Synchronization with fibers increases the power consumption, complicates the structure, and increases the cost of the radar system. Additionally, it is terribly difficult to carry out experiments with complicate and heavy low-frequency GPR in polar glacier and high tableland [9, 10].

In order to simplify the low-frequency GPR structure and reduce the cost of hardware system, a simple and efficient wireless automatic synchronization method is proposed. Firstly, our developed low-frequency coded GPR will be described briefly. Secondly, the wireless automatic synchronization method will be discussed in detail, and the estimation error of synchronization time will be analyzed. Thirdly, the performance of the proposed method of wireless automatic synchronization will be validated with experiment.

2. Description of Our Developed Low-Frequency Coded GPR

Our developed low-frequency coded GPR consists of a pair of resistively loaded wire dipole antennas with a length of 12 m, a transmitter with a peak power of 200 W, and a receiver with two sampling channels, which is shown in Figure 1. In

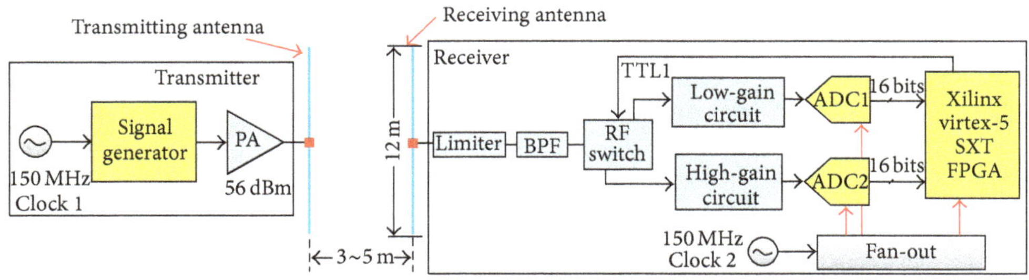

FIGURE 1: The architecture of our developed low-frequency coded GPR.

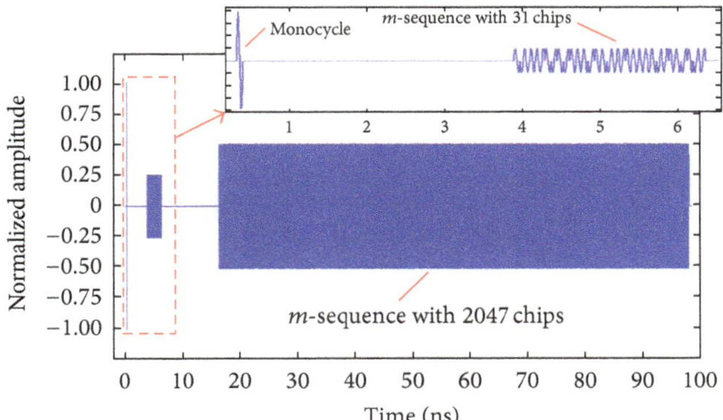

FIGURE 2: The transmit signal from the signal generator.

TABLE 1: Parameters of the low-frequency coded GPR.

Parameter	Value
Width of the monocycle	80 ns
Width of each chip	80 ns
Length of the long m-sequence	2047 chips
Peak power	50 dBm
Receiver sensitivity	–115 dBm
Real-time sampling rate of ADC	150 Mbps
The linear average factor	512
The scanning rate	8 scans/s
Pulse repeated period	200 us
Maximal detection time window	4 us

practical applications, the distance between the two antennas is from 3 m to 5 m.

2.1. Transmitter. One steady-state 150 MHz clock source is the reference clock of the signal generator. The coded signals produced by the signal generator are amplified by a power amplifier (PA) with a maximal peak power of 56 dBm to feed the transmitting antenna. In order to achieve good synchronization without affecting the radar echo, the transmit signals consist of a synchronization preamble and m-sequence of 2047 chips at a time interval of 10 us as shown in Figure 2. The synchronization preamble includes a monocycle and m-sequence of 31 chips at a time interval of 3.5 us. The monocycle is used to estimate coarsely the time of arrival of the direct wave of the m-sequence with 31 chips, which is used to determine accurately the synchronization time. The m-sequence of 2047 chips is used to obtain the impulse response of the deep targets with low side-lobes.

Additionally, the peak power of the monocycle, m-sequence with 31 chips, and m-sequence of 2047 chips are 56 dBm, 44 dBm, and 50 dBm, respectively. Both the two m-sequences are modulated with a 12 MHz sine wave, and the width of each chip of m-sequence is about 80 ns.

2.2. Receiver. Another steady-state 150 MHz clock source is used as the main clock of the receiver. As shown in Figure 1, the 150 MHz clock is sent to the fan-out chip to generate three synchronized clocks, which are used as the clocks of the two ADCs and FPGA. The limiter is used to prevent the receiver from damage. During the estimation of synchronization time, the synchronization preamble is switched to the low-gain circuit with 3 dB gain and sampled by the first ADC at the sampling rate of 150 MHz, and FPGA estimates the synchronization time. Once the synchronization time has been estimated, the radar echo is switched to high-gain circuit with 30 dB and is sampled by the second ADC at the same sampling rate of 150 MHz. Then the FPGA performs the cross-correlation between the sampled radar echo and the responding reference signal to obtain the impulse response of the detection scenarios.

Additionally, the main parameters of the proposed low-frequency coded GPR are summarized in Table 1.

3. Design of Wireless Automatic Synchronization

The transmitter omits the coded signals at a pulse repeated period (PRP) of 200 us, and the receiver estimates accurately the synchronization time and receives the radar echoes. The responding reference signals of the two m-sequences can be obtained in close-loop situation as shown in Figure 3, and the two reference signals are stored in two on-chip RAMs of the FPGA. The reference signal of the m-sequence with 31 chips is denoted as $m(i)$, and the reference signal of the m-sequence with 2047 chips is denoted as $m_l(i)$. The architecture

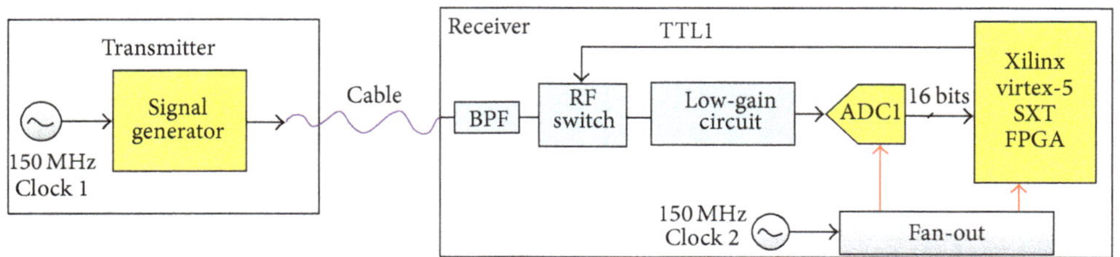

FIGURE 3: Obtain the reference signal of the m-sequence in close-loop situation.

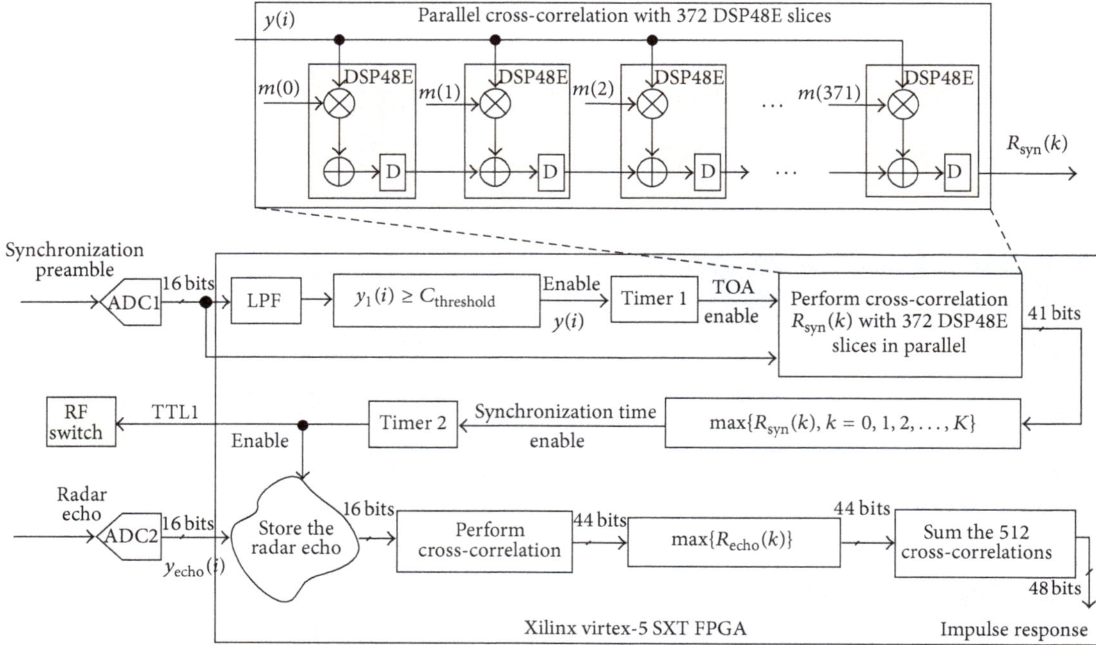

FIGURE 4: Estimation of the synchronization time based on FPGA.

of wireless automatic synchronization in the receiver is shown in Figure 4.

3.1. Estimation of the Synchronization Time. Firstly, the direct wave of the monocycle sampled by the first ADC, which is denoted as $y(i)$, is filtered by the digital low pass filter (LPF) to remove the noise and compared with the threshold $C_{\text{threshold}}$. The threshold $C_{\text{threshold}}$ should be chosen according to the amplitude of the direct wave and the level of the noise or clutter. For our low-frequency GPR system, the maximum clutter or noise is about 42 mV, and the amplitude of the direct wave is about 610 mV when the distance between the two antennas is 3 m. When the distance is increased to 5 m, the amplitude of the direct wave is decreased to about 220 mV. Thus the threshold is fixed at 150 mV in this work, and the distance between the two antennas should be less than 5 m in practical applications.

Secondly, when the filtered signal $y_l(i)$ is more than the threshold, the first timer Timer 1 is enabled to count. The TOA of the m-sequence with 31 chips comes when the timer Timer 1 counts to 3.5 us. At this time, the sampled signal $y(i)$ could be considered coarsely as the direct wave of the m-sequence of 31 chips and sent to the parallel cross-correlation module to obtain the cross-correlation $R_{\text{syn}}(k)$ as

$$R_{\text{syn}}(k) = \sum_{i=0}^{371} m(i) \cdot y(k+i), \quad k = 0, 1, \dots, K, \tag{1}$$

where $m(i)$ is the reference signal of the m-sequence of 31 chips and stored in a RAM of FPGA. M-sequence with 31 chips lasts for about 2.48 us and the sampling clock of ADC is 150 MHz, and the length of the reference signal $m(i)$ is 372. The clock of the parallel cross-correlation module is 150 MHz synchronized with the sampling clock of ADCs. The pipeline delay and throughput rate of the parallel cross-correlation module are 372 clock periods and 150 Msps, respectively. Then FPGA can find the maximum value of the cross-correlation by comparison between the two adjacent cross-correlations in a time window of 2.48 us. Ultimately, the synchronization time can be determined by the maximal value of cross-correlation.

Thirdly, when the synchronization time is estimated, the timer Timer 2 is enabled to count and the radar echo is switched to the high-gain circuit and sampled by the second

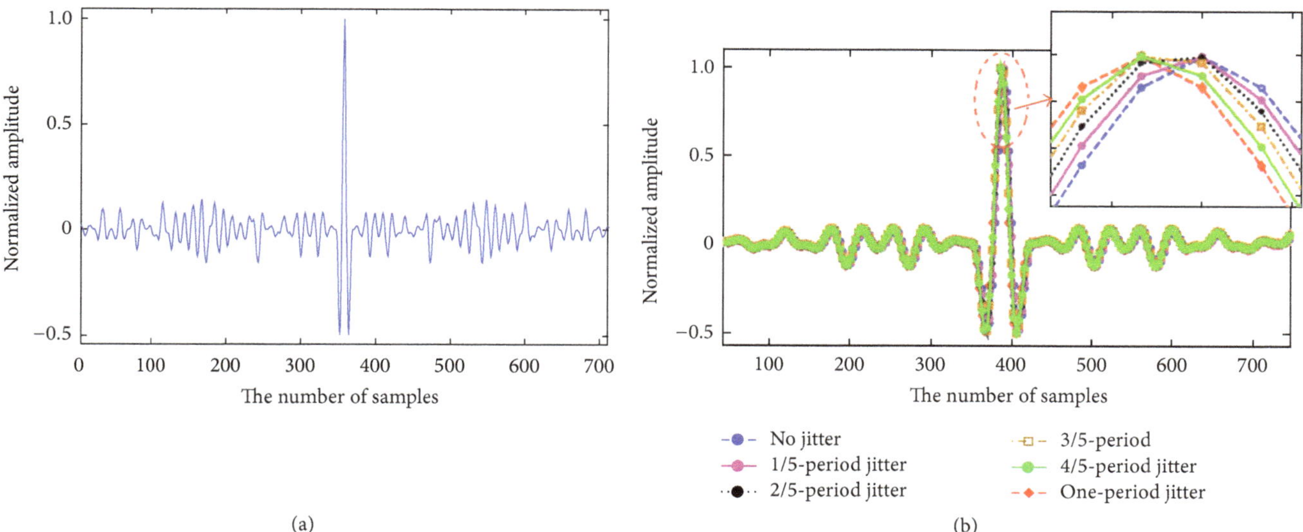

FIGURE 5: Cross-correlation of the m-sequence with 31 chips in close-loop situation. (a) Cross-correlation of the sampled m-sequence with 31 chips. (b) Cross-correlations of the sampled m-sequence with unstable initial sampling time.

ADC when the timer Timer 2 counts to 10 us. Then the cross-correlation $R_{\text{echo}}(k)$ of the echo is performed to obtain the impulse response of detection scenarios. In order to perform correctly the linear averaging to improve the SNR of the radar echoes, the maximal value of the cross-correlation $R_{\text{echo}}(k)$ in each returned signal should be found and then performs addition among the adjacent 512 traces to obtain the ultimate impulse response of the underground targets.

3.2. Analysis of the Estimation Error. As shown in Figure 1, the two reference clocks, Clock 1 and Clock 2, are not synchronized, which induces the estimation error of synchronization time. It is equivalent to the fact that the estimation error comes from the random jitter of the initial sampling time of the receiver, and the maximal jitter is one period of Clock 2. In close-loop situation as shown in Figure 3, all the reference signals of the cross-correlation have been sampled at a sampling clock of Clock 2 and stored in FPGA beforehand. According to (1), the cross-correlation of the sampled m-sequence with 31 chips in close-loop situation is shown in Figure 5(a). Figure 5(b) shows the estimation errors of synchronization time come from different jitter of the initial sampling time of the receiver, which indicates that the maximal estimation error is also one period of Clock 2 when the jitter is more than a half period of Clock 2. Then the estimation error will affect the initial acquisition time for the radar echo.

Figure 6 shows the comparison between the cross-correlation of the received m-sequence of 2047 chips with one-period estimation error of synchronization time and that of no estimation error. It can be seen that the main-lobe to side-lobe ratio of the two cross-correlations is almost the same because of the high sampling rate relative to the low-frequency m-sequence. However, the location of the maximal value of the cross-correlation with one-period error is ahead one period of that of the cross-correlation without error.

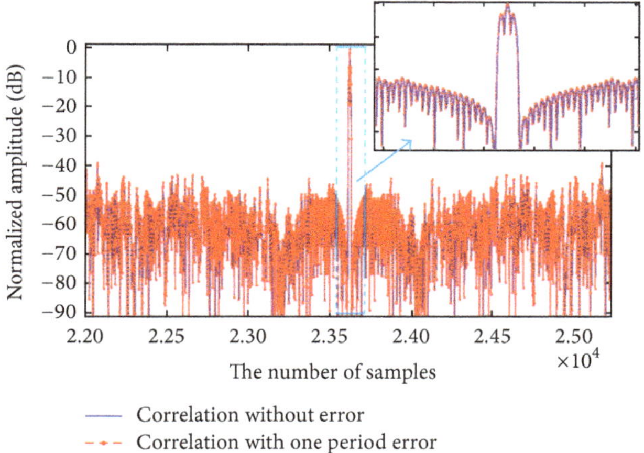

FIGURE 6: Comparison between the cross-correlation of the m-sequence with 2047 chips with one-period estimation error and that of m-sequence with no estimation error.

In order to perform correctly the linear averaging with 512 traces, the maximal value of the cross-correlation should be found firstly, and the 512 additions are performed from the maximal value of the 512 cross-correlations.

4. Experiments and Results

The typical configuration for the low-frequency GPR is shown in Figure 7. The distance between the two wire antennas is about 3.8 m, and the synchronization is implemented with two methods: high-performance ORTEL's fiber models (including OTS-1RefR-100 and OTS-1RefT-100 [11]) and the proposed wireless automatic synchronization method. For this wireless automatic synchronization, the received direct wave of the monocycle as the blue line is shown in

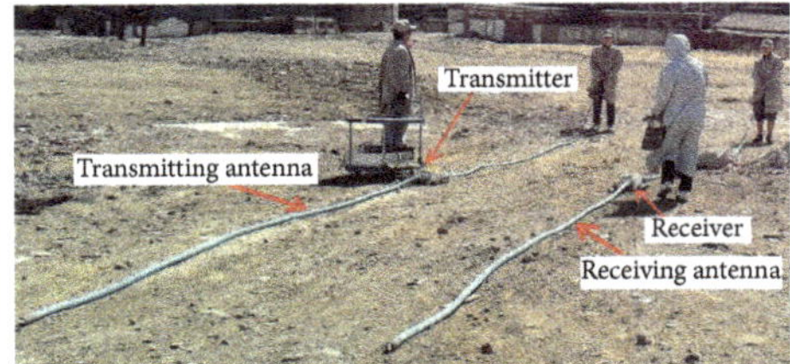

FIGURE 7: The typical configuration for the low-frequency coded GPR.

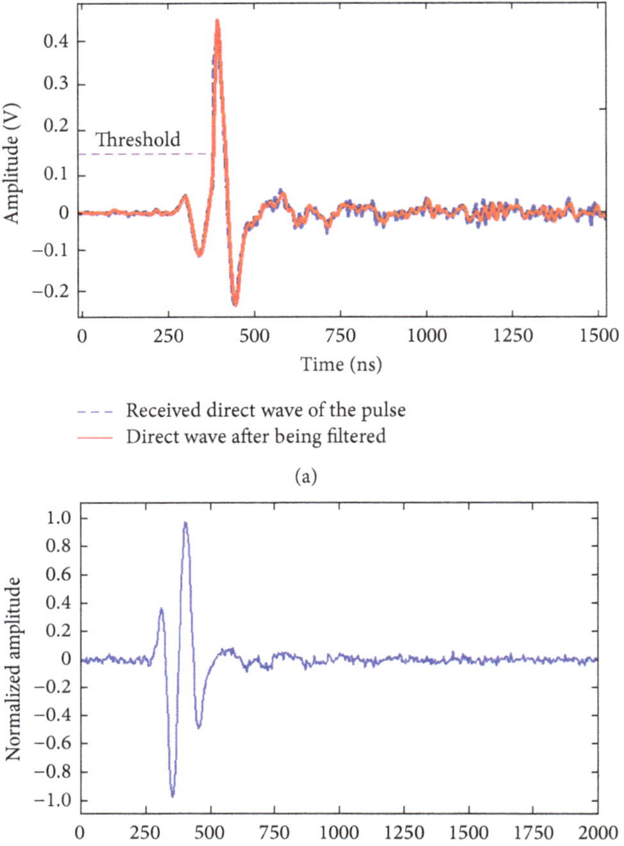

FIGURE 8: Received synchronization preamble. (a) The direct wave of the monocycle. (b) The cross-correlation of the direct wave of the m-sequence with 31 chips.

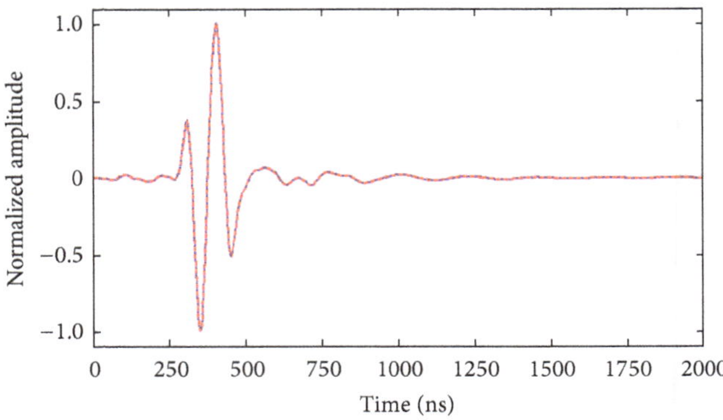

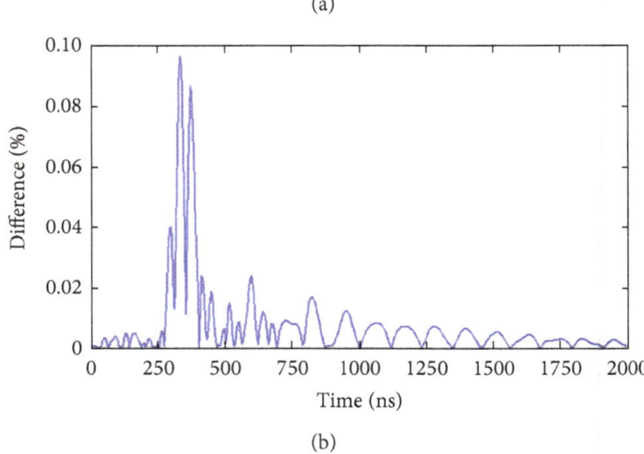

FIGURE 9: The obtained impulse responses with the two methods. (a) Two impulse responses. (b) The difference between them.

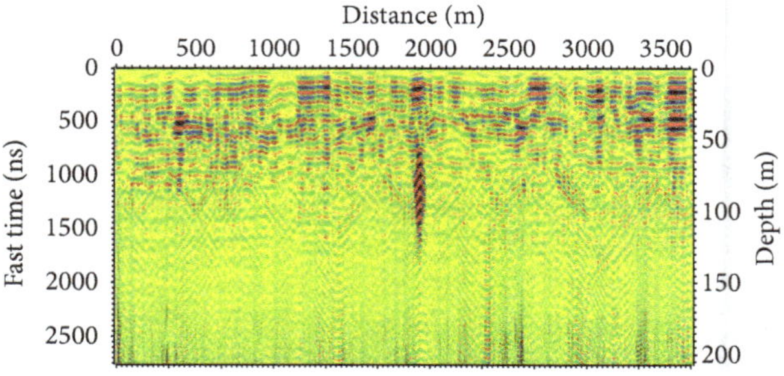

FIGURE 10: The detection result with the low-frequency coded GPR, and the dielectric constant is 8.

Figure 8(a), while the filtered signal as the red line is shown in Figure 8(a). The direct wave of the m-sequence with 31 chips is received, and its cross-correlation can be obtained as shown in Figure 8(b). The synchronization time can be estimated by the maximal value of the cross-correlation.

The red line as shown in Figure 9(a) is the obtained impulse response with the proposed wireless automatic synchronization method, while the blue line in Figure 9(a) is the obtained impulse response with the fiber modules. The difference between the two impulse responses is shown in Figure 9(b), which indicates that the two impulse responses are almost same, while different clutter and noise have been received in different time. Figure 10 shows the responding detection result with the proposed low-frequency coded GPR in desert. It can be seen that some clear layers in depth of 45 m and 70 m can be detected.

5. Conclusion

This paper presents a wireless automatic synchronization method for the low-frequency coded GPR without any other synchronization chips or modules, which simplify the architecture of the GPR system and reduce the hardware cost. Although the maximum synchronization error with the proposed method is one sampling period of ADC-clock, it degrades hardly the main-lobe to side-lobe ratio of pulse compression due to the high sampling rate relative to the low-frequency m-sequence. In order to reduce the effect from the synchronization error farther, the ADC-clock can be increased to an appropriate value to satisfy the requirement of different applications, such as 400 MHz or 1 GHz.

Conflict of Interests

The authors declare that there is no conflict of interests regarding the publication of this paper.

Acknowledgment

This work was supported by the National High Technology Research and Development Program of China (863 Program: 2012AA121901).

References

[1] V. Utsi, "Design of a GPR for deep investigations," in *Proceedings of the 4th International Workshop on Advanced Ground Penetrating Radar*, pp. 222–225, Aula Magna Partenope, June 2007.

[2] L. Fu, S. Liu, L. Liu, and L. Lei, "Development of an airborne ground penetrating radar system: antenna design, laboratory experiment, and numerical simulation," *IEEE Journal of Selected Topics in Applied Earth Observations and Remote Sensing*, vol. 7, no. 3, pp. 761–766, 2014.

[3] Z. Xia, G. Fang, S. Ye, Q. Zhang, C. Chen, and H. Yin, "A novel handheld pseudo random coded UWB radar for human sensing applications," *IEICE Electronics Express*, vol. 11, no. 23, pp. 1–7, 2014.

[4] S. R. J. Axelsson, "Noise radar using random phase and frequency modulation," *IEEE Transactions on Geoscience and Remote Sensing*, vol. 42, no. 11, pp. 2370–2384, 2004.

[5] A. Nezirović, A. G. Yarovoy, and L. P. Ligthart, "Signal processing for improved detection of trapped victims using UWB radar," *IEEE Transactions on Geoscience and Remote Sensing*, vol. 48, no. 4, pp. 2005–2014, 2010.

[6] Z. Xia, G. Fang, S. Ye et al., "Design of modulated m-sequence ultrawideband radar for life detection," in *Proceedings of the 15th International Conference on Ground Penetrating Radar (GPR '14)*, pp. 960–963, IEEE, Brussels, Belgium, July 2014.

[7] F. Nicollin, Y. Barbin, W. Kofman et al., "An HF bi-phase shift keying radar: Application to ice sounding in Western Alps and Spitsbergen glaciers," *IEEE Transactions on Geoscience and Remote Sensing*, vol. 30, no. 5, pp. 1025–1033, 1992.

[8] G. Fang and M. Pipan, "Design of a low frequency ultra-wideband (UWB) antenna and its applications in ground penetrating radar (GPR) system," in *Proceedings of the tenth International Conference on Ground penetrating Radar*, pp. 109–111, Delft, The Netherlands, 2004.

[9] M. Carnevale and J. Hager, "Low frequency GPR in difficult terrain," in *Proceedings of the 4th International Workshop on Advanced Ground Penetrating Radar*, pp. 68–73, Aula Magna Partenope, Italy, June 2007.

[10] M. Tallini, D. Ranalli, M. Scozzafava, and G. Manacorda, "Testing a new low-frequency GPR antenna on karst environments of central Italy," in *Proceedings of the Tenth International Conference Ground Penetrating Radar (GPR '04)*, pp. 133–135, June 2004.

[11] http://www.emcore.com/.

14

Time-Frequency Analysis of Clinical Percussion Signals using Matrix Pencil Method

Moinuddin Bhuiyan,[1] Eugene V. Malyarenko,[2] Mircea A. Pantea,[3] Dante Capaldi,[3] Alfred E. Baylor,[4] and Roman Gr. Maev[1,2,3]

[1]*Department of Physics, University of Windsor, 401 Sunset Avenue, Windsor, ON, Canada N9B 3P4*
[2]*Tessonics Corp., 2019 Hazel Street, Birmingham, MI 48009, USA*
[3]*Institute for Diagnostic Imaging Research, University of Windsor, 401 Sunset Avenue, Windsor, ON, Canada N9B 3P4*
[4]*Detroit Medical Center, 4201 St. Antoine Street, Detroit, MI 48201, USA*

Correspondence should be addressed to Moinuddin Bhuiyan; moni.bit.ctg@gmail.com

Academic Editor: Peter Jung

This paper discusses time-frequency analysis of clinical percussion signals produced by tapping over human chest or abdomen with a neurological hammer and recorded with an air microphone. The analysis of short, highly damped percussion signals using conventional time-frequency distributions (TFDs) meets certain difficulties, such as poor time-frequency localization, cross terms, and masking of the lower energy features by the higher energy ones. The above shortcomings lead to inaccurate and ambiguous representation of the signal behavior in the time-frequency plane. This work describes an attempt to construct a TF representation specifically tailored to clinical percussion signals to achieve better resolution of individual components corresponding to physical oscillation modes. Matrix Pencil Method (MPM) is used to decompose the signal into a set of exponentially damped sinusoids, which are then plotted in the time-frequency plane. Such representation provides better visualization of the signal structure than the commonly used frequency-amplitude plots and facilitates tracking subtle changes in the signal for diagnostic purposes. The performance of our approach has been verified on both ideal and real percussion signals. The MPM-based time-frequency analysis appears to be a better choice for clinical percussion signals than conventional TFDs, while its ability to visualize damping has immediate practical applications.

1. Introduction

Clinical percussion is a centuries-old bedside diagnostic technique used to identify various conditions of the thorax and abdomen in health and disease [1, 2]. It is a method of eliciting sounds by striking body parts with fingertips or a percussion hammer [1, 2]. The purpose of percussion is to provide the physician with qualitative information about the size, consistency, and borders of the vital organ or pathology under the percussed area. Trained physicians are able to recognize many different kinds of percussion sounds, the main three of which are widely known as "resonant," "tympanic," and "dull" [1, 2]. Resonant sounds are low-pitch, hollow sounds heard over normal lung tissue. Abnormal lungs may be hyperresonant, dull, or stony dull. Dullness is expected over the liver and over the heart [2]. Typical tympanic sound is drum-like and is usually heard over the air cavities in the abdomen as well as over pneumothorax [2, 3]. In the latter case, percussion response of the chest changes from resonant to tympanic due to acoustic impedance mismatch at the air cavity boundary [1]. Due to disrupted acoustic coupling between the chest wall and the lung parenchyma (normally provided by the wet contact between visceral and parietal pleurae), the transfer of the acoustical energy to the lungs (where it subsequently dissipates) becomes less efficient [1]. Hence the energy dissipates less rapidly and stays within the chest wall for a longer period of time, leading to the extended signal duration and decreased damping [1]. The goal of this work is to come up with a practical way to visualize this effect of changed damping of certain signal components in the time-frequency plane.

The majority of preceding work attempting the objective classification of percussion sounds is based on the time-domain and Fourier spectral analysis [2, 4]. Spectral peaks correspond to individual oscillation modes, while their width is a direct measure of damping. When closely spaced oscillation modes have high damping, their broad spectral peaks overlap and often cannot be resolved with required accuracy. Time-frequency analysis combines the advantages of the pure time-domain and spectral methods and brings important additional benefits, such as the ability to trace the evolution of the spectral content with time, thus providing a convenient tool to dissect, analyze, and interpret signals [5, 6].

A clinical percussion signal represents free transient response of the human body to a brief mechanical impact. Due to damping, the amplitude of these signals changes with time, and hence they can be considered nonstationary in the broad sense of this definition [7]. Since the impact usually excites oscillations of more than one anatomical subsystem, percussion signals usually represent more than one independent oscillation mode with different frequencies and hence can be considered multicomponent, again, in the broad sense of this term [7]. Representing such signals in the time-frequency domain is not always easy due to their short duration and possible coexistence of multiple frequency components with very different amplitudes. When dealing with multicomponent signals, most conventional TFDs demonstrate poor time-frequency localization, lack of accuracy, insufficient cross terms suppression, and masking of the lower energy transient components by the higher energy ones [3, 8, 9]. Such drawbacks severely distort the time-frequency portrait of percussion signals, limiting practical applications. The most extensively exploited linear time-frequency analysis methods are the Short Time Fourier Transform (STFT) and Wavelet Transforms [5, 10]. Although STFT is immune to cross terms, its biggest limitation is the tradeoff between the time and frequency resolution due to the Heisenberg uncertainty principle [8, 11]. In case of the Wavelet Transform, the frequency resolution decreases with increasing frequency [10]. If more than one frequency component exists in the signal, satisfactory resolution for all modes cannot be obtained. This is an inherent characteristic of the wavelet analysis [10]. On the other hand, the quadratic (Cohen's class) time-frequency distributions are based on estimation of instantaneous energy using bilinear operation on the signal [12]. Among the most commonly used time-frequency distributions from Cohen's class are Wigner-Ville Distribution (WVD), Choi-Williams Distribution (CWD), and Zhao-Atlas Marks (ZAM) Distribution [5, 6, 8]. WVD has excellent time-frequency resolution due to the absence of averaging over any finite time interval [11, 13]. While performing well on the single-component signals, WVD's performance degrades dramatically as the number of components increases, because of the spurious peaks arising from the cross terms and confusing the visual interpretation of its time-frequency spectrum [8, 14]. CWD is an improvement of the WVD intended to suppress the cross terms by multiplying it by a kernel function [6]. Unfortunately, increased suppression of the cross terms invariably leads to smearing or loss of resolution of the autoterms in the time-frequency plane.

Another possible way to construct a time-frequency representation of a signal is by representing it as a linear superposition of elementary functions with well-behaved TFDs and then assembling such elementary TFDs on a compound plot. The majority of methods for performing linear signal decomposition involve overcomplete waveform dictionaries. By selecting the optimal (in certain sense) set of available waveforms from the dictionary, one can obtain a sparse model of the signal. In this work, we chose to represent clinical percussion signals as a linear combination of exponentially damped sinusoids (EDS). Such decomposition seems compatible with the underlying physics. Indeed, the percussed human body represents a passive viscoelastic system with multiple degrees of freedom. A percussion impact excites several such degrees at once, and then the oscillations fade out with their respective rates. If the impact is weak, the transient response can be considered linear, and the damping of all normal modes can be considered exponential [4].

In this work, the decomposition of medical percussion signals into damped harmonics was accomplished using Matrix Pencil Method (MPM). The MPM was chosen because it is reportedly the best performing of all damped harmonic analysis methods, especially in the presence of noise. The MPM decomposes the signal into a number of EDS components [15], detecting the exact location of the dominant frequencies and accurately distinguishing closely spaced frequency components [15, 16]. Each EDS is fully characterized by its amplitude, frequency, damping, and initial phase. The first two parameters can be visualized in the frequency domain as vertical line segments of certain height (amplitude). This representation allows for direct comparison with the Fourier amplitude spectrum [17]. However, without seeing the damping parameter it is difficult to judge the relative contribution of individual EDS to the signal. Visualizing the damping parameter would help better capture the behavior of each EDS and show their time evolution in the time-frequency plane.

2. Materials and Methods

The percussion signals analyzed in this paper were collected from normal volunteers at the Detroit Medical Center by trained medical personnel. The testing of human subjects was performed according to Protocol # 0710005340 (Portable Pulmonary Injury Diagnostic Device) approved by the Human Investigation Committee for the Wayne State University International Review Board (M1) for the period from 25 September 2008 to 24 September 2009. The signals were produced by gently tapping with a neurological hammer over a mediator plate (plessimeter) placed on the subject's chest or abdomen. The signals were received with a tripod-based omnidirectional electret condenser air microphone. In each test, the microphone was placed 150–300 mm away from the percussion spot.

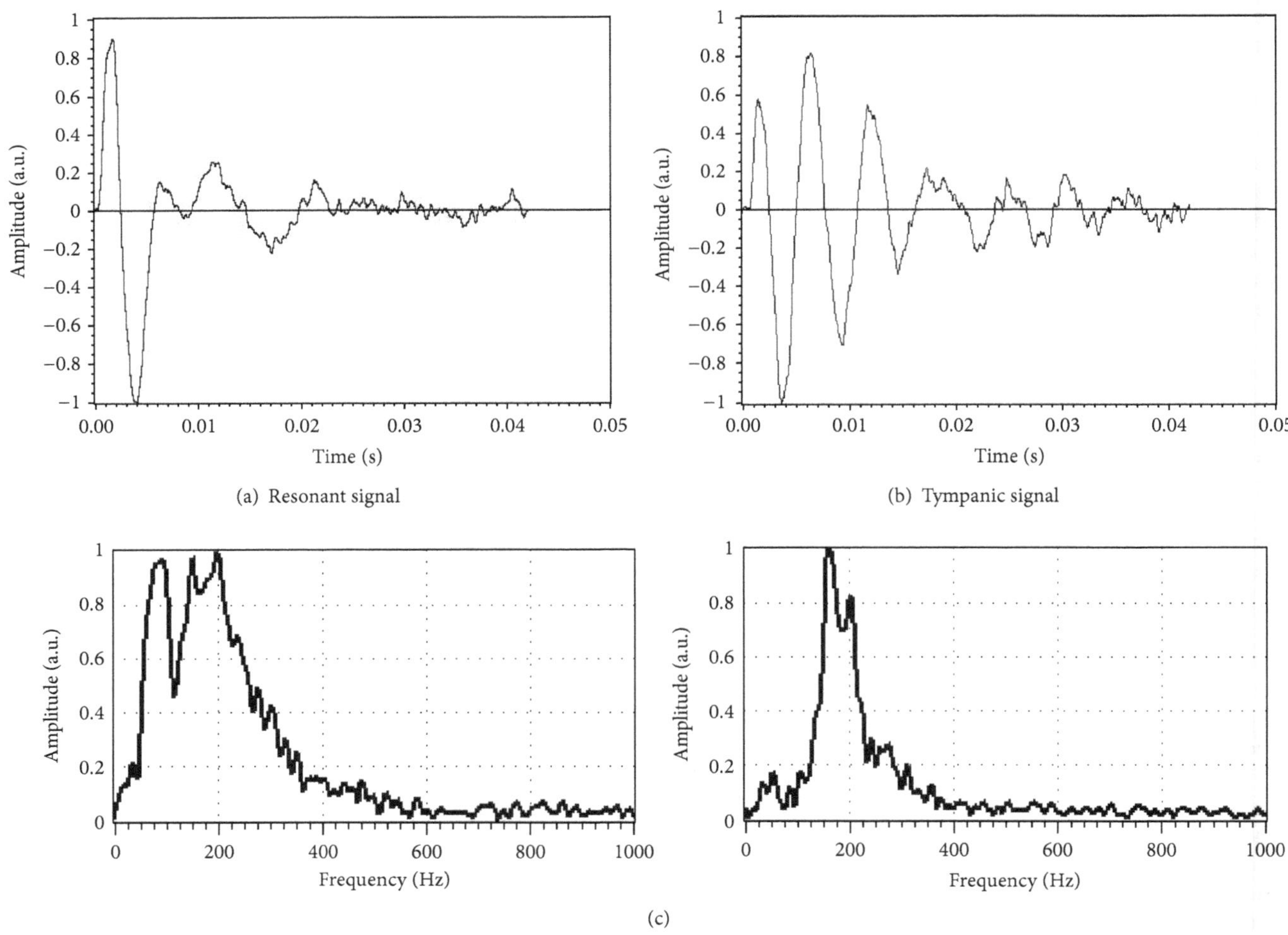

FIGURE 1: Two examples of audible percussion signals collected from healthy volunteers: (a) Signal from the left subclavicular area ("resonant" character); (b) signal from the left abdominal area ("tympanic" character). The normalized spectra of both signals are shown in (c).

Then the signals were amplified and digitized at 48 KHz sampling rate using a 24-bit computer sound card. Additional descriptions of our testing setup can be found in [4, 17, 18]. Two examples of percussion signal waveforms are shown in Figure 1. The "resonant" signal (Figure 1(a)) was acquired over the upper chest in the subclavicular area and the "tympanic" signal (Figure 1(b)) was acquired over the left portion of the abdomen. The signals recorded in our experiments are quite repeatable, hardware-independent, and, in fact, similar in shape to those recorded by Murray and Neilson in 1975 [2]. The amplitude spectra of both signals are also shown in Figure 1. Although the fine spectral structure may differ significantly from signal to signal, in general, "tympanic" spectra tend to be narrower and with fewer major subpeaks than "resonant" ones.

3. Damped Harmonic Model

In this model, the observed percussion response $y(t) = x(t) + n(t)$ is represented as a sum of noise $n(t)$ and the actual signal composed of M EDS with frequencies f_i, amplitudes A_i, initial phases φ_i, and damping factors d_i (total $4M$ parameters). In the discrete case,

$$y(kT_s) = x(kT_s) + n(kT_s) = \sum_{i}^{M} R_i z_i^k + n(kT_s). \quad (1)$$

Here, $k = 0, \ldots, N-1$, $i = 1, \ldots, M$, T_s is the sampling period, $R_i = A_i e^{-j\varphi_i}$ are complex amplitudes, and $z_i = e^{(j2\pi f_i - d_i)T_s}$ are signal poles [16, 19].

Decomposition (1) is carried out using MPM, which is described in more detail in [15–20]. One of the major MPM advantages is that it does not require any prior knowledge or guessing about the EDS present in the signal. Another advantage is the built-in signal denoising process. After performing singular value decomposition (SVD) of the Hankel data matrix constructed from signal samples, the individual singular values σ_i are arranged on the main diagonal of the resulting matrix in the descending order: $\sigma_1 \geq \sigma_2 \geq \cdots \geq \sigma_{\min}$. Some noise can be filtered out by discarding singular values σ_i smaller than the empirically chosen threshold β, for example, those for which $(\sigma_i/\sigma_{\max}) < \beta$ (in this work

$\beta = 10^{-3}$) [15]. This built-in thresholding process is very useful, but it does not always completely eradicate the noise. The remaining noise can be further reduced using physical reasons. Indeed, in typical situations, some of the EDS components produced by the MPM may have too low amplitude, zero or negative damping, or unrealistically high frequency. Such EDS can be discarded, as they are artificially introduced by the MPM to better fit the noise or discontinuities in the signal. If the noise level is too high, the number of such artificial EDS can grow uncontrollably, as the signal no longer obeys the damped harmonic model. Therefore, prior to applying MPM, it is necessary to estimate the Signal to Noise Ratio (SNR), which in clinical percussion signals is typically not constant. It is highest at the onset and then decreases as the signal fades out due to damping. A detailed analysis of the effects of the noise on the MPM performance can be found in [17]. The most optimal conditions for MPM application to clinical percussion signals estimated in [17] suggest that the SNR at the end of the signal window should not exceed 20 dB.

4. Results and Discussion

4.1. MPM Decomposition of Percussion Signals. After filtering both signals shown in Figure 1 with a 50–1000 Hz band-pass Butterworth filter and digitally finding the approximate location of the signal onset, 1500 samples (~30 ms duration) of each signal were processed with the MPM algorithm without downsampling. The Butterworth filter was chosen as it has a reasonably sharp frequency response but does not induce substantial passband ripple. The four parameters of all resulting EDS (only those with positive frequency and damping values were selected) are listed in Tables 1 and 2. The EDS are sorted by the normalized amplitude.

4.2. Analysis of Clinical Percussion Signals

4.2.1. Comparison of the Conventional and MPM-Based TFD. The time-frequency analysis of resonant and tympanic clinical percussion signals of Figure 1 is presented in Figure 2. Figures 2(a) and 2(b) show the results of applying WVD and CWD (with the value of the cross term suppression parameter $\alpha = 0.8$). These two distributions were chosen because they are the most popular quadratic TFDs and because one of them uses a special kernel function for reducing the cross terms and the other one does not. WWD (Figure 2(a)) is the most blurred and the most difficult to interpret of all TFDs presented. Multiple cross term artifacts (not suppressed by the WVD) indirectly confirm the multicomponent character of clinical percussion signals.

The CWD (Figure 2(b)) appears to be much cleaner than WVD. Beyond the blurred 0–6 ms interval for the resonant signal and 0–12 ms for the tympanic signal, CWD shows several components (three for the resonant signal and two for the tympanic signal) with different frequencies, which do not change with time. The amplitudes of these components appear to decay with time. The frequency values of these components are in qualitative agreement with the frequencies of the high-amplitude EDS from Tables 1 and 2. However,

Table 1: Parameters of all EDS extracted by the MPM algorithm from the resonant percussion signal of Figure 1(a).

A	f [Hz]	d [s^{-1}]	φ [rad]
1.0	197.39	226.164	−1.221
0.368	75.937	67.235	2.225
0.207	299.127	251.635	3.062
0.066	142.49	27.705	0.73
0.057	492.73	145.182	1.183
0.046	550.313	152.348	1.654
0.032	927.405	245.756	2.109
0.01	831.218	11.861	1.539
0.009	724.028	40.647	−2.372
0.007	358.181	58.512	1.24
0.003	642.492	38.446	−1.26
0.003	989.675	18.548	0.693

Table 2: Parameters of all EDS extracted by the MPM algorithm from the tympanic percussion signal of Figure 1(b).

A	f [Hz]	d [s^{-1}]	φ [rad]
1.0	159.629	79.883	0.167
0.586	210.907	86.856	−2.813
0.505	277.04	206.404	2.895
0.051	849.697	189.087	−0.921
0.02	370.45	74.406	1.742
0.018	552.295	6.767	2.383
0.008	486.105	13.611	2.98

not all of the EDS can be identified from the CWD plot. For example, the 299.1 Hz mode of the resonant signal does not extend beyond the blurred 0–6 ms interval due to relatively high damping and therefore cannot be detected. The blurring within these time intervals can be explained by the high intensity of cross terms, which CWD suppresses at the expense of the resolution.

The frequency resolution of the STFT, another conventional non-model-based TFD, can be adjusted by choosing appropriate length of the sliding window, apodization, zero padding, and other common Fourier domain techniques. Figure 2(c) shows the results of applying a reassigned spectrogram (STFT) with a 1400-point apodization window (almost as long as the signal itself). This method is able to resolve same frequency components as CWD. Also, similar to CWD, the separation is not satisfactory in the beginning ($t < 5$–9 ms). This can be explained by noting that the spectrogram is essentially a Fourier transform that fits each damped oscillation mode with infinite sine waves. Hence, each damped mode appears to be broadband and contains multiple Fourier harmonics at the same time, making the regular STFT (not shown here) highly blurred and not practically useful. The reassignment procedure improves the resolution of the spectrogram by artificially highlighting its ridges. This helps visualize the Fourier bandwidth of each EDS and even single out the main three frequency components. However,

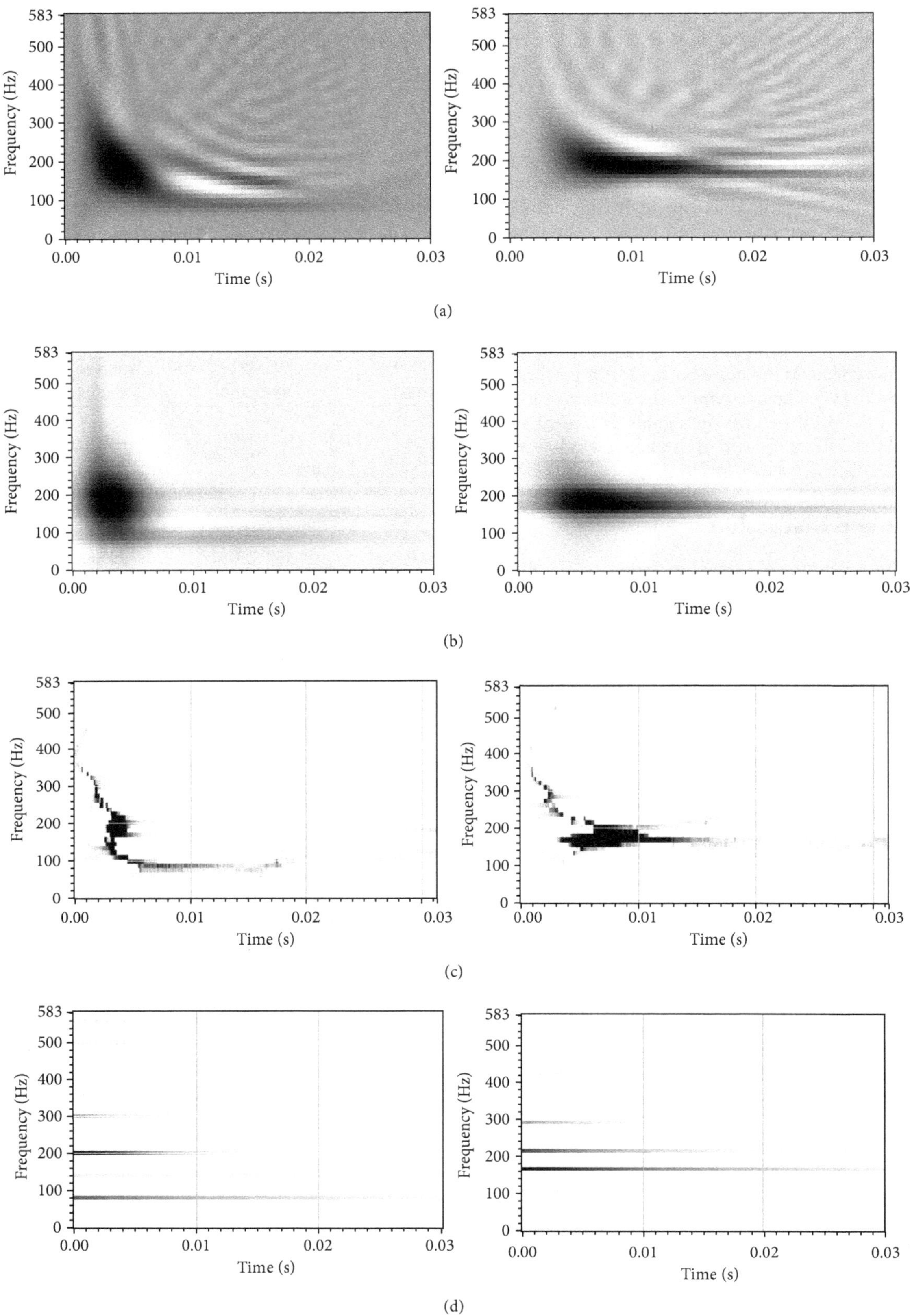

FIGURE 2: Time-frequency representations of the chest and abdominal percussion signals of Figure 1. (a) WVD; (b) CWD ($\alpha = 0.8$); (c) STFT (spectrograms) taken with 1400-point Gaussian window; (d) MPM TFR (amplitudes of all EDS extracted from both signals).

even after reassignment, the STFT of percussion signals still contains too many artifacts to be practically useful.

Figure 2(d) shows the EDS-based time-frequency representation of the chest and abdominal percussion signals. Since the EDS decomposition is performed using MPM, this representation is named MPM TFR. Each EDS (defined by its frequency f, amplitude A, damping d, and phase φ) is plotted as a line parallel to the time axis, having ordinate value of f and colored according to the value of Ae^{-bt}. This representation is very intuitive and readily reveals the main features of the signal. The damping of each EDS component can be estimated from its visible duration, while the color indicates the instantaneous amplitude (darker color means higher amplitude). Each component is associated with a particular frequency and experiences exponential decay with time. Components with high starting amplitude and low damping are dominant and determine the overall character of the signal. Weak components or those with very high damping are less important, as they are often artifacts produced by the MPM to fit the noise. It should be mentioned that the graphical representation of the MPM results is not completely equivalent to that of the results obtained by the other methods. The frequencies on the vertical axis correspond to frequency of the EDS components and not to Fourier components. Even though they are represented by the same parameter, the frequency, the set of EDS represent a different basis for the decomposition of the signals. Damping of a harmonic component leads to widening of its spectral peak. This may result in overlapping of various components in the spectrum. In the EDS representation, the damping influences the apparent length of the horizontal lines. There is no overlapping along the frequency axis in the EDS representation, no matter how close in frequency and how damped the components are. This feature should be considered when one decides which representation is the most suitable for a specific application.

The MPM TFR images have no cross term artifacts, infinite frequency resolution, and provide convenient visualization of damping. However, MPM TFR follows from the EDS-based model, which is only an empirical suggestion and needs to be justified. Indeed, decomposition into a number of EDS is just one possible type of representing the signal. Other basis functions can be selected to fit the same signal, and the resulting TFR may look different than the MPM-based one. One argument in favor of the EDS-based model comes from physical reasons, as described in the Introduction. Another independent proof could be obtained by demonstrating similarity of the time-frequency portraits produced by the MPM TFR and by non-model-based distributions. Indeed, the time-frequency portrait is an intrinsic property of the signal, and it should not depend on the choice of any particular analysis tool. Various TFDs can visualize this portrait only approximately, while introducing their specific artifacts. If certain TFDs introduce less severe artifacts than others for a given signal type, they are expected to produce more or less similar looking time-frequency portraits (approaching the actual portrait). On the other hand, showing that a model-based distribution produces time-frequency portrait that looks similar to that produced by a non-model-based distribution could be considered as an argument in favor of the model.

The STFT and CWD belong to different classes (Fourier transform-based and bilinear) yet show very similar time-frequency portraits of the same clinical percussion signal. Notably, all dominant features present in the spectrogram and CWD are also present in the MPM TFR plot. Based on the above reasoning, this could be considered an independent proof of the validity of the EDS-based model of clinical percussion signals. On the other hand, MPM TFR reveals additional components, not resolved by conventional distributions. Those components do not play critical role in the signal reconstruction as they have relatively low amplitudes and/or significantly higher damping compared to the dominant ones, determining the signal behavior. The physical origin of these minor components is unclear. The EDS with low amplitude, too high frequency, negative or excessive damping, and so forth could be artificially introduced by MPM to fit the noise or edge discontinuities in the signal. Alternatively, they may represent real (but weak) oscillation modes of the human body and be of practical importance. On the other hand, the noise (which is always present in percussion signals) could be the major cause of the poor resolution of the WVD, CWD, and reassigned spectrogram. Without the noise, these distributions could possibly show much better resolution. In order to answer those questions and to better understand the effect of noise on the behavior of the MPM TFR and other conventional distributions, we constructed two idealized, noise-free signals closely resembling the actual resonant and tympanic percussion waveforms. These simulated signals are analyzed below.

4.3. Analysis of Simulated Percussion Signals. Two simulated noise-free percussion signals were constructed by adding four highest-amplitude EDS from Table 1 and three such EDS from Table 2, respectively, to model resonant and tympanic signals. The resulting signals (1500 points sampled at 48000 Hz) and their spectra are shown in Figure 3.

The simulated "resonant" signal contains three EDS with frequencies of 75.9, 142.5, 197.4, and 299.1 Hz. The first three EDS have well-resolved spectral peaks, while the fourth one is masked by the broad (highly damped) and higher-amplitude third peak. The simulated "tympanic" signal contains three EDS with frequencies of 159.6, 210.9, and 277.0 Hz. All three spectral peaks are very closely spaced, but still resolvable, as the first (and the strongest) two of them have relatively low damping.

The time-frequency analysis of both simulated signals is presented in Figure 4. Figures 4(a) and 4(b) show the results of applying WVD and CWD (with $\alpha = 0.8$). Visual examination of the WVD plots reveals many additional components, which were not present in the original signal. These are undesirable cross term artifacts produced by the WVD and are making the image difficult to interpret. Still, the WVDs of ideal signals are cleaner than those of real signals (Figure 2). As expected, the level of cross term artifacts in the CWD images is significantly reduced. CWD allows resolving, respectively, three and two dominant components

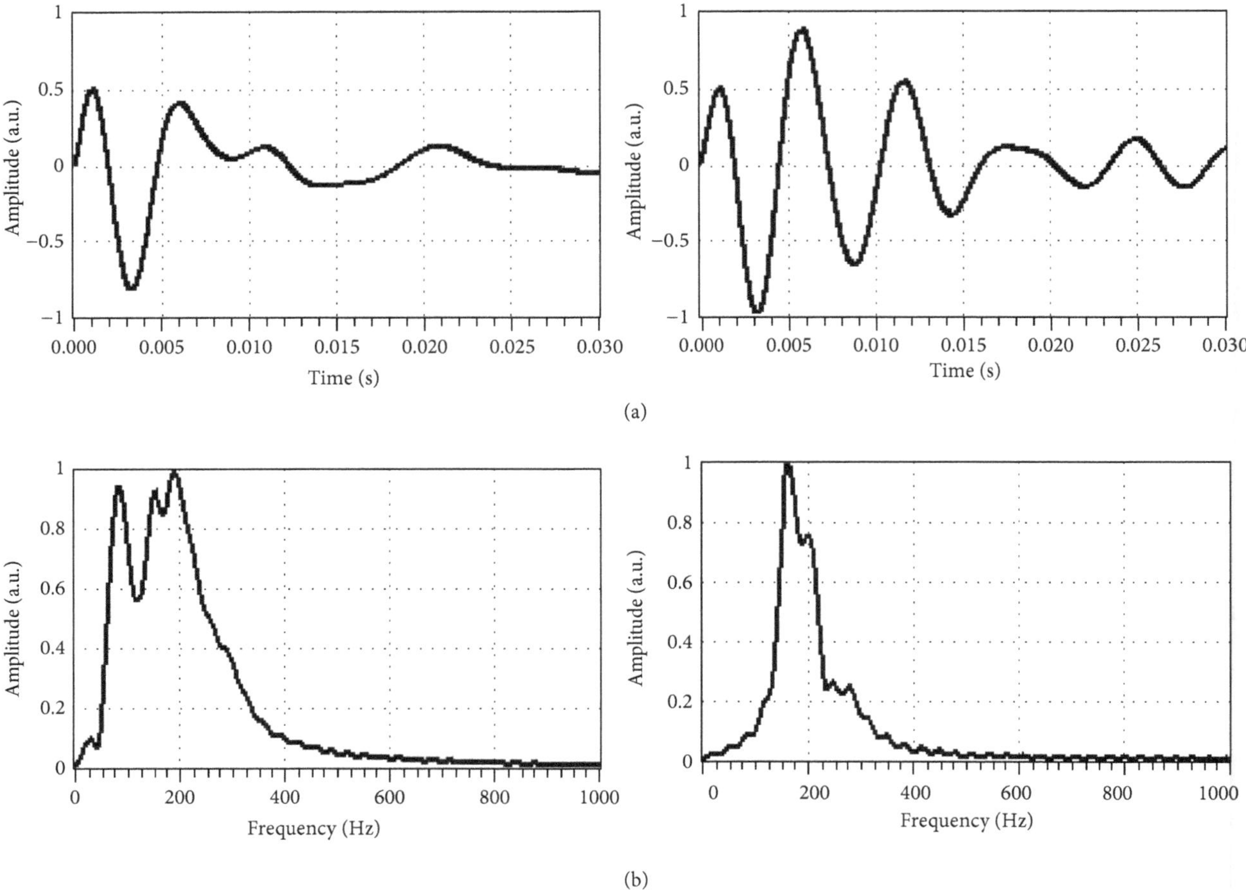

FIGURE 3: Noise-free "resonant" and "tympanic" percussion signals synthesized with four and three highest-amplitude EDS extracted from the real signals shown in Figure 1. The spectra of both signals are shown in (b).

in both images and noticing that their frequencies remain constant while the amplitudes decay with time. The only issue with CWD is its poor frequency resolution in the beginning ($t < 6$ and $t < 12$ ms), where all components have high amplitudes and the strongest interference. This happens because CWD suppresses the cross terms at the expense of the resolution. Therefore, even for these simplistic simulated signals neither WVD nor CWD can simultaneously resolve the cross terms problem and increase the resolution. This problem is exacerbated in practical applications, as the real percussion signals always contain noise. Figure 4(c) shows the results of applying a reassigned spectrogram (STFT) with a 1400-point apodization window. This method shows resolution similar to CWD, only with slightly reduced low-resolution zone ($t < 5$–9 ms). Figure 4(d) contains time-frequency representations of the original EDS components used to construct both synthetic signals (MPM TFR). Both images have no cross term artifacts, infinite frequency resolution, and provide convenient visualization of damping. They adequately represent the noise-free resonant and tympanic signals and are clearly superior to WVD, CWD, and spectrogram. By comparing the noise-free plots of Figure 4 with their real-world counterparts of Figure 2 it can be observed that the extent of the low-resolution zone in CWD and reassigned spectrogram is approximately the same for the real and noise-free percussion signals. Therefore, the low-resolution zone in the CWD and STFT cannot be explained by the noise in the signal. In another simulated experiment, a signal was synthesized from the same top three EDS of Table 2, but with frequency values of the second and third EDS changed to 310 and 580 Hz. The resulting CWD and reassigned spectrogram (not shown here) successfully resolved these frequencies starting from 3 ms. Further spreading of the EDS frequencies or reducing the damping (less spectral overlapping) leads to further shrinkage of the blurred zone. Therefore, the blurred areas are caused by the spectral overlapping and not by the noise. The observed "blurring" is then a real feature of the noiseless signal. It is due to the attempt to decompose a signal made from EDS into harmonic modes (analogous to decomposition of a vector into a nonorthogonal basis). For these cases, the MPM TFR correctly identified all constituent EDS including weak and highly damped modes, not resolvable by conventional distributions, and eliminates the overlapping along the frequency axis.

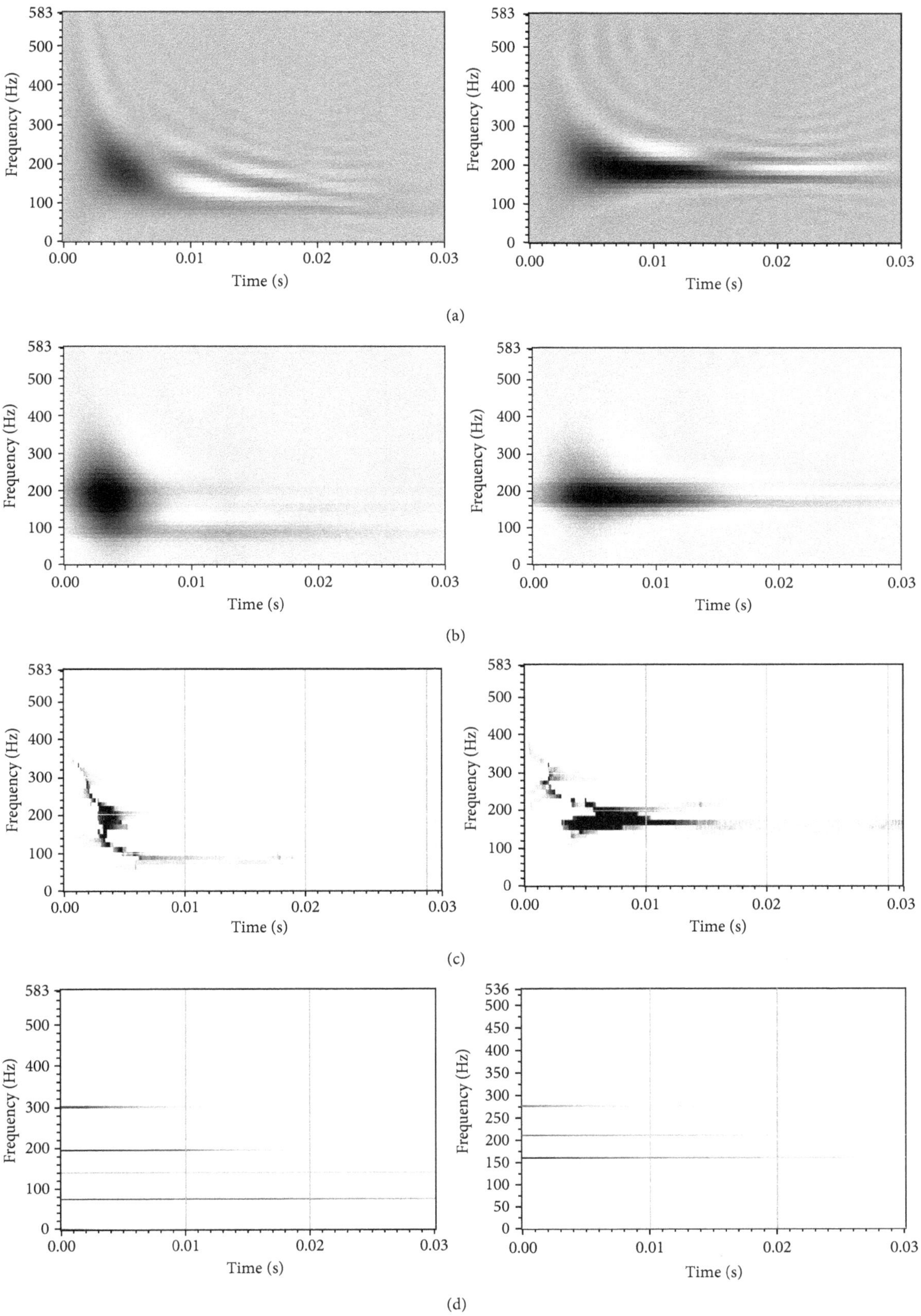

FIGURE 4: Time-frequency representations of the ideal, noise-free "resonant" (left) and "tympanic" (right) percussion signals shown in Figure 3. The "resonant" signal is a sum of four EDS; the "tympanic" signal is a sum of three EDS. (a) WVD; (b) CWD (α = 0.8); (c) STFT (spectrograms) taken with 1400-point Gaussian window; (d) MPM TFR (amplitudes of the original EDS comprising both signals).

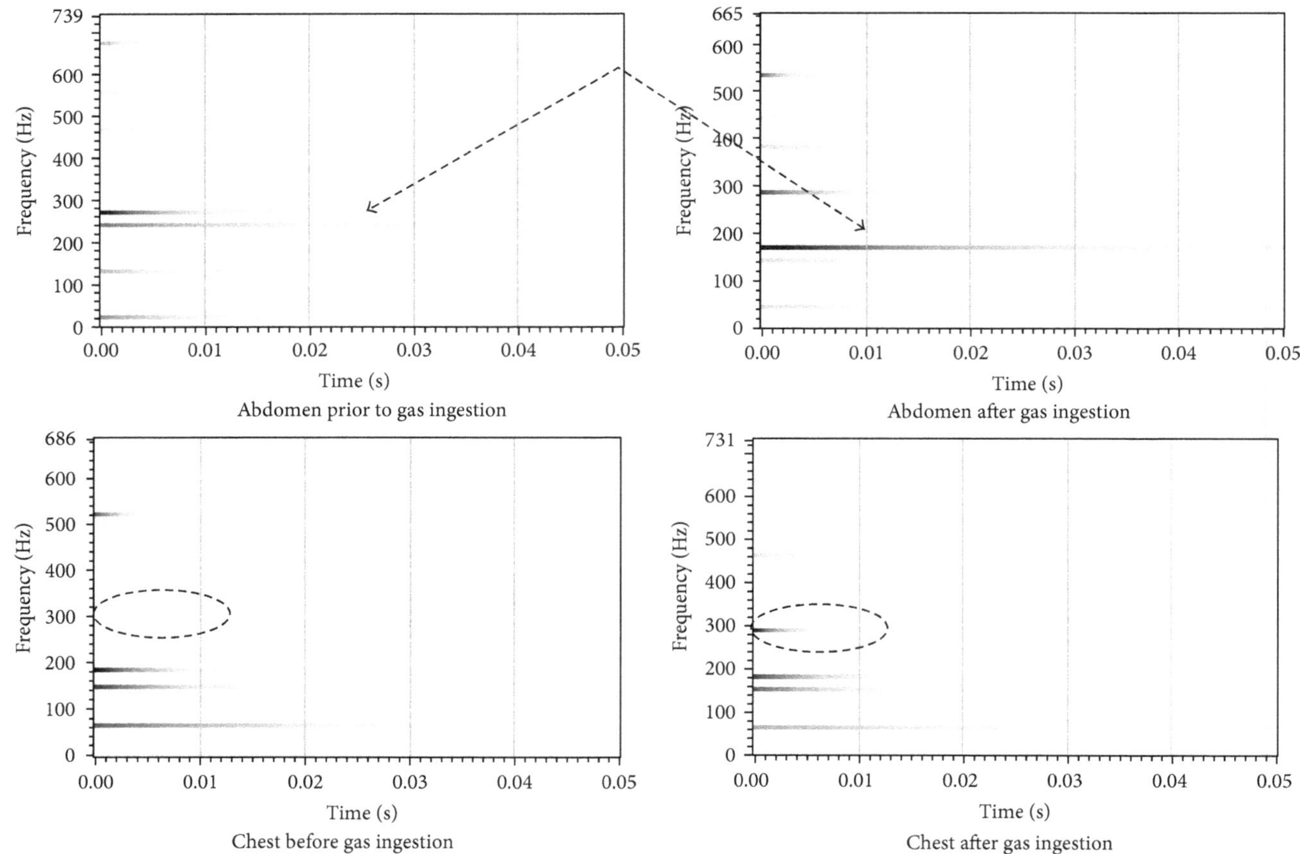

Figure 5: MPM TFR of percussion signals from a normal volunteer before and after intake of a carbonated drink. The percussion signals were recorded both over the abdomen and the chest.

4.4. Using MPM TFR to Detect Changes in Physical Condition. Up to this point, we have demonstrated the superiority of the MPM-based TFR over WVD, CWD, and spectrogram when applied to clinical percussion signals. Below we discuss some of its practical applications. The rates of decay (damping), the frequencies, and the relative amplitudes of dominant EDS are important parameters, possibly reflecting pathophysiological properties of the percussed organ.

An illustrative example of how the MPM-based time-frequency analysis can be practically deployed to detect changes in the signal is shown in Figure 5. In this experiment, a normal volunteer did not eat for 15 hours and then consumed a carbonated drink. The percussion signals were recorded over his abdomen and chest before and after consuming the drink. The MPM TFRs of these signals were constructed and compared. It can be seen that before consuming the drink, the least damped EDS at ~240 Hz responsible for the overall tympanic character of the abdominal signal lasted for only ~20 ms (see Figure 5, top row). After taking the drink, the intestinal gas pattern changed, so that the frequency of the least damped EDS shifted to ~190 Hz, and its duration increased beyond 30 ms. Changes in the modal composition of the chest signal can be observed as well (see Figure 5, bottom row). For example, an additional strong but highly damped mode has appeared at ~300 Hz, as marked by the dashed oval. Other signal modes have remained almost unchanged. One possible explanation of this result is that gas-filled stomach has freed one of the chest vibration modes indirectly through diaphragm.

It should be noted that none of the observed effects can be reliably detected using conventional TFRs examined in this paper due to their inferior resolution in the 0–5 or 0–12 ms interval and due to their insensitivity to slight changes in the damping.

5. Conclusions

The proposed MPM TFR yields a clear and efficient graphical representation of both resonant and tympanic percussion signal types. The ability to resolve oscillation modes with closely spaced frequencies and high damping makes MPM TFR a preferable method for the analysis of clinical percussion signals. Its superiority becomes especially evident when analyzing the 5 ms time interval following the signal onset, where such commonly used TFDs as STFT, WVD, and CWD all fail to resolve individual frequency components.

We have successfully demonstrated the usefulness of MPM TFR in detecting subtle percussion signal changes caused by physiological processes. We were able to detect the appearance of a new oscillation mode or slight changes in damping of an existing mode. Such effects cannot be reliably detected using conventional TFRs examined in this paper.

Decomposing the signal into a sum of elementary functions with known time-frequency representations and then combining those elementary TFRs produce the image free of cross terms and other common artifacts. MPM TFR is based on the EDS signal model, which can be justified both from physical principles and through direct comparison with CWD and STWT results. Besides clinical percussion, the proposed MPM TFR can be used in many other areas, including music and speech processing, where the EDS decomposition model is applicable.

Conflict of Interests

The authors declare that there is no conflict of interests regarding the publication of this paper.

Acknowledgments

The authors express sincere gratitude to the personnel of Tessonics Corp., the Institute for Diagnostic Imaging Research, the Detroit Medical Center, and Departments of Electrical & Computer Engineering and Physics at the University of Windsor.

References

[1] J. C. Yernault and A. B. Bohadana, "Chest percussion," *European Respiratory Journal*, vol. 8, no. 10, pp. 1756–1760, 1995.

[2] A. Murray and J. M. M. Neilson, "Diagnostic percussion sounds: 1. A qualitative analysis," *Medical and Biological Engineering*, vol. 13, no. 1, pp. 19–28, 1975.

[3] A. Kacha, F. Grenez, and K. Benmahammed, "Time-frequency analysis and instantaneous frequency estimation using two-sided linear prediction," *Signal Processing*, vol. 85, no. 3, pp. 491–503, 2005.

[4] M. A. Pantea, E. V. Malyarenko, A. E. Baylor, and R. G. Maev, "A physical approach to the automated classification of clinical percussion sounds," *The Journal of the Acoustical Society of America*, vol. 131, no. 1, pp. 608–619, 2012.

[5] F. Hlawatsch and G. F. Boudreaux-Bartels, "Linear and quadratic time-frequency signal representations," *IEEE Signal Processing Magazine*, vol. 9, no. 2, pp. 21–67, 1992.

[6] H.-I. Choi and W. J. Williams, "Improved time-frequency representation of multicomponent signals using exponential kernels," *IEEE Transactions on Acoustics, Speech, and Signal Processing*, vol. 37, no. 6, pp. 862–871, 1989.

[7] G. Putland and B. Boashash, "Can a signal be both monocomponent and multicomponent?" in *Proceedings of the 3rd Australasian Workshop on Signal Processing Applications (WoSPA '00)*, Brisbane, Australia, December 2000.

[8] B. Barkat and B. Boashash, "A high-resolution quadratic time-frequency distribution for multicomponent signals analysis," *IEEE Transactions on Signal Processing*, vol. 49, no. 10, pp. 2232–2239, 2001.

[9] L. A. Escobar-Moreira, "Ultrasonic fault machinery monitoring by using the wigner-ville and Choi-Williams distributions," in *Proceedings of the 11th International Conference on Electrical Machines and Systems (ICEMS '08)*, pp. 741–745, IEEE, Wuhan, China, October 2008.

[10] Z. Li and M. J. Crocker, "A study of joint time-frequency analysis-based modal analysis," *IEEE Transactions on Instrumentation and Measurement*, vol. 55, no. 6, pp. 2335–2342, 2006.

[11] M. D. Davidović and V. Vojisavljevic, "Time-frequency analysis of nonstationary optical signals using Husimi type function," *Acta Physica Polonica A*, vol. 116, no. 4, pp. 675–677, 2009.

[12] C. Griffin, "A comparison study on the Wigner and Choi-Williams distributions for detection," in *Proceedings of the International Conference on Acoustics, Speech, and Signal Processing (ICASSP '91)*, pp. 1485–1488, May 1991.

[13] Y. Noguchi, K. Watanabe, E.-I. Kashiwagi et al., "Time-frequency analysis with eight-figure kernel," in *Proceedings of the 19th Annual International Conference of the IEEE Engineering in Medicine and Biology Society*, pp. 1324–1327, November 1997.

[14] A. Goli, D. M. McNamara, and A. K. Ziarani, "A novel method for decomposition of multicomponent nonstationary signals," in *Proceedings of the IEEE Workshop on Applications of Signal Processing to Audio and Acoustics (WASPAA '07)*, pp. 255–258, New Paltz, NY, USA, October 2007.

[15] T. K. Sarkar and O. Pereira, "Using the matrix pencil method to estimate the parameters of a sum of complex exponentials," *IEEE Antennas and Propagation Magazine*, vol. 37, no. 1, pp. 48–55, 1995.

[16] J. Laroche, "The use of the matrix pencil method for the spectrum analysis of musical signals," *Journal of the Acoustical Society of America*, vol. 94, no. 4, pp. 1958–1965, 1993.

[17] M. Bhuiyan, E. V. Malyarenko, M. A. Pantea, F. M. Seviaryn, and R. Gr. Maev, "Advantages and limitations of using matrix pencil method for the modal analysis of medical percussion signals," *IEEE Transactions on Biomedical Engineering*, vol. 60, no. 2, pp. 417–426, 2013.

[18] M. Bhuiyan, E. V. Malyarenko, M. A. Pantea, R. G. Maev, and A. E. Baylor, "Estimating the parameters of audible clinical percussion signals by fitting exponentially damped harmonics," *The Journal of the Acoustical Society of America*, vol. 131, no. 6, pp. 4690–4698, 2012.

[19] H. Fengduo, T. Sarkar, and H. Yingbo, "The spectral parameter estimation by using pre filtering and matrix pencil method," in *Proceedings of the 5th ASSP Workshop on Spectrum Estimation and Modeling*, pp. 45–49, Rochester, NY, USA, October 1990.

[20] Y. Hua and T. K. Sarkar, "Matrix pencil method for estimating parameters of exponentially damped/undamped sinusoids in noise," *IEEE Transactions on Acoustics, Speech, and Signal Processing*, vol. 38, no. 5, pp. 814–824, 1990.

15

Two Improved Cancellation Techniques for Direct-Conversion Receivers

Xueyuan Hao and Xiaohong Yan

Nanjing University of Posts and Telecommunications, Nanjing 210003, China

Correspondence should be addressed to Xueyuan Hao; haoxy@njupt.edu.cn

Academic Editor: Ahmed M. Soliman

To solve the problems of carrier leakage and DC offset in direct-conversion receiver (DCR) system, the paper proposed two kinds of improved technology to overcome the problems in DCR system. One is the RF carrier cancellation technology; the traditional cancellation technology based on lumped parameter filter can be easily influenced by distribution parameters, the improved circuits use a 3 db bridge to realize a 180-degree phase shifter, and the method can adapt to a wider range of RF frequency. Another is DC offset cancellation technique; a novel DC servo loop circuit is proposed to replace the traditional AC-coupled amplifier circuit. It can improve the integrity of the baseband signal and reduces the complexity of the subsequent software algorithm. Experimental results show that two kinds of improved technology can improve the performance of DCR and expand its scope of application.

1. Introduction

The direct-conversion receiver (DCR), also known as zero-IF receiver, is a radio receiver design that realizes the RF signal one-time conversion to baseband signal. Compared to the super heterodyne structure, the DCR system has no mirror frequency interference and can be easily realized with low cost [1].

In the single antenna DCR system, a special component, the circulator, is generally used to realize the isolation between sending signals and receiving signal; it is shown in Figure 1.

Circulator is a three-port device, including the launch port, the antenna port, and a receive port. Due to the properties of the magnetic material, the circulator is not easy to be designed with high isolation. Measured by the vector network analyzer in good match case, the isolation can only reach 26 dB. In actual application, due to wiring length in circuit board and the antenna impedance mismatch, the isolation is often lower than 20 dB [2]. The low isolation will cause some problems, the first one is linearity issue; the transmitted signal can be leaked into the receive port by several ways as shown in Figure 2; one way is through the antenna port directly leaking into the receive port, and the other one is from the launch port circulator leakage to the receiving port. The power of leakage signal in receiving port is far greater than the received signal power; it is easy to cause the receiver front-end circuit (LNA and mixer) saturation.

Another issue of the DCR system is the DC offset problem. Any leakage between LO and RF ports of the mixer will produce an undesired DC component; the large DC signal will stature the subsequent DC amplifier. Because of the existence of the DC voltage, the wideband amplifier cannot work in DC coupling mode and can only work in the AC coupling mode. The AC capacitor not only limits the data rate but also influences the pulse width of the baseband signal, and the uncertain baseband signal pulse width will seriously affect the subsequent decoding algorithm.

2. The Traditional Workaround Method Used in DCR System

To solve the above problem, people put forward some improvement measures [3, 4]. One method is to try to improve the impedance matching of the circulator port, using Voltage Standing Wave Ratio (VSWR) better antenna switch to increase compensation link in the actual circuit. However, practice shows that the effects of the method are rather limited because of the circulator own property. The second

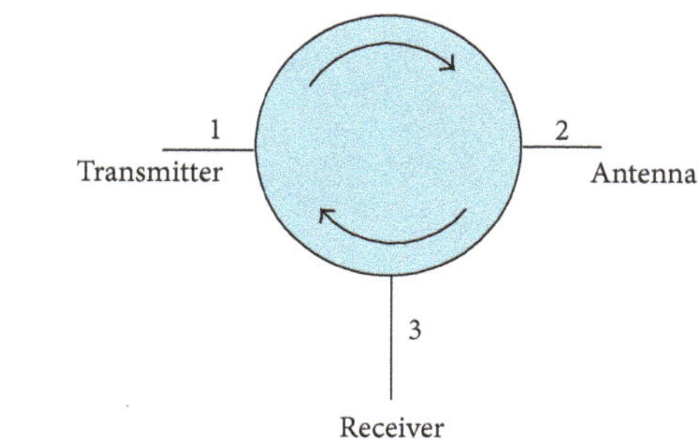

FIGURE 1: Circulator.

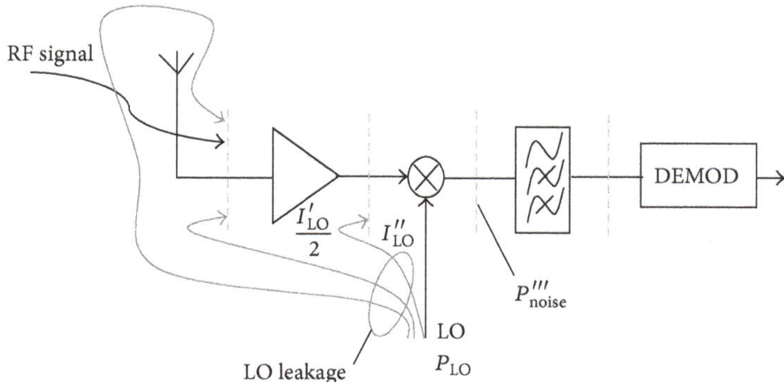

FIGURE 2: Paths of carrier leakage.

method is to reduce the power of the transmitted signal leaks to the receive terminal in order to reduce the transmit signal power level; the result is a reduction in communication distance. The third method, high IP3 double balanced mixers, is used at the receiver; the result is reducing the receiver SNR and also affecting the communication distance and the communication quality at the receiver.

The abovementioned methods are not significantly improving the mentioned problems, and they limit the DCR system scope of the application system.

3. A Novel Carrier Cancellation Technique

Carrier cancellation technology is a reverse power synthesis technology [5] that can be used to solve the nonlinear distortion of the power amplifier and extended the measurement frequency limit in super heterodyne spectrum analyzer. Some articles also proposed design methods of carrier cancellation. However, such design methods commonly use lumped parameter approach, and the cancellation signal is generated by the RLC filter, which is generally used in occasion of less than 900 M band. When applied to the higher RF band, the lumped parameter circuit will be seriously affected by distribution parameters, and therefore carrier cancellation function cannot be well implemented.

In order to eliminate the influence of the lumped parameters, a new carrier cancellation method is proposed as shown in Figure 3.

The improved circuit increases the directional coupler, microstrip 180° phase shifter, and Electric Regulating Control (ERC) attenuator; the ERC 180° phase shifter is a key circuit. The following describes the role of each circuit.

(1) Directional Couplers A and B. Two circuits are identical in structure, but different in functions. The role of directional coupler *A*, used as a power divider, is coupled out from a certain energy of RF signal. The role of directional coupler *B*, used as a power combiner, is superposition of the electrical output of the RF modulated signal attenuator and the leakage signal from the circulator and antenna port. Its coupling degree is about −6 dB, and the center frequency is 2.45 GHz in the paper.

(2) ERC 180° Phase Shift. This is a key circuit, which not only achieves near 180° phase shift but also has a very fine phase adjustment capability. Since the carrier phase offset technique is very sensitive in RF frequency, precise matching degree of phase offset will seriously affect the results. In actual implementation, both the leakage signal phase shift and the phase offset cancellation should be considered. Based on the above considerations, we propose using microstrip electronic bridge to realize the phase shifter. To solve the problem of fine-tuning phase, two PIN diodes through an impedance transformation network are placed in the bridge's two arms as voltage control devices. Phase shifter has center frequency of 2.45 GHz, 3 dB bandwidth of about 600 MHz, and the insertion loss at 2.45 GHz frequency of 3.5 dB. It is shown in Figure 4.

(3) ERC Attenuator. The common PIN diode can be used at lower frequencies, but for high-frequency applications, the

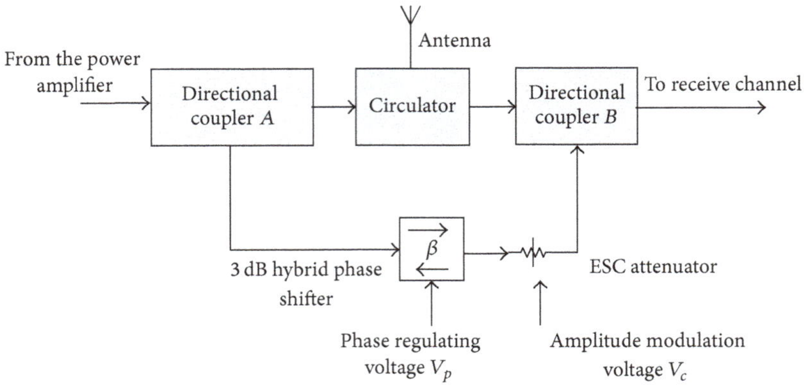

FIGURE 3: Carrier cancellation circuit.

FIGURE 4: ERC 180° phase shifter.

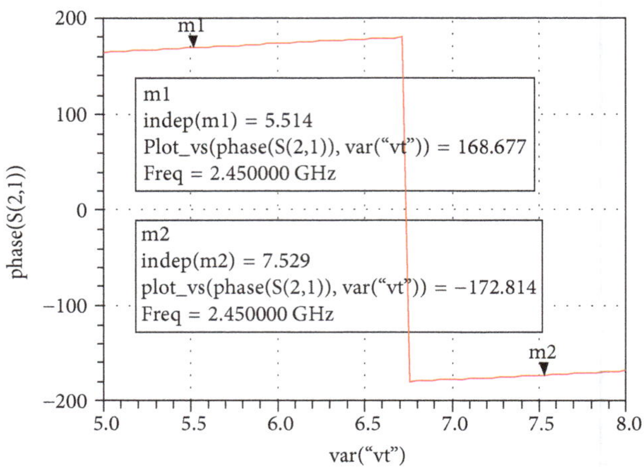

FIGURE 5: The voltage control phase shifter; when the phase-controlled voltage V_p changes at 5.5 V~7.5 V, the electrical phase shifter adjusts a phase shift variation of 169.2°~187.4°; taking into account the additional phase shift of the circulator and directional coupler itself, the phase control range of the voltage V_p may also be larger than the theoretical results, which can meet the overall requirements of the circuit to compensate the phase shift. By carefully adjusting the control voltage V_p, the exact 180° phase shift can be obtained between the transmitter port of the circulator and the cancellation signal.

capacitance of PIN diode junction will reduce the amount of power attenuation. In this paper, the attenuator chip of Skyworks is used to minimize errors. When V_c changes 0 V~1.2 V, amplitude adjustment range is −22 dB~−2 dB. ERC Attenuator cannot only find the best amplitude offset points and is able to adapt to different isolation difference within the circulator.

4. The ADS Simulation of Carrier Leakage Circuit and Experiment

Two ports are coupled to each other by the coupling coefficient M; the scattering matrix is

$$\begin{bmatrix} b_1 \\ b_2 \end{bmatrix} = \frac{1}{2}\begin{bmatrix} s_{11} & s_{12} \\ s_{21} & s_{22} \end{bmatrix}\begin{bmatrix} a_1 \\ a_2 \end{bmatrix}, \tag{1}$$

where

$$\begin{aligned} s_{11} &= T^2\left(\alpha_A^2\Gamma_A - \alpha_B^2\Gamma_B\right) - 2jT^2M\alpha_A\alpha_B, \\ s_{12} &= TC\left(\alpha_A\Gamma_A - \alpha_B\Gamma_B\right) + I\left(\alpha_A - \alpha_B\right) \\ &\quad - jTCM\left(\alpha_A + \alpha_B\right), \\ s_{21} &= TC\left(\alpha_A\Gamma_A - \alpha_B\Gamma_B\right) + I\left(\alpha_A - \alpha_B\right) \\ &\quad - jTCM\left(\alpha_A + \alpha_B\right), \\ s_{22} &= C^2\left(\Gamma_A - \Gamma_B\right) - 2jC^2M. \end{aligned} \tag{2}$$

Tx leakage can be expressed as follows:

$$\begin{aligned} \frac{b_2}{a_1} &= \frac{1}{2}\left(TC\left(\alpha_A\Gamma_A - \alpha_B\Gamma_B\right) + I\left(\alpha_A + \alpha_B\right)\right. \\ &\quad \left. - jTCM\left(\alpha_A + \alpha_B\right)\right). \end{aligned} \tag{3}$$

The relationship between the voltage control phase shifter is shown in Figure 5.

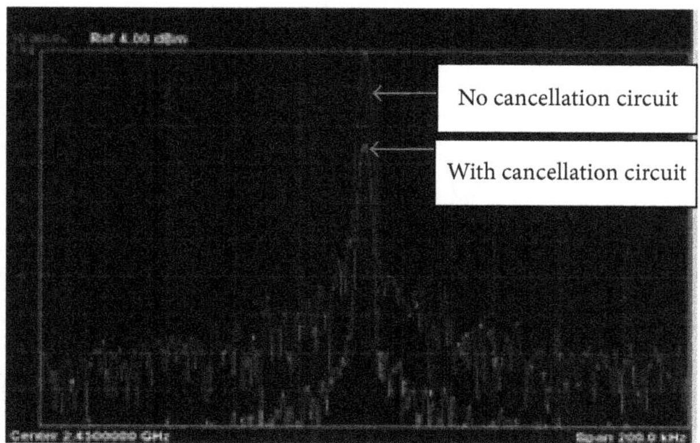

FIGURE 6: Comparison results after adding cancellation circuit.

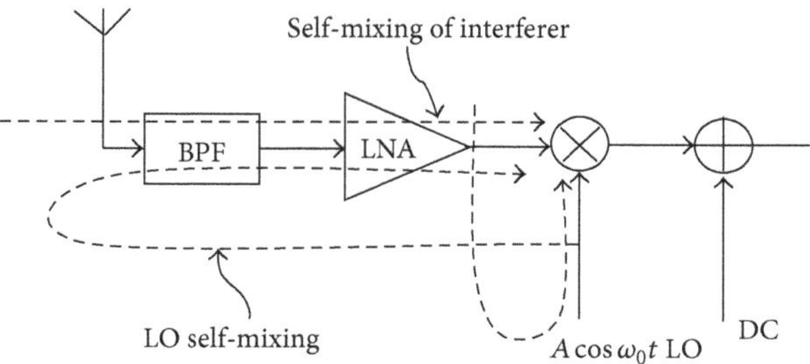

FIGURE 7: Three main sources of leakages give rise to DC offset.

The real result is tested by the spectrum analyzer as shown in Figure 6. Experimental results show that the system can be obtained at 30 db attenuation.

5. DC Offset Problems and Solutions in DCR System

DCR receiver will bring unwanted DC component; it is much larger than the useful baseband signal amplitude. Three main sources of leakages give rise to DC offset as shown in Figure 7. For example, if the receiver front-end mixer LO level is +17 dBm, transmit power is 1 W and circulator isolation is about 20 dB. When not using the carrier cancellation circuit, the leakage to the mixer local oscillator signal power is up to 10 mW.

The DC offset voltage is attained:

$$\begin{aligned} V_m &= k_m \times (V_{\mathrm{DC}I} + V_I) \times (V_{\mathrm{DCQ}} + V_Q) = k_m \times [V_{\mathrm{DC}I} \\ &\quad \times V_{\mathrm{DCQ}} + (V_I \times V_{\mathrm{DCQ}} + V_Q \times V_{\mathrm{DC}I} + V_I \times V_Q)] \\ &= U + W, \end{aligned} \tag{4}$$

where k_m *denotes* the amplifying coefficient, and

$$\begin{aligned} U &= k_m \times V_{\mathrm{DC}I} \times V_{\mathrm{DCQ}}, \\ W &= k_m \times (V_I \times V_{\mathrm{DCQ}} + V_Q \times V_{\mathrm{DC}I} + V_I \times V_Q), \end{aligned} \tag{5}$$

where U is DC signal and W is signal including both DC and AC components.

Taking into account the conversion 6 db loss of the mixer, the DC voltage output from the mixer is about 447 mV. With the carrier cancellation circuit, the DC voltage output of the mixer is about 30 mV, and the DC voltage is larger than the amplitude of the baseband signal.

Due to the presence of the DC voltage, an AC-coupled amplifier is generally used to amplify the weak baseband signal. Theoretical analysis shows that AC coupling will have an impact on the quality of the digital baseband signal. Figure 8 shows the result of the different data rate using multistage AC amplifier.

For a fixed coupling constant AC amplifier, the special data rate is appropriate for the special data rate, but when the rate of the baseband signal is higher or lower than the frequency, amplified signal will produce integral or differential effects, and the baseband signal integrity is damaged, so some complex software algorithms were designed to correct the distorted baseband signals [6–9].

6. A New Wideband DC Amplifier Design with DC Feedback Loop

To solve these problems, we proposed a new novel wideband and high gain amplifier with low frequency servo loop. The basic circuit is shown in Figure 9.

The circuits contain two low-pass filters, a multistage wideband DC amplifier, an integrator, and an active threshold decision circuit. The high-frequency signal of mixer is filtered by the low-pass filter 1, and baseband and DC signal are retained. The multistage DC amplifiers have large dynamic scope; the gain is set at 100 db in the simulation circuit, and the full-power bandwidth of amplifier satisfies the maximum rate requirements of DCR system. The low-pass filter 2 and the integrator circuit are the core components to cancel the effect of DC and $1/f$ noise; the circuits filter out the AC component from the DC amplifier while retaining its DC and low frequency components; the active integrator accumulates the DC voltage and feed back to the inverse input of the high-speed DC amplifier, because the integrator samples the signal from the last amplifier. The DC component produced by the front mixer or by the high-speed operational amplifier self will be limited to a very small level. Taking into account $1/f$ noise in wideband amplifier, the low-pass filter 2 and integrator must be set at a higher cut-off frequency, but in order to try to maintain the integrity of the baseband signal, the cut-off frequency should be set lower.

The baseband signal is series rectangle pulse, and according to Fourier transform, the rectangular pulse can be expressed as

$$\begin{aligned} |F_n| &= \frac{1}{2}\sqrt{a_n^2 + b_n^2} = \frac{1}{2}\sqrt{\frac{4\sin^2(n\omega_0\tau/2)}{n^2\pi^2}} \\ &= \frac{1}{2} \cdot \frac{2\sin(n\omega_0\tau/2)}{n\pi} = \frac{\sin(n\omega_0\tau/2)}{n\pi}. \end{aligned} \tag{6}$$

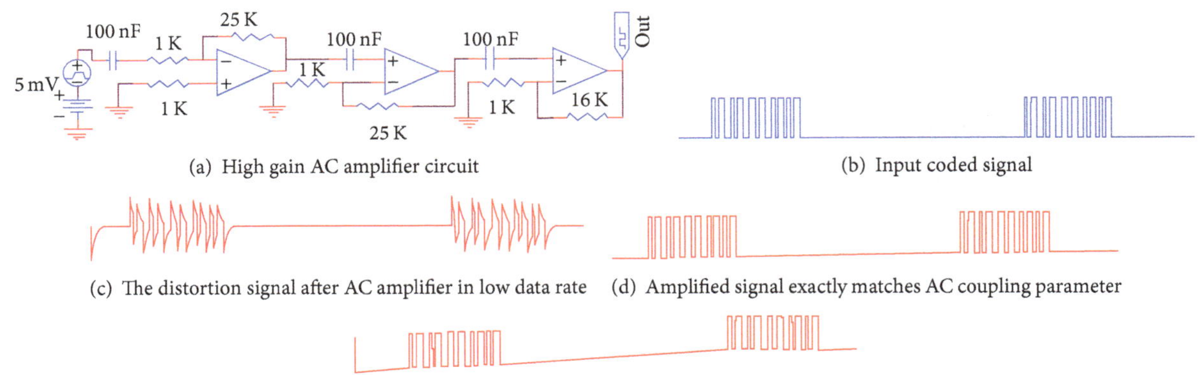

(a) High gain AC amplifier circuit

(b) Input coded signal

(c) The distortion signal after AC amplifier in low data rate

(d) Amplified signal exactly matches AC coupling parameter

(e) The distortion signal after AC amplifier in high data rate

FIGURE 8: AC amplification distortions.

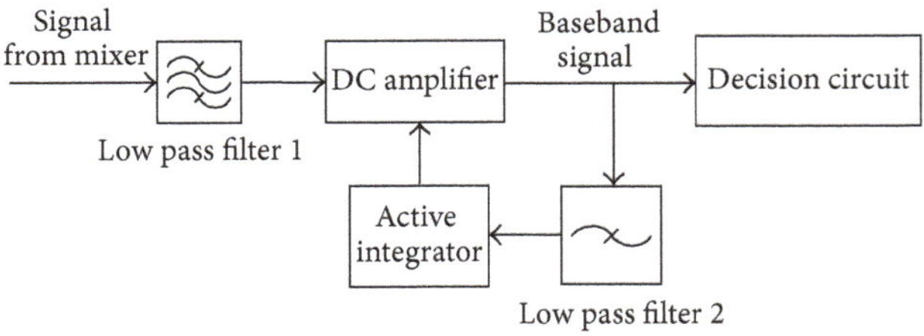

FIGURE 9: DC amplifier circuit with low frequency servo loop.

Power spectrum is expressed as

$$\begin{aligned}\varphi(\omega) &= 2\pi \cdot \sum_{n=-\infty}^{\infty} \frac{\sin^2(n\omega_0\tau/2)}{n^2\pi^2}\delta(\omega - n\omega_0) \\ &= 2\pi \sum_{n=-\infty}^{\infty} \frac{\sin^2(n\pi \cdot (\tau/T))}{n^2\pi^2}\delta(\omega - n\omega_0).\end{aligned} \tag{7}$$

The average normalized power spectral density of a series of rectangular pulses representing n data bits is thus

$$\begin{aligned}G(f) &= \frac{nA^2\tau^2 \sin c^2(\pi f\tau)}{n\tau} \\ &= A^2\tau \sin c^2(\pi f\tau) \text{ volts}^2/\text{Hz}.\end{aligned} \tag{8}$$

Generally, the percentage of signal's power within a frequency band is a good indication of the signal distortion. By the circuit simulation result, it is appropriate that the cut-off frequency is set at 0.1% of the baseband data rate; for example, if the data rate is 32 kHz, the cut-off frequency of low-pass filter 2 can be set at 32 Hz.

At the same time, the cut-off frequency determines the offset $1/f$ noise; according to the typical MOSFET amplifiers model, the $1/f$ noise in the vicinity of 1 KHz can be expressed as

$$\frac{K}{WLC_{\text{OX}}} \cdot \frac{1}{f}\bigg|_{f\approx 1\,\text{KHz}} \approx 4kT\frac{2}{3g_m}. \tag{9}$$

The $1/f$ noise power in a bandwidth from 0 Hz to 32 Hz can be calculated:

$$\begin{aligned}P_{n1} &= \int_{0\,\text{Hz}}^{32\,\text{Hz}} \frac{K}{WLC_{\text{OX}}} \cdot \frac{1}{f}df = \frac{K}{WLC_{\text{OX}}} \ln 32 \\ &= 4kT\frac{2}{3g_m}(1\ \text{KHz}) \ln 32.\end{aligned} \tag{10}$$

For the thermal noise, we have

$$\begin{aligned}P_{n2} &= \int_{0\,\text{Hz}}^{32\,\text{Hz}} \frac{K}{WLC_{\text{OX}}}df = \int_{0\,\text{Hz}}^{32\,\text{Hz}} 4kT\frac{2}{3g_m}df \\ &= 4kT\frac{2}{3g_m} \cdot 32.\end{aligned} \tag{11}$$

Thus, the cancellation noise power is

$$\frac{P_{n1}}{P_{n2}} = 108.1 = 20.3\,\text{dB}. \tag{12}$$

Detailed circuits are shown in Figure 10; compared to the AC coupling model (Figure 8), the improved circuit is DC coupling mode with low frequency feedback servo loop. The multistage amplifier gain is set at 100 db in the simulation, and the Manchester coded data with 5 mV DC offset are applied to both circuits.

The simulation results (Figure 11) show that both circuits can remove the DC offset, but the AC coupling amplifier cannot adapt various data rate and causes signal distortion at high data rate and low data rate; while DC coupling amplifier with low frequency servo loop can keep the signal integrity at any data speed, the signal has no significant distortion.

This circuit is applied in DCR RFID system, and the test result is shown in Figure 12; the wave of baseband signal reflected from the tag can be very well maintained in different data rate.

7. Conclusion

This paper presents the design of microstrip circuit carrier cancellation technology for DCR system; experimental

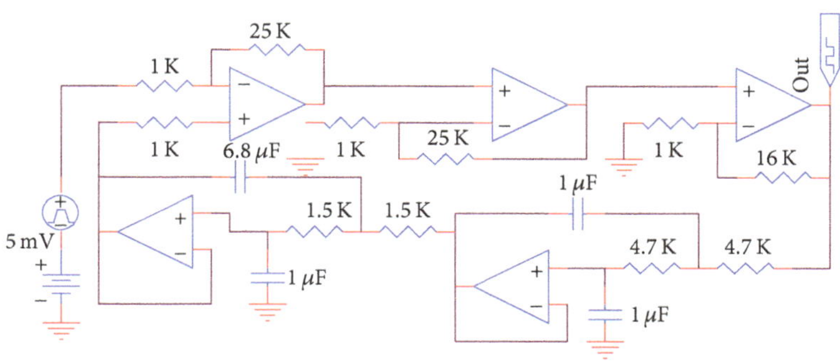

FIGURE 10: Improved DC coupling circuit with low frequency servo loop based on discrete components.

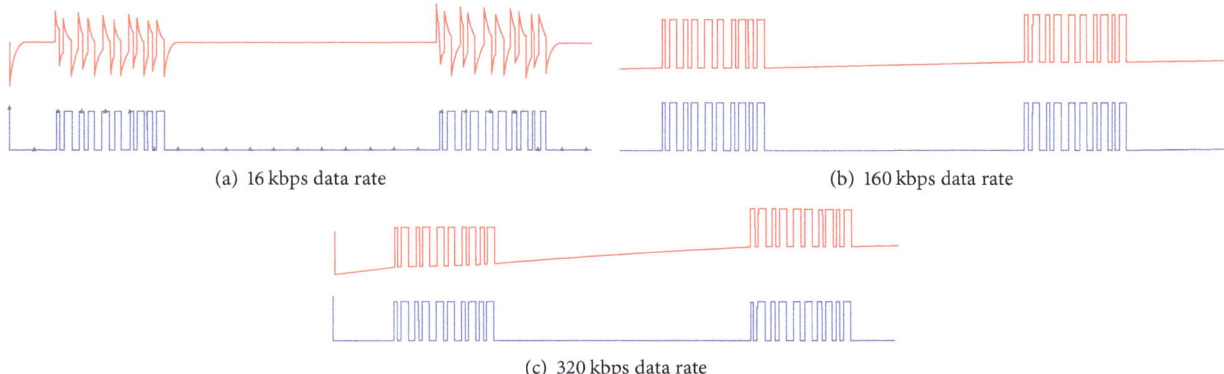

(a) 16 kbps data rate

(b) 160 kbps data rate

(c) 320 kbps data rate

FIGURE 11: (a), (b), and (c) show the comparison results of AC coupling (red color) and improved DC coupling amplifier (blue color) in 16 kbps, 160 kbps, and 320 kbps data rate.

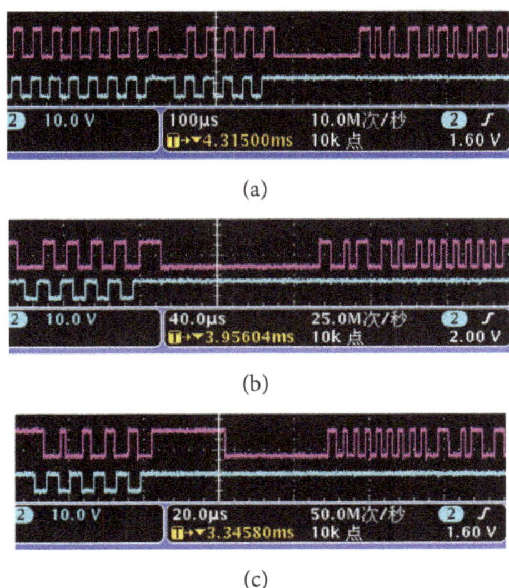

(a)

(b)

(c)

FIGURE 12: Measured waveform of the reader and tag: pink color (top) is tag reflected signal, and blue color (bottom) is reader command. (a), (b), and (c) are 40 kbps, 80 kbps, and 160 kbps data rate.

results show that the use of this technology solution can reduce the carrier leakage power 25 dB, equivalent to the circulator isolation up to 45 dB. This approach ensures the linearity of the receiver, so that more low noise preamplifiers can be used in the front-end circuit of the receiver to increase the communication distance of DCR system. Further, for the problem of DC offset, a new wideband amplifier circuit with DC stereo loop is proposed; the simulation and experimental results show that the novel circuits can solve the DC offset problem and keep the integrity of baseband signal.

Competing Interests

The authors declare that they have no competing interests.

Acknowledgments

This work was supported by the National Research Foundation of China grant funded by the China government (11374162) and University Natural Science Project (TJ215009, NY215162). Thanks are due for the help of Senior Engineer Jingqing Cheng and Yufeng Guo.

References

[1] B. Razavi, "Design considerations for direct-conversion receivers," *IEEE Transactions on Circuits and Systems II: Analog and Digital Signal Processing*, vol. 44, no. 6, pp. 428–435, 1997.

[2] W.-G. Lim and J.-W. Yu, "Balanced circulator structure with enhanced isolation characteristics," *Microwave and Optical Technology Letters*, vol. 50, no. 9, pp. 2389–2391, 2008.

[3] A. Yoshizawa and Y. P. Tsividis, "Anti-blocker design techniques for MOSFET-C filters for direct conversion receivers," *IEEE Journal of Solid-State Circuits*, vol. 37, no. 3, pp. 357–364, 2002.

[4] K.-J. Cho, J.-H. Kim, and S. P. Stapleton, "A highly efficient doherty feedforward linear power amplifier for W-CDMA base-station applications," *IEEE Transactions on Microwave Theory and Techniques*, vol. 53, no. 1, pp. 292–300, 2005.

[5] M. Grimm, M. Allén, J. Marttila, M. Valkama, and R. Thomä, "Joint mitigation of nonlinear RF and baseband distortions in wideband direct-conversion receivers," *IEEE Transactions on Microwave Theory and Techniques*, vol. 62, no. 1, pp. 166–182, 2014.

[6] G. Byeon, S. Oh, and G. Jang, "A new DC offset removal algorithm using an iterative method for real-time simulation," *IEEE Transactions on Power Delivery*, vol. 26, no. 4, pp. 2277–2286, 2011.

[7] L. Yu and W. M. Snelgrove, "A novel adaptive mismatch cancellation system for quadrature if radio receivers," *IEEE Transactions on Circuits and Systems II: Analog and Digital Signal Processing*, vol. 46, no. 6, pp. 789–801, 1999.

[8] H. Yoshida, H. Tsurumi, and Y. Suzuki, "DC offset canceller in a direct conversion receiver for QPSK signal reception," in *Proceedings of the 9th IEEE International Symposium on Personal, Indoor and Mobile Radio Communications*, vol. 3, pp. 1314–1318, IEEE, Boston, Mass, USA, September 1998.

[9] J.-Y. Jung, C.-W. Park, and K.-W. Yeom, "A novel carrier leakage suppression front-end for UHF RFID reader," *IEEE Transactions on Microwave Theory and Techniques*, vol. 60, no. 5, pp. 1468–1477, 2012.

16

DOA Estimation based on Sparse Signal Recovery Utilizing Double-Threshold Sigmoid Penalty

Hanbing Wang, Hui Li, and Bin Li

Department of Electronics and Information, Northwestern Polytechnical University, Xi'an, Shaanxi 710129, China

Correspondence should be addressed to Hanbing Wang; xiaosuperhan@gmail.com

Academic Editor: John N. Sahalos

This paper proposes a new algorithm based on sparse signal recovery for estimating the direction of arrival (DOA) of multiple sources. The problem model we build is about the sample covariance matrix fitting by unknown source powers. We enhance the sparsity by the double-threshold sigmoid penalty function which can approximate the l_0 norm accurately. Our method can distinguish closely spaced sources and does not need the knowledge of the number of the sources. In addition, our method can also perform well in low SNR. Besides, our method can handle more sources accurately than other methods. Simulations are done to certify the great performance of the proposed method.

1. Introduction

The estimation of the direction of arrival (DOA) of multiple sources plays a key role in many applications including radar, sonar, and wireless communication. So far, amounts of superresolution algorithms for DOA estimation have been developed. The nonparametric methods include Capon method [1] and subspace-based methods. The traditional subspace-based algorithms, like MUSIC which firstly exploits the orthogonality between the signal space and the noise subspace [2] and ESPRIT which utilizes the rotational invariance of the signal subspace [3], can achieve excellent performance in high SNR, specially, when the snapshots are long. The maximum likelihood (ML) methods including the deterministic maximum likelihood (DML) and stochastic maximum likelihood (SML) possess good statistical properties [4–7] but require a large number of samples. All the above methods need the knowledge of the number of sources.

Sparse representation of signals and compressed sensing have become a hot topic in many fields [8, 9], and the DOA estimation methods based on sparse reconstruction have already been paid more attention by researchers. The well-known l_1-SVD is a pretty good algorithm [10]; it combines the sparse signal recovery method based on l_1 norm with the singular value decomposition (SVD). The l_1-SVD algorithm can handle closely spaced correlated sources when the number of sources is known, while its performance is degraded without knowing the number of sources. In [11], the joint l_0 approximation DOA (JLZA-DOA) algorithm is proposed. This algorithm processes with a mixed $l_{2,0}$ approximation approach; it can acquire high resolution without the knowledge of the number of sources and with only a few snapshots. However, the JLZA-DOA algorithm may fail to make a good estimation in low SNR.

The algorithms we mentioned in previous paragraph are all based on the multiple measurement vectors (MMV) problem [12]. Recently, the coprime array technique is proposed [13, 14]. This technique can enhance the degrees of freedom (DOFs) of array. With vectorizing the sample covariance matrix, it deals with the sparse covariance fitting problem eventually when signals are sparsely represented. That is to say, the MMV problem is transformed into the single measurement vector (SMV) problem. In the SMV case, DOA estimation can be implemented using many techniques, such as the orthogonal matching pursuit (OMP) and the improved smoothed l_0 approximation algorithm (ISL0) [15, 16]. The OMP algorithm enjoys a short computational time. However, it needs to know the sparsity in advance; that is, it requires the knowledge of the number of sources. The ISL0 performs better than OMP, but it costs longer computational time. And its performance is degraded in low SNR.

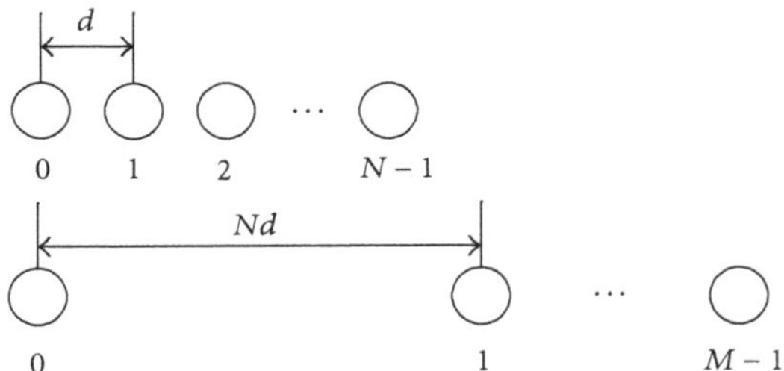

FIGURE 1: Coprime array configuration.

In this paper, we adopt the coprime array technique and propose a new Newton-like algorithm based on double-threshold sigmoid penalty for handling the sparse covariance fitting problem. We know that the direct l_0 norm optimization problem is NP-hard. Many algorithms approximate the l_0 norm by l_1 norm, but the estimation errors increase when the magnitudes of the nonzero elements to be estimated are greater than one. In addition, the l_1 norm method is not robust to noise and takes lots of iteration to converge. Instead of replacing the l_0 norm with l_1 norm, we utilize the double-threshold sigmoid penalty function to approach the l_0 norm [17]. And by adjusting the upper threshold at each step on the iteration, our algorithm can preserve most of the advantages of l_0 norm; this contributes to improving the estimation performance. Besides it also can accelerate the speed of convergence by adjusting the upper threshold. Numerical simulations demonstrate that our proposed method can achieve high resolution without the knowledge of the number of sources and perform well in low SNR.

Throughout the paper, the lower-case and upper-case bold letters are used to denote the vectors and matrices, respectively. $(\cdot)^T$, $(\cdot)^H$, $(\cdot)^*$, and $(\cdot)^{-1}$ present transpose, conjugate transpose, complex conjugate, and inverse, respectively. $\otimes$ denotes the Kronecker product, and $E[\cdot]$ is the statistical expectation operator.

2. Problem Model

Consider a coprime array with compressed interelement spacing, as illustrated in Figure 1. The two subarrays consist of M and N sensors, respectively, where $N = M + 1$. d is usually set to $\lambda/2$, where λ denotes the wavelength. Consequently, the array sensors are positioned at

$$\mathrm{P} = \{nd \mid 0 \le n \le N-1\} \cup \{Nmd \mid 0 \le m \le M-1\}; \quad (1)$$

the two subarrays share the first sensor at zeroth position; denote $\mathbf{p} = [p_1, p_2, \ldots, p_{M+N-1}]^T$ as the positions of the array sensors, where $p_i \in \mathrm{P}$, $\forall i$.

Supposing that K uncorrelated narrow-band signals impinge on the array from angles $\boldsymbol{\Theta} = [\theta_1, \theta_2, \ldots, \theta_K]$, the tth observation can be represented as

$$\mathbf{y}(t) = \mathbf{A}\mathbf{s}(t) + \mathbf{n}(t), \quad (2)$$

where $\mathbf{s}(t) = [s_1(t), s_2(t), \ldots, s_K(t)]^T$ is the vector of unknown signals and $\mathbf{n}(t) \in \mathrm{C}^{M+N-1\times 1}$ denotes the additive white Gaussian noises. $\mathbf{A} \in \mathrm{C}^{M+N-1\times K}$ is the manifold matrix, the column of which is corresponding to the directions of the source

$$\mathbf{A} = [\mathbf{a}(\theta_1), \mathbf{a}(\theta_2), \ldots, \mathbf{a}(\theta_K)], \quad (3)$$

where $\mathbf{a}(\theta_k) = [e^{-j2\pi p_1 \sin(\theta_k)/\lambda}, e^{-j2\pi p_2 \sin(\theta_k)/\lambda}, \ldots, e^{-j2\pi p_{M+N-1} \sin(\theta_k)/\lambda}]^T$ $(k = 1, \ldots, K)$.

The covariance matrix of receive data vector $\mathbf{y}(t)$ can be obtained as

$$\mathbf{R} = E[\mathbf{y}(t)\mathbf{y}^H(t)] = \mathbf{A}\mathbf{R}_\mathbf{s}\mathbf{A}^H + \sigma_n^2 \mathbf{I}_{M+N-1}, \quad (4)$$

where $\mathbf{R}_\mathbf{s} = E[\mathbf{s}(t)\mathbf{s}^H(t)]$ and $\mathbf{I}_{M+N-1}$ represents an identity matrix.

The classical DOA estimation problem can be reformulated as a sparse representation problem. We consider a gird of G equally interval angles $\boldsymbol{\Phi} = [\varphi_1, \ldots, \varphi_G]$ with $G \gg K$. Assuming $\boldsymbol{\Theta}$ is a subset of $\boldsymbol{\Phi}$, we construct an overcomplete matrix $\overline{\mathbf{A}}$ by collecting the steering vectors corresponding to all the potential source locations. Accordingly, received signal model (2) can be represented as

$$\mathbf{y}(t) = \overline{\mathbf{A}}\mathbf{x}(t) + \mathbf{n}(t), \quad (5)$$

where $\overline{\mathbf{A}} = [\mathbf{a}(\varphi_1), \mathbf{a}(\varphi_2), \ldots, \mathbf{a}(\varphi_G)]$, $\mathbf{x}(t) \in \mathrm{C}^{G\times 1}$ is a sparse vector, and the nonzero entries of $\mathbf{x}(t)$ are the positions which correspond to the source locations. That is to say, the qth component $x_q(t)$ of $\mathbf{x}(t)$ is nonzero only if $\varphi_q = \theta_k$.

Consequently, we can obtain a spatial covariance matrix in terms of $\overline{\mathbf{A}}$; it takes the following form:

$$\mathbf{R} = E[\mathbf{y}(t)\mathbf{y}^H(t)] = \overline{\mathbf{A}}\mathbf{R}_\mathbf{x}\overline{\mathbf{A}}^H + \sigma_n^2 \mathbf{I}_{M+N-1}, \quad (6)$$

where $\mathbf{R}_\mathbf{x} = E[\mathbf{x}(t)\mathbf{x}^H(t)]$. In practice, the spatial covariance matrix is replaced by the sample covariance matrix $\widehat{\mathbf{R}} = (1/T)\sum_{t=1}^{T} \mathbf{y}(t)\mathbf{y}^H(t)$, where T denotes the sample snapshots.

Under the assumption that sources are uncorrelated, $\mathbf{R}_\mathbf{x} \in \mathrm{C}^{G\times G}$ is a sparse diagonal matrix, and only K entries are nonzero. Denote $\mathrm{diag}(\mathbf{R}_\mathbf{x}) = [b_1, b_2, \ldots, b_G]^T = \mathbf{b}$, and it is obvious that $\mathbf{b}$ is a $G \times 1$ sparse vector with nonzero entries at positions which correspond to source locations. Moreover, the elements of $\mathbf{b}$ are real valued and negative; that is, $\mathbf{b} \in \mathrm{R}_+^{G\times 1}$.

Applying vectorization to (6) yields

$$\mathbf{z} = \mathrm{vec}(\mathbf{R}) = \widetilde{\mathbf{A}}\mathbf{b} + \sigma_n^2 \widetilde{\mathbf{I}}, \quad (7)$$

where $\widetilde{\mathbf{A}} = [\widetilde{\mathbf{a}}(\varphi_1), \widetilde{\mathbf{a}}(\varphi_2), \ldots, \widetilde{\mathbf{a}}(\varphi_G)]$, $\widetilde{\mathbf{a}}(\varphi_g) = \mathbf{a}^*(\varphi_g) \otimes \mathbf{a}(\varphi_g)$, and $\widetilde{\mathbf{I}} = \mathrm{vec}(\mathbf{I}_{M+N-1})$. The distinct rows of $\widetilde{\mathbf{A}}$ behave like the manifold of an array whose sensor locations are given by the values in the set of cross differences $\{(Nm - n)d,\ 0 \le m \le M-1,\ 1 \le n \le N-1\}$ and the self-differences $\{(n_1 - n_2)d,\ 1 \le n_1, n_2 \le N-1\}$ and $\{(Nm_1 - Nm_2)d,\ 0 \le m_1, m_2 \le M-1\}$. The set of these differences include all the $2N(M-1)+1$ differences continuously from $-N(M-1)$ to $N(M-1)$. As such, $\mathbf{z}$ is similar to the observation data from a long array which includes actual elements and virtual elements.

To avoid the calculation of complex number and reduce the computational cost, we can separate the real and image part. Then (7) can be reformulated as

$$\begin{bmatrix} \mathbf{z}_r \\ \mathbf{z}_i \end{bmatrix} = \begin{bmatrix} \widetilde{\mathbf{A}}_r \\ \widetilde{\mathbf{A}}_i \end{bmatrix} \mathbf{b} + \sigma_n^2 \begin{bmatrix} \widetilde{\mathbf{I}} \\ \mathbf{0}_{(M+N-1)^2\times 1} \end{bmatrix}, \tag{8}$$

where $\mathbf{z}_r = \mathrm{Re}(\mathbf{z})$, $\mathbf{z}_i = \mathrm{Im}(\mathbf{z})$, $\widetilde{\mathbf{A}}_r = \mathrm{Re}(\widetilde{\mathbf{A}})$, and $\widetilde{\mathbf{A}}_i = \mathrm{Im}(\widetilde{\mathbf{A}})$.

Our proposed method can work on both real valued case and complex valued case; we just take the complex valued case into account in the following discussion.

3. Proposed Method

3.1. Proposed L0A Algorithm. The problem expressed in (7) can be solved by the l_1 regularization method whose model can be written as

$$\mathbf{b}_*(h) = \arg\min_{\mathbf{b}} L(\mathbf{b}, h), \tag{9}$$

$$L(\mathbf{b}, h) = \frac{1}{2}\left\|\widetilde{\mathbf{A}}\mathbf{b} - \mathbf{z}\right\|_2^2 + h\|\mathbf{b}\|_1, \tag{10}$$

where $\|\mathbf{b}\|_1$ is the penalty term which approximates the l_0 norm. h denotes the regularization parameter; it controls the tradeoff between the sparsity of the signal and the residual energy. In the following, we derive a gradient-based method with double-threshold sigmoid (DTHS) penalty.

Before deriving our method, we first deal with the problem that the derivation of $|b_g|$ is not defined at zero, which always happens when handling the sparse problem. Some approximation techniques for this problem are adopted in practice [18]. Now we make an approximation to $|b_g|$ as $|b_g| \approx \sqrt{b_g^2 + \delta}$, where δ is a small positive constant, and $\sqrt{b_g^2 + \delta}$ approaches $|b_g|$ when $\delta \to 0^+$. It is set to 0.0001 in this paper. Then

$$\frac{d|b_g|}{db_g} \approx \frac{b_g}{\sqrt{|b_g|^2 + \delta}}. \tag{11}$$

Obviously, this equation is also held in the complex valued case.

In order to approximate the l_0 norm more accurately, some thresholding penalties have been utilized. Then (10) can be rewritten as

$$L(\mathbf{b}, h) = \frac{1}{2}\left\|\widetilde{\mathbf{A}}\mathbf{b} - \mathbf{z}\right\|_2^2 + h\sum_{g=1}^{G}\mathrm{Th}\left(|b_g|\right), \tag{12}$$

where $\mathrm{Th}(\cdot)$ denotes the thresholding function and $\sum_{g=1}^{G}\mathrm{Th}(|b_g|)$ is an approximation of $\|\mathbf{b}\|_0$.

Now, we adopt a DTHS function in our method. The ideal one is divided into three parts by the upper and lower threshold points, which is shown in Figure 2. Denote the upper threshold as τ_u and the lower one as τ_l. The values smaller than τ_l are set to zero; the values between τ_l and τ_u are multiplied by a particular constant, and the values larger than τ_u are set to a particular positive value. However, the derivative in the two threshold points does not exist, so we replace the ideal DTHS function by the flowing function given by

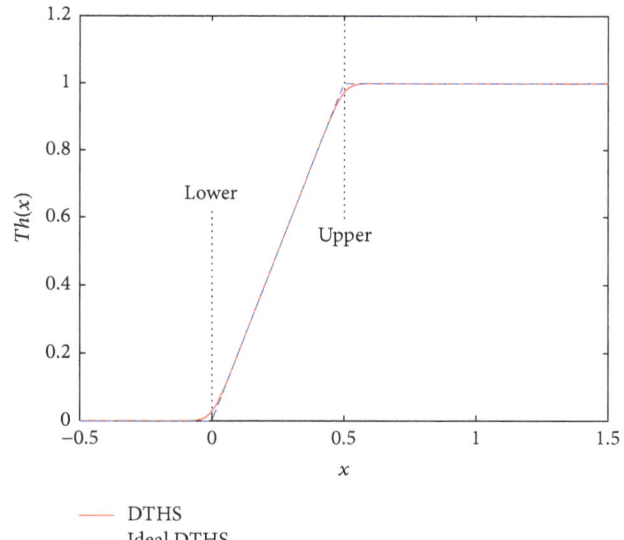

FIGURE 2: The ideal DTHS function and DTHS function.

$$\mathrm{Th}(x; \tau_l, \tau_u) = \frac{1}{\alpha(\tau_u - \tau_l)} \ln \frac{1 + e^{\alpha(x-\tau_l)}}{1 + e^{\alpha(x-\tau_u)}} \tag{13}$$

which can approximate the ideal DTHS function greatly when $\alpha \to +\infty$; we name this function the DTHS function and set α to 50 in this paper. Substituting the thresholding function in (12) by (13), we obtain a new problem:

$$L(\mathbf{b}, h) = \frac{1}{2}\left\|\widetilde{\mathbf{A}}\mathbf{b} - \mathbf{z}\right\|_2^2 + hF_{\tau_l,\tau_u}(\mathbf{b}), \tag{14}$$

$$F_{\tau_l,\tau_u}(\mathbf{b}) = \sum_{g=1}^{G}\mathrm{Th}\left(|b_g|; \tau_l, \tau_u\right). \tag{15}$$

Figure 3 shows the graph of the DTHS function with four different upper thresholds. For any given $b > 0$, we have

$$\lim_{\tau_u \to 0}\mathrm{Th}(b) = 1. \tag{16}$$

Consequently, the function $\sum_{i=1}^{N}\mathrm{Th}(b_i)$ behaves like $\|b\|_0$. That is to say, the DTHS penalty will approach l_0 norm when a smaller τ_u is used. However, while we choose a smaller τ_u, the function $F_{\tau_l,\tau_u}(b_g)$ might have many local minima. As τ_u increases, $F_{\tau_l,\tau_u}(b_g)$ becomes smoother. Then, we can handle our problem by solving a sequence optimization problem. Start solving (14) with a larger τ_u; subsequently, we reduce τ_u by a small factor ρ and solve (14) again for $\tau_u = \rho\tau_u$.

We can obtain the gradient of (11) as

$$\begin{aligned}\nabla_{\mathbf{b}}L(\mathbf{b}, h) &= \widetilde{\mathbf{A}}^H\left(\widetilde{\mathbf{A}}\mathbf{b} - \mathbf{z}\right) + h\mathbf{W}\mathbf{b} \\ &= \left(\widetilde{\mathbf{A}}^H\widetilde{\mathbf{A}} + h\mathbf{W}\right)\mathbf{b} - \widetilde{\mathbf{A}}^H\mathbf{z},\end{aligned} \tag{17}$$

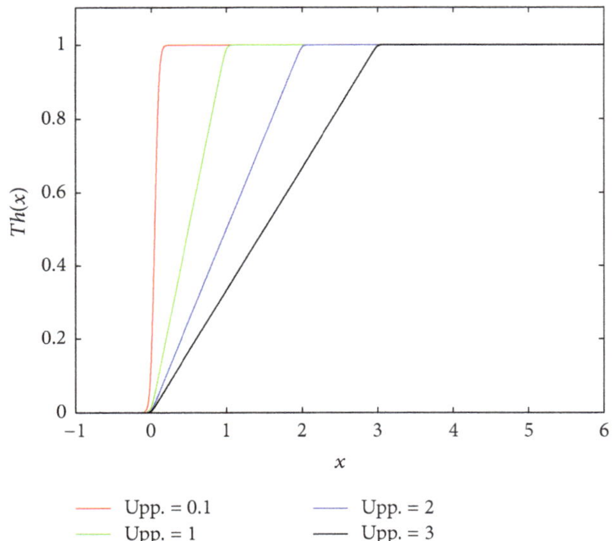

FIGURE 3: DTHS function with four upper thresholds.

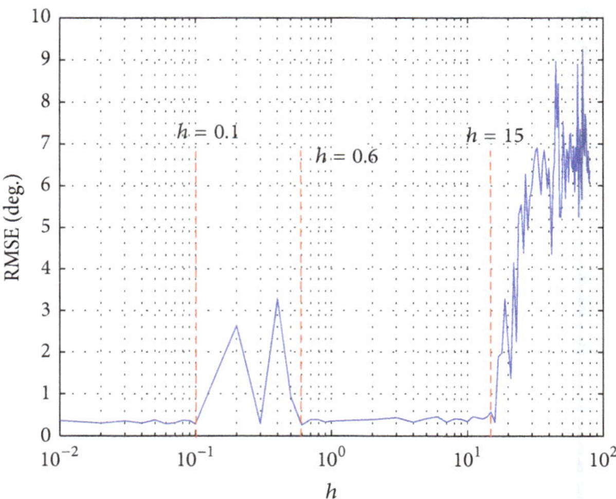

FIGURE 4: RMSE as a function of h.

where $\mathbf{W}$ is a diagonal matrix:

$$\mathbf{W} = \begin{bmatrix} \frac{\mathrm{Th}'(|b_1|;\tau_l,\tau_u)}{\sqrt{|b_1|^2+\delta}} & \cdots & 0 \\ \vdots & \ddots & \vdots \\ 0 & \cdots & \frac{\mathrm{Th}'(|b_G|;\tau_l,\tau_u)}{\sqrt{|b_G|^2+\delta}} \end{bmatrix}, \tag{18}$$

$$\mathrm{Th}'(b_g;\tau_l,\tau_u) = \frac{1}{\tau_u-\tau_l}\left(\frac{1}{1+e^{-\alpha(|b_g|-\tau_l)}} - \frac{1}{1+e^{-\alpha(|b_g|-\tau_u)}}\right).$$

Instead of calculating the Hessian matrix of (12), we adopt $\mathbf{H} = \widetilde{\mathbf{A}}^H\widetilde{\mathbf{A}} + h\mathbf{W}$ to approximate it; that is, $\nabla^2_{\mathbf{b}} L(\mathbf{b},h) \approx \mathbf{H}$. Then the iteration process can be accelerated, and we obtain the qth quasi-Newton iteration formula of problem (8), which is written as

$$\begin{aligned}\mathbf{b}^{(q+1)} &= \mathbf{b}^{(q)} - \beta\mathbf{H}^{-1}\left(\mathbf{H}\mathbf{b}^{(q)} - \widetilde{\mathbf{A}}^H\mathbf{z}\right) \\ &= (1-\beta)\,\mathbf{b}^{(q)} + \beta\mathbf{H}^{-1}\widetilde{\mathbf{A}}^H\mathbf{z},\end{aligned} \tag{19}$$

where β denotes the step size. The whole algorithm is summarized in Algorithm 1; we call our method the l_0 approximation (L0A) algorithm.

Algorithm 1 (L0A algorithm).

Initialization. $\mathbf{b}^0 = \mathbf{b}_0$, $\beta = 1$, $\mu,\eta,\rho \in (0,1)$, $\tau_u = \tau_{u0}$, $\tau_{\text{stop}} \in [0.1, 0.01]$, $\lambda \in \mathbb{R}$.

Main Iteration

(1) Calculate $\mathbf{H}^{(q)}$: $\mathbf{H}^{(q)} = \widetilde{\mathbf{A}}^H\widetilde{\mathbf{A}} + h\mathbf{W}^{(q)}$, $\mathbf{W}^{(q)} = \operatorname{diag}(\mathrm{Th}'(\mathbf{b}^{(q)};\tau_l,\tau_u)/\sqrt{|\mathbf{b}^{(q)}|^2+\delta})$.

(2) Update $\mathbf{b}$: $\mathbf{b}^{(q+1)} = (1-\beta)\mathbf{b}^{(q)} + \beta\mathbf{H}^{-1}\widetilde{\mathbf{A}}^H\mathbf{z}$.

(3) When $L(\mathbf{b}^{(q+1)},h) > L(\mathbf{b}^{(q)},h)$, $\beta = \eta\beta$.

(4) If $\|\mathbf{b}^{(q+1)} - \mathbf{b}^{(q)}\|_2 \le \mu\tau_u$, then $\tau_u = \rho\tau_u$.

(5) If $\tau_u < \tau_{\text{stop}}$, stop the iteration.

Output. The solution is $\mathbf{b}^{(q)}$ after q iterations.

We initialize $\mathbf{b}^0$ by $\mathbf{b}_0 = \widetilde{\mathbf{A}}^+\mathbf{z}$ for accelerating the iteration, where $\widetilde{\mathbf{A}}^+$ denotes the Moore-Penrose pseudoinverse of $\widetilde{\mathbf{A}}$. As discussed above, if τ_{u0} is too small, function (15) may have many local minima. This will degrade the estimation performance of our method. So τ_{u0} should be initialized by a larger value. To preserve most of the advantages of l_0 norm, τ_{u0} also should not be chosen to be too large. Consequently, the value of τ_{u0} should be selected moderately according to the signal magnitude. Usually, we set $\tau_{u0} = 5\max(\text{abs}(\mathbf{b}_0))$, where abs$(\cdot)$ denotes the absolute value function. As for the lower threshold τ_l, some entries close to zero will be picked into the nonzero components of $\mathbf{b}^{(q)}$ when $\tau_l > 0$. Consequently, we fixed it as 0. According to numerous experiments, we suggest that μ should be set from 0.3 to 0.5, and ρ is set from 0.6 to 0.8. In this paper we set τ_{stop} to 0.01.

Now, we talk about the selection of the regularization parameter h. Firstly, we have estimated an approximate range by calculating the root mean square error (RMSE) of our proposed method as a function of h, as shown in Figure 4. It demonstrates that the low RMSE can be achieved when $0.01 < h < 0.1$ and $0.6 < h < 15$. However, if $h < 5$, there will be false peaks. And it is going to worsen as h decreases; the speed of convergence also becomes slower simultaneously. When $h > 10$, some true peaks may disappear, and the performance will be more terrible as h increases. Consequently, we set h from 5 to 10. Some more simulations about how h affects the estimation performance are conducted in Section 4.

3.2. Applying MUSIC Algorithm. Now we make a brief introduction about how to apply MUSIC algorithm correctly for DOA estimation under the problem model $\mathbf{z} = \text{vec}(\mathbf{R}) = \widetilde{\mathbf{A}}\mathbf{b} + \sigma_n^2\widetilde{\mathbf{I}}$ which is obtained by vectorizing $\mathbf{R}$ in (4); more details can be seen in [19]. It is obvious that the virtual source signal becomes a single snapshot of $\mathbf{b}$. And the rank of the covariance matrix of $\mathbf{z}$, $\mathbf{R}_\mathbf{z} = \mathbf{z}\mathbf{z}^H$, is one in the noise-free case. Then the problem resembles dealing with fully coherent sources. Consequently, MUSIC will fail to work when multiple sources are impinging the array. As described in [19], the spatial smoothing technique can be used to overcome this problem. Since spatial smoothing requires a continuous set of differences, we construct a new matrix $\widetilde{\mathbf{A}}_1$ of size $2N(M-1)+1 \times K$ where we have extracted precisely those rows from $\widetilde{\mathbf{A}}$ which correspond to the $2N(M-1)+1$ successive differences and also sorted them. This is equivalent to removing the corresponding rows from the observation vector $\mathbf{z}$ and sorting them to get a new vector $\mathbf{z}_1$ expressed as

$$\mathbf{z}_1 = \widetilde{\mathbf{A}}_1\mathbf{b} + \sigma_n^2\widetilde{\mathbf{I}}_1. \tag{20}$$

We divide this virtual array into $N(M-1)+1$ overlapping subarrays and construct a full-rank covariance matrix so that the MUSIC algorithm can be applied for DOA estimation.

4. Simulation Results

In this section, we illustrate the simulation results of our proposed method. We consider the coprime arrays talked about in Section 2 consisting of 12 array sensors, $M = 6$ and $N = 7$.

We assume four uncorrelated sources impinging on the array; their locations are 10°, 14°, 50°, and 61°. The SNR is set to 5 dB and the number of snapshots is 200. The estimated spectrum is shown in Figure 5 with above conditions, where the grid resolution is 0.5°. It demonstrates that L0A method performs well in low SNR and can distinguish closely spaced sources. We also consider two correlated sources at 21° and 30° based on the same conditions. As it is shown in Figure 6, the estimated results are close to the actual angles with small errors.

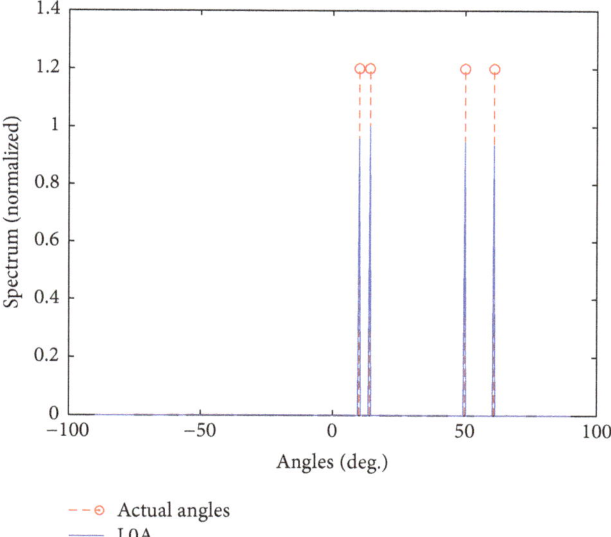

FIGURE 5: Spectrum of four uncorrelated sources.

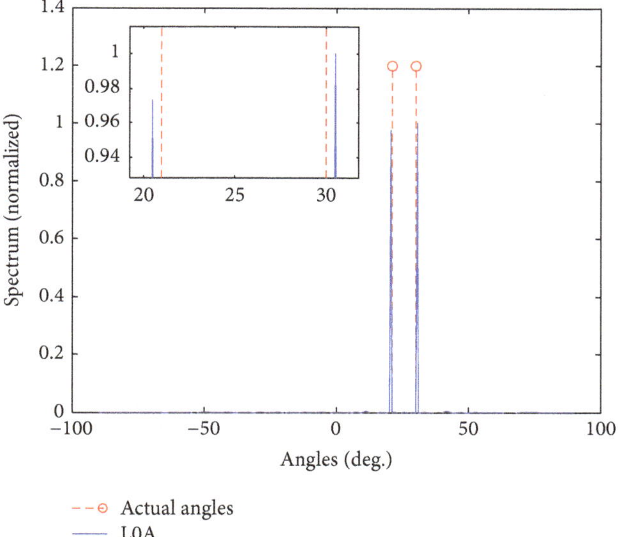

FIGURE 6: Spectrum of two correlated sources.

By adopting the coprime arrays, higher degrees of freedom can be achieved. Now we consider 25 uncorrelated sources with uniform space between −60° and 60°. We set the number of snapshots to 500 and the SNR is 5 dB, and the grid resolution is set to 0.2°. We compare the performance of L0A and MUSIC method under the same condition. As shown in Figures 7 and 8, the L0A method apparently performs better than MUSIC method in low SNR; it can recognize all the closely spaced sources greatly, while the MUSIC method fails to solve some angles of sources.

Figure 9 shows the RMSEs of the L0A, OMP, ISL0, and MUSIC as a function of the SNR. Twenty uncorrelated sources with 500 snapshots are taken into account in the simulations, and the grid resolution is set to 0.5°. It manifests that the estimation performance of these algorithms degrades with SNR decreasing. Our proposed L0A enjoys a better performance than others, especially in low SNR. In Figure 10, we make a comparison among their performance of DOA estimation under different snapshots in 5 dB. With the snapshots increasing, the performance becomes better. However, to achieve the same RMSE, our proposed method just needs smaller snapshots.

Our simulations were run on the computer with a 3.40 GHz Intel (R) Core (TM) i7-2600 CPU, where the operating system is Microsoft Windows XP with 32 bits. Table 1 shows the computation time for different algorithms. Here, we consider the same case used in Figure 9. We can see that SNR has little effect on the computation time of OMP and MUSIC. As SNR becomes low, the computation time of L0A and ISL0 increases. However, our proposed L0A costs shorter computation time than ISL0.

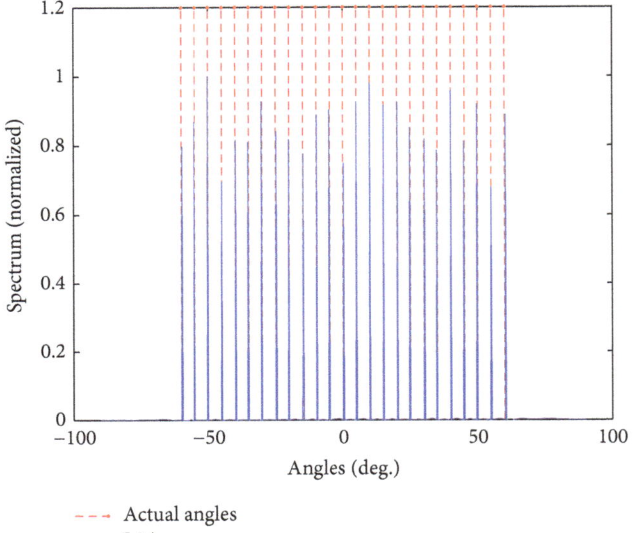

FIGURE 7: Spectrum estimated by L0A-DOA.

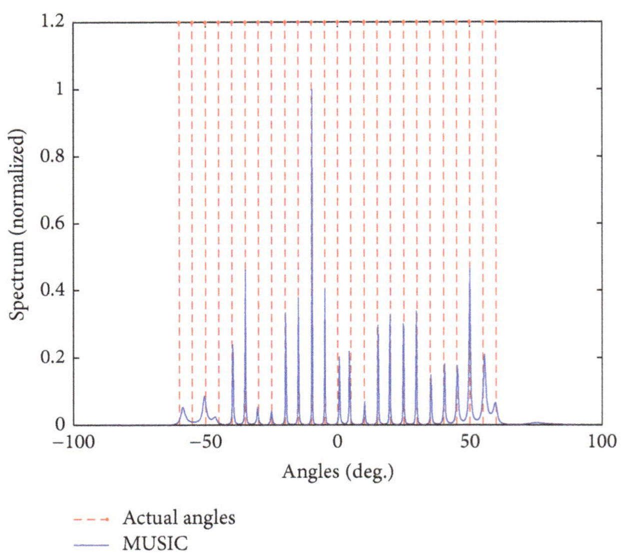

FIGURE 8: Spectrum estimated by MUSIC.

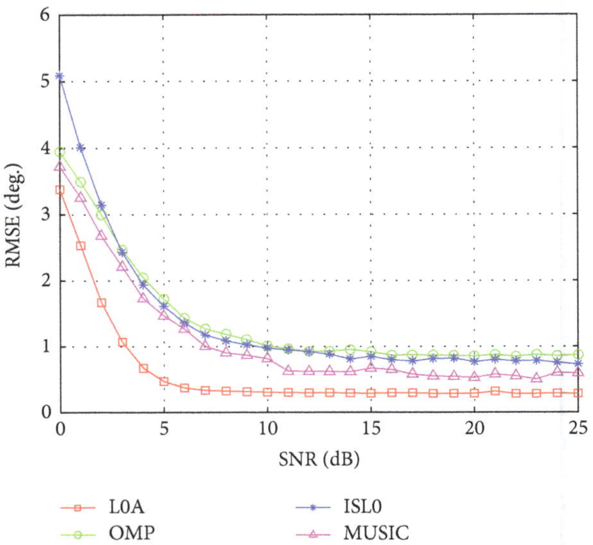

FIGURE 9: RMSE as a function of SNR.

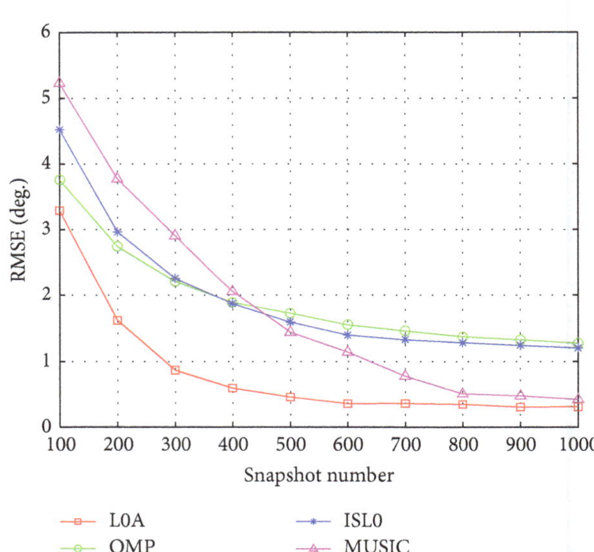

FIGURE 10: RMSE as a function of the snapshot number.

TABLE 1: Computation time comparison.

SNR (dB)	Time (seconds)			
	L0A	OMP	ISL0	MUSIC
5	0.4404	0.0098	0.6362	0.0104
10	0.3637	0.0099	0.4586	0.0105
15	0.3346	0.0098	0.4303	0.0103

Finally, we show the effect of h on the performance of DOA estimation from Figures 11 and 12. We consider the same case used in Figure 7. Figure 11 shows the case where $h = 1$. Obviously, there are some false peaks in the spectrum. When $h = 15$, true peaks disappear at the position of -50° and 55° as shown in Figure 12. Consequently, an appropriate value of h is important to the DOA estimation.

5. Conclusion

In this paper, we propose a new method based on sparse reconstruction for finding the directions of sources impinging on a coprime array. By approximating the l_0 norm with the DTHS function and adjusting the upper threshold dynamically, our method achieves an excellent performance. The proposed method not only can perform well without the knowledge of the number of sources but also could work on correlated sources with small bias. Furthermore, it can also distinguish the closely spaced sources. With the high DOFs, this method could resolve much more the number of sources

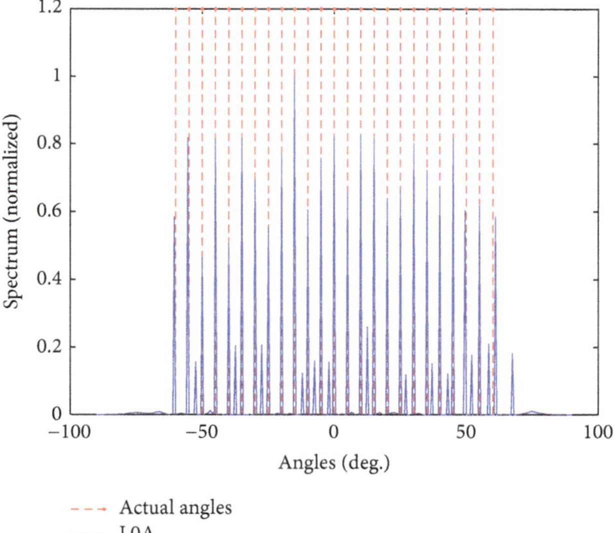

FIGURE 11: Spectrum of L0A-DOA with $h = 1$.

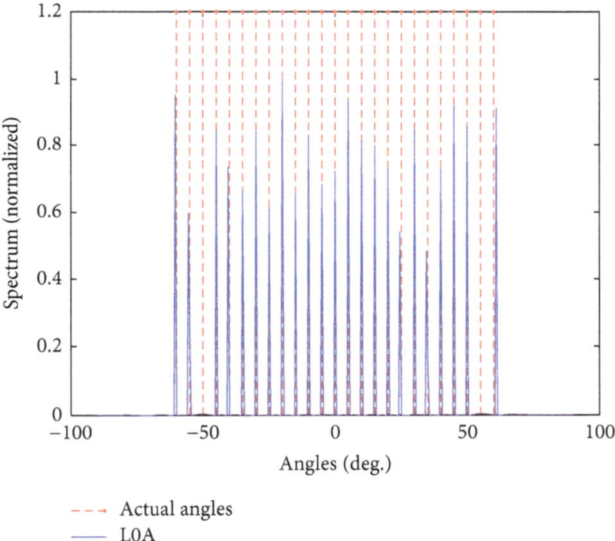

FIGURE 12: Spectrum of L0A-DOA with $h = 15$.

and have a better performance in low SNR and with smaller snapshots.

Conflict of Interests

The authors declare that there is no conflict of interests regarding the publication of this paper.

Acknowledgment

This work was supported by the National Natural Science Foundation of China (61171155 and 61571364).

References

[1] R. J. Weber and Y. Huang, "Analysis for Capon and MUSIC DOA estimation algorithms," in *Proceedings of the IEEE Antennas and Propagation Society International Symposium (APSURSI '09)*, pp. 1–4, IEEE, Charleston, SC, USA, June 2009.

[2] X. Li, G. Yang, and Y. Gu, "Simulation analysis of music algorithm of array signal proccessing DOA," in *Proceedings of the International Conference on Automatic Control and Artificial Intelligence (ACAI '12)*, pp. 1838–1841, March 2012.

[3] A. T. Y. Lok, P. Davoodian, R. C. Chin, J. Bermudez, Z. Aliyazicioglu, and H. K. Hwang, "Sensitivity analysis of DOA estimation using the ESPRIT algorithm," in *Proceedings of the IEEE Aerospace Conference*, pp. 1–7, Big Sky, Mont, USA, March 2010.

[4] P. Stoica and K. C. Sharman, "Maximum likelihood methods for direction-of-arrival estimation," *IEEE Transactions on Acoustics, Speech, and Signal Processing*, vol. 38, no. 7, pp. 1132–1143, 1990.

[5] X. Mestre, P. Vallet, and P. Loubaton, "On the resolution probability of conditional and unconditional maximum likelihood DOA estimation," in *Proceedings of the 21st European Signal Processing Conference (EUSIPCO '13)*, pp. 1–5, Marrakech, Morocco, September 2013.

[6] J.-W. Shin, Y.-J. Lee, and H.-N. Kim, "Reduced-complexity maximum likelihood direction-of-arrival estimation based on spatial aliasing," *IEEE Transactions on Signal Processing*, vol. 62, no. 24, pp. 6568–6581, 2014.

[7] C. E. Chen, F. Lorenzelli, R. E. Hudson, and K. Yao, "Stochastic maximum-likelihood DOA estimation in the presence of unknown nonuniform noise," *IEEE Transactions on Signal Processing*, vol. 56, no. 7, pp. 3038–3044, 2008.

[8] D. L. Donoho, M. Elad, and V. N. Temlyakov, "Stable recovery of sparse overcomplete representations in the presence of noise," *IEEE Transactions on Information Theory*, vol. 52, no. 1, pp. 6–18, 2006.

[9] D. L. Donoho, "Compressed sensing," *IEEE Transactions on Information Theory*, vol. 52, no. 4, pp. 1289–1306, 2006.

[10] D. Malioutov, M. Cetin, and A. S. Willsky, "A sparse signal reconstruction perspective for source localization with sensor arrays," *IEEE Transactions on Signal Processing*, vol. 53, no. 8, pp. 3010–3022, 2005.

[11] M. M. Hyder and K. Mahata, "Direction-of-arrival estimation using a mixed approximation," *IEEE Transactions on Signal Processing*, vol. 58, no. 9, pp. 4646–4655, 2010.

[12] S. F. Cotter, B. D. Rao, K. Engan, and K. Kreutz-Delgado, "Sparse solutions to linear inverse problems with multiple measurement vectors," *IEEE Transactions on Signal Processing*, vol. 53, no. 7, pp. 2477–2488, 2005.

[13] S. Qin, Y. D. Zhang, and M. G. Amin, "Generalized coprime array configurations for direction-of-arrival estimation," *IEEE Transactions on Signal Processing*, vol. 63, no. 6, pp. 1377–1390, 2015.

[14] K. Han and A. Nehorai, "Improved source number detection and direction estimation with nested arrays and ULAs using jackknifing," *IEEE Transactions on Signal Processing*, vol. 61, no. 23, pp. 6118–6128, 2013.

[15] J. A. Tropp and A. C. Gilbert, "Signal recovery from random measurements via orthogonal matching pursuit," *IEEE Transactions on Information Theory*, vol. 53, no. 12, pp. 4655–4666, 2007.

[16] M. M. Hyder and K. Mahata, "An improved smoothed l^0 approximation algorithm for sparse representation," *IEEE Transactions on Signal Processing*, vol. 58, no. 4, pp. 2194–2205, 2010.

[17] J. Shi, R. H. Ding, G. Xiang, and X. L. Zhang, "Complex-valued sparse recovery via double-threshold sigmoid penalty," *Signal Processing*, vol. 114, pp. 231–244, 2015.

[18] M. Schmidt, G. Fung, and R. Rosales, "Optimization methods for l_1-regularization," Tech. Rep. TR-2009-19, University of British Columbia, 2009.

[19] P. Pal and P. P. Vaidyanathan, "Coprime sampling and the music algorithm," in *Proceedings of the IEEE Digital Signal Processing Workshop and IEEE Signal Processing Education Workshop (DSP/SPE '11)*, pp. 289–294, IEEE, Sedona, Ariz, USA, January 2011.

17

Signal Processing based Remote Sensing Data Simulation in Radar System

Renxuan Hao[1] and Tan Guo[2]

[1] *School of Electronic Engineering, University of Electronic Science and Technology of China, Chengdu, China*
[2] *College of Communication Engineering, Chongqing University, Chongqing, China*

Correspondence should be addressed to Renxuan Hao; jszs.zl23@163.com

Academic Editor: John N. Sahalos

Range cell migration has a serious impact on the precision of image formation, especially for high-resolution and large-scale imaging. To get the full resolution and high quality of the image, the range cell migration correction in the azimuth time domain must be considered. For tackling this problem, this paper presents a novel and efficient range cell migration correction method based on curve fitting and signal processing. By emulating a curve to approximate the range-compressed echo of a strong point, the range location indexes of the strong point along the azimuth direction can be obtained under the least squares criterion. The merits of the proposed method are twofold: (1) the proposed method is robust to the uncertainty of system parameters (strong tolerance) under real flights and (2) the generalization of the proposed method is better and can be easily adapted to different synthetic aperture radar (SAR) modes (e.g., monostatic and bistatic). The experimental results on real remote sensing data from both the monostatic and the bistatic SAR demonstrate the effectiveness. The regressed distance curve is completely coincident with the trajectory of strong points of the echo. Finally, the imaging focus results also validate the efficiency of the proposed method.

1. Introduction

Synthetic aperture radar (SAR) is an important remote sensing technique, which has attracted more and more attention since it was proposed [1]. The most significant characteristic of SAR is its ability to obtain high-resolution microwave images day and night under all weather conditions [2]. For this reason, SAR has been widely applied to many fields, such as generation of digital elevation maps, observation of volcanic activities and flood disasters, land and sea traffic monitoring, observation of vegetation growth, monitoring of ocean currents and traveling icebergs, and detection of oil spills in the ocean [3–8].

One of the most fundamental and important techniques for SAR application is the imaging algorithm [9]. By now, many efficient imaging algorithms have been put forward, such as range-Doppler (RD) algorithm [10], chirp-scaling (CS) algorithm [11], wave number domain algorithm (Omega-K) [12], and back projection (BP) algorithm [13]. Although all these algorithms are effective, RD algorithm is the most popular algorithm since it is simple, highly efficient, and intuitive [14].

A key problem of the RD algorithm is the range cell migration correction (RCMC) [15]. SAR obtains low-quality images of the target due to the relative motion between the radar and the target [16]. Therefore, to obtain high-resolution images, the signals from the same scatter should be situated in one range cell. Nevertheless, it is not practical in view of the platform movement relative to the target, which is called range cell migration [15]. Generally, the movement of the range cell will result in coupling between the range direction and the azimuth direction. That is why the range cell migration must be corrected before further processing in RD algorithms [17–19]. The basic objective of range cell migration correction (RCMC) is to adjust the received echo of the same target back to the same range cell [20]. Traditionally, range cell migration can be corrected through time-domain interpolation or frequency-domain compensation [21]. The main problems of these methods are the excessive data operation and the narrow compatibility for different SAR configurations. Therefore, there remains a need for an efficient method that can solve the problems.

In this paper, a curve fitting based range cell migration correction method is proposed to solve the foregoing problem. According to the Taylor expansion theorem, the ranges between the radar and targets can be expressed as polynomials for both monostatic radar and bistatic radar [22–25]. A polynomial fitting method based on the least squares principle [26] is presented to compensate the range migration. The main idea is to approximate the trajectory of a strong point in range-compressed echo by making use of a simulation of the curve, which is fast and robust for real SAR data application. With the proposed method, the operation of RCMC will become easy, flexible, and efficient.

The following parts of this paper are organized as follows. In Section 2, the basic formulation of the range cell migration and the RCMC problem is briefly presented. The basic principle of curve fitting and the proposed CF-RCMC method is described in Section 3. In Section 4, experiments on the real monostatic and bistatic SAR data for remote sensing by using the proposed CF-RCMC method are presented and analyzed. Finally, in Section 5, a concluding remark of the present work is given.

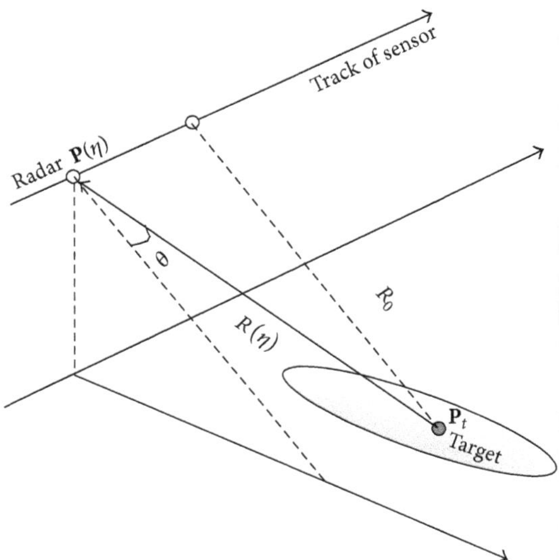

FIGURE 1: SAR geometry. The radar travels along its course at an altitude with a constant velocity V_T and the radar transmits and receives pulses with a squint angle of θ. The center of the imaging area is P_t, and the nearest slant distance between P_t and the radar is R_0. The motion of the radar relative to the target results in RCM, which is the intrinsic feature of SAR and, however, must be corrected.

2. Range Cell Migration

The simplified data acquisition geometry of the monostatic SAR is illustrated in Figure 1. The radar is carried on a platform moving along a straight line at a constant velocity V_T and a constant attitude. The instantaneous position of the sensor is $\mathbf{P}(\eta)$, where η denotes the slow time. The slant range from target $\mathbf{P}_t$ to the radar at η is denoted by $R(\eta)$. R_0 denotes the reference range, which is defined as the nearest slant range here and is perpendicular to the velocity.

With the well-known Pythagorean Theorem, the range between the target and the radar can be given by (bistatic)

$$\begin{aligned} R(\eta) &= R_T(\eta) + R_R(\eta) \\ &= \sqrt{R_{0T}{}^2 + V_T{}^2\eta^2} + \sqrt{R_{0R}{}^2 + V_R{}^2\eta^2}, \end{aligned} \tag{1}$$

where it can be approximated by the following polynomials according to the Taylor expansion theorem:

$$\begin{aligned} R(\eta) = {} & R_T(\eta_c) + R_R(\eta_c) \\ & + \left(\frac{V_T{}^2\eta_c}{R_T(\eta_c)} + \frac{V_R{}^2\eta_c}{R_R(\eta_c)}\right)(\eta - \eta_c) \\ & + \frac{1}{2}\left(\frac{V_T{}^2\cos^2\theta_{\mathrm{Tr},c}}{R_T(\eta_c)} + \frac{V_R{}^2\cos^2\theta_{\mathrm{Rr},c}}{R_R(\eta_c)}\right)(\eta - \eta_c)^2 \\ & + \cdots, \end{aligned} \tag{2}$$

where η_c denotes the azimuth center time when the beam center passes through the target; $\theta_{\mathrm{Tr},c}$ and $\theta_{\mathrm{Rr},c}$ denote the squint angles for the transmitter and receiver with respect to η_c; R_{0T} and R_{0R} denote the nearest distance at the initial time t_0; V_T and V_R denote the velocity of the radar and the target; $R_T(\eta)$ and $R_R(\eta)$ denote the slant distance between the radar and the target at time η. Hence,

$$\theta_{jr,c} = \arcsin\left(-\frac{V_j\eta c}{R_j(\eta c)}\right), \tag{3}$$

where j denotes the subscript T or R.

It is not hard to realize that the range history of the monostatic SAR can be also expressed like (2). However, the expression is much more complex than that of the bistatic SAR.

Compared with the size of the range cell, the higher order terms are so small that they can be ignored:

$$\begin{aligned} R(\eta) & \\ = {} & R_T(\eta_c) + R_R(\eta_c) \\ & + \left(\frac{V_T{}^2\eta_c}{R_T(\eta_c)} + \frac{V_R{}^2\eta_c}{R_R(\eta_c)}\right)(\eta - \eta_c) \\ & + \frac{1}{2}\left(\frac{V_T{}^2\cos^2\theta_{\mathrm{Tr},c}}{R_T(\eta_c)} + \frac{V_R{}^2\cos^2\theta_{\mathrm{Rr},c}}{R_R(\eta_c)}\right)(\eta - \eta_c)^2. \end{aligned} \tag{4}$$

The linear and quadratic components in (4) denote the range walk and curve, respectively, and they are collectively known as range cell migration (RCM). RCM results from the motion of the radar relative to the target results. This relative motion is the base of SAR and thus RCM is the intrinsic feature of SAR. However, RCM must be corrected in the RD algorithm.

Under the condition of small squint angle, the beam center approximates the zero Doppler direction and $\eta_c \approx 0$,

$\theta_{r,c} \approx 0$. So, the range history is simplified as follows (take the monostatic situation as an example):

$$R(\eta) = R_0 + \frac{V^2\eta^2}{2R_0}, \quad (5)$$

where there is only range curve.

In the RD algorithm, the range cell migration is corrected in the range-Doppler domain. The corresponding range cell migration under the condition of side looking is [9]

$$R(f_\eta) = R_0 + \frac{\lambda^2 R_0 f_\eta^{\ 2}}{8V^2}, \quad (6)$$

where f_η is the Doppler frequency and λ is the wavelength.

In the range-Doppler domain, the targets with the same range have the same range cell migration and they can be corrected simultaneously, which is also the main advantage of RD. The objective of RCMC is to remove the migrated actual track to the nearest slant range location.

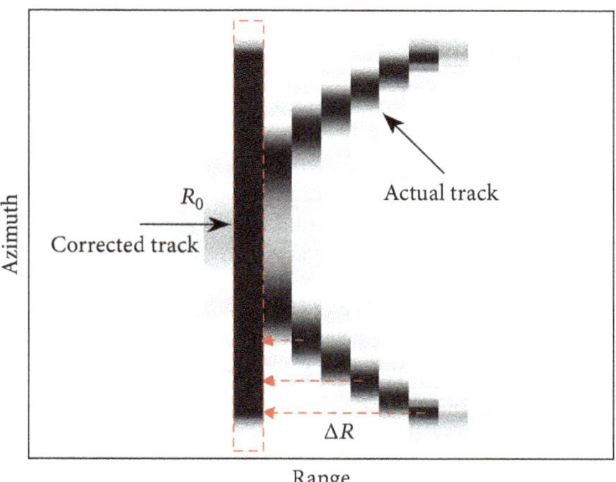

FIGURE 2: RCMC sketch. The actual curve track must be removed to the location of the nearest slant range R_0.

3. The Proposed Curve Fitting Based Range Cell Migration Correction (CF-RCMC) Method

In traditional methods, RCMC is accomplished through time-domain interpolation or frequency-domain compensation. However, the main problems of these methods are the excessive data operation and the narrow compatibility for different SAR configurations.

From Figure 2, we can see that the actual track of the target is a parabola, which is in accordance with the range history in (5) or (6). If the range history curve can be known, we can use this curve to correct the range cell migration. With this idea, we apply the polynomial curve fitting to obtain this range curve through simulation.

3.1. Curve Fitting. The curve fitting method [26] is a common method that can be applied to construct the approximate expression of the function $y = f(x)$ according to a set of experimental data points (x_k, y_k), where x_k and y_k denote the values of independent and dependent variables, respectively. Polynomial fitting is one of the most common fitting models, which can be expressed as

$$f(x) = a_0 + a_1 x + a_2 x^2 + \cdots + a_n x^n. \quad (7)$$

The fitting error between the polynomial and the measured value is

$$e_k = f(x_k) - y_k. \quad (8)$$

There are many kinds of criterions to minimize the fitting error. The least squares criterion is widely used and the squares error is formulated as

$$Q = \sum_{k=1}^{m} e_k^{\ 2} = \sum_{k=1}^{m} (f(x_k) - y_k)^2 = \sum_{k=1}^{m} \left(a_0 + a_1 x_k + a_2 x_k^{\ 2} + \cdots + a_n x_k^{\ n} - y_k\right)^2. \quad (9)$$

One can solve the equation set or utilize the optimization method by minimizing the error in (9) to obtain the fitting coefficients in (7).

3.2. The Proposed CF-RCMC. On the occasion of real data processing, the motion parameters may not be accurately obtained and the achieved curve might not fit the RCM well. In the range-compressed range-Doppler domain, the track of a strong point target can reflect RCM phenomena. However, the RCM curve of the strong point target cannot be directly obtained through the motion parameters since the strong point target is arbitrarily located. Therefore, we could adjust the motion parameters to obtain the coefficients of the quadratic polynomial. Then, we can correct the range cell migration easily with a totally known expression of the RCM curve.

The range-compressed echo is a two-dimensional data matrix with range cell migration. Figure 2 shows the range migration phenomena of a strong point target in the range-Doppler domain. ΔR denotes the range migration of the strong point target.

The slant ranges are then transformed into range cells for data processing. In this case, the range index of a strong point target at the azimuth time i is given as

$$\mathrm{ID}(i) = \mathrm{round}\left(\frac{(R(i) - R_0) F_s}{c}\right), \quad (10)$$

where $R(i)$ is the slant range between the scene center and the radar position; R_0 is the reference range; c denotes the wave propagation speed; F_s denotes the sampling frequency. The symbol round$(\cdot)$ denotes the rounding operation.

Once all the ranges are received, they will be employed by using curve fitting under the least squares criterion. To obtain appropriate coefficients of the quadratic polynomial, we can adjust the scene center, the reference range, or the platform position, such that the simulation curve and the range migration curve of a strong point are basically coincident. Then, the obtained curve function can be used

```
while 1
  (1) Initialize the scene center P_c and the reference range R_0;
  (2) Select N azimuth times; i = 0;
      while 1
          i++;
          Compute the slant range R(i) by using Equation (12);
          Compute the location ID(i) of the strong point target in the echo by using Equation (13);
          if i > N
             break;
          end if
      end while
  (3) Perform curve fitting on ID vector and get the simulation curve C.
  (4) if the simulation curve C coincides with the range migration line of the strong point;
          break;
      end if
end while
  (5) Remove the target echo of range cells of the nearest slant range location in frequency domain.
```

PSEUDOCODE 1: Pseudocode of the proposed CF-RCMC method.

to compute the offsets. Therefore, the corresponding number of range cells where the echo should be removed can be calculated by using the simulation curve, as follows:

$$S(i) = \mathrm{round}\left(f(i) - f(n)\right), \tag{11}$$

where n denotes slow time of the nearest slant range.

After that, it is easy to convert the range migration line of the strong point into a line by using the properties of Fourier transformation. That is, removal in the fast time domain can be achieved through the multiplication operation in frequency domain. This operation will be done for every range cell. Finally, the RD algorithm can be used for SAR imaging.

The implementation steps of the CF-RCMC method are listed as follows.

Step 1. Initialize the scene center $\mathbf{P}_c$ and the reference range R_0.

Step 2. Select one azimuth time and compute the slant range between the scene center and the radar positions:

$$R(i) = \sqrt{\left(P_1(i) - P_c\right)^2} + \sqrt{\left(P_2(i) - P_c\right)^2}, \tag{12}$$

where $P_1(i)$ and $P_2(i)$ denote the positions of the receiver and the transmitter, respectively. For monostatic SAR, $P_1(i) = P_2(i)$.

Step 3. Compute the location of the strong point target in the echo

$$\mathrm{ID}(i) = \mathrm{round}\left(\frac{\left(R(i) - R_0\right)F_s}{c}\right), \tag{13}$$

where the round function denotes the nearest integer operator.

Step 4. Change the azimuth times one by one, and repeat Steps 2 and 3 until the IDs at all the azimuth times are computed.

Step 5. Based on the calculated IDs, utilize the curve fitting method to obtain the quadratic polynomial coefficients.

Step 6. Adjust P_c and R_0 and repeat Steps 1–5 until the simulation curve and range migration line of the strong point are basically coincident.

Step 7. Remove the target echo of the corresponding range cells with respect to the nearest slant range location in the range frequency domain, by making use of the properties of Fourier transformation.

Specifically, the pseudocode of the CF-RCMC method is summarized in Pseudocode 1.

Further, the flow chart of the CF-RCMC method is summarized in Figure 3.

4. Experimental Results

In this section, the proposed CF-RCMC method is applied to real data for both the monostatic airborne SAR and the bistatic airborne SAR. The velocities and the trajectories are not exactly known because only GPS was used to measure the antenna attitude, and hence measurement precision is not very high.

4.1. A Brief Introduction of the Radar System. The mentioned radar system in this paper is an air force military radar. The referred band is x-band with chirp signal and the bandwidth is 300 MHz. The effective distance of the transmitter is 30~60 km with a transmission power of 4000 watts (w). The effective distance of the receiver is 10~20 km, and the data sampling frequency is 1.8 GHz.

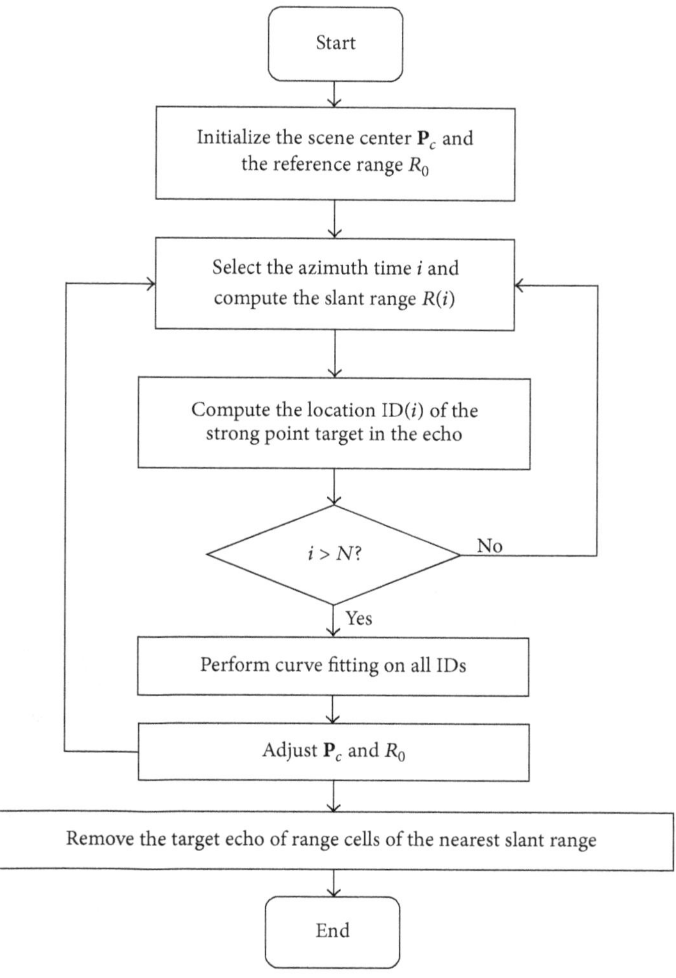

FIGURE 3: The flow chart of the proposed method, where N is the total azimuth times.

4.2. Monostatic SAR Experiment. The track of a strong point target in the range-compressed echo domain is shown in Figure 4, from which we can see the range cell migration phenomenon clearly. However, the nearest slant range location does not appear in this figure. When we correct the RCM, the range at the first PRF (pulse recurrence frequency) is selected as the nearest slant range. The range cell migration is extremely serious, which is nearly 90 range cells along 5000 azimuth cells.

After adjusting the scene center and the reference range, the fitted curve (red solid line) is illustrated in Figure 5. We can see that this fitted curve is exactly coincident with the RCM curve of the strong point target. Then, we use the fitted quadratic curve expression and the properties of Fourier transformation to correct the RCM in the range frequency domain. Figure 6 shows the result of CF-RCMC. The track of the strong point target has been corrected to a horizontal straight line along the azimuth direction, which states the efficiency of the proposed CF-RCMC method. However, it is worth noting that this CF-RCMC method can only correct the RCM for partial targets around the selected one located at "the nearest slant range." In fact, all the other RCMC methods also cannot correct the RCM for every target.

After azimuth compression, the imaging results based on conventional RCMC used in [19, 20] and the proposed CF-RCMC are presented in Figures 7 and 8, respectively. In Figure 7, the whole scene is seriously defocused, especially in the azimuth direction. In Figure 8, obviously, the image is well focused and the targets features can be clearly distinguished. Note that the focusing effect becomes better when the target is closer to the strong point, which is referred to correct the RCM.

4.3. Bistatic SAR Experiment. For bistatic SAR, the range history is much more complex than monostatic SAR, which thus results in more complicated RCMC. However, the proposed RCMC method based on curve fitting can be directly applied to the bistatic SAR. The processing method for bistatic SAR is totally the same as that for monostatic SAR. The results for bistatic SAR are concisely provided below.

The track of a strong point target in the range-compressed echo domain is shown in Figure 9. The range cell migration amount is about 400 range cells along 5000 azimuth cells,

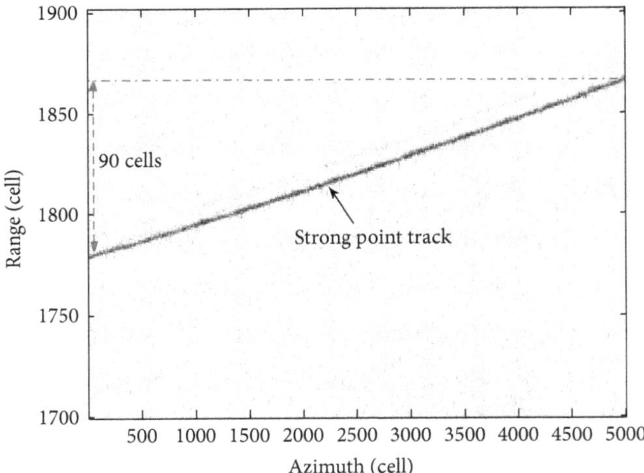

FIGURE 4: Track of a strong point in the range-compressed echo (monostatic SAR). The RCM of 90 cells is very serious.

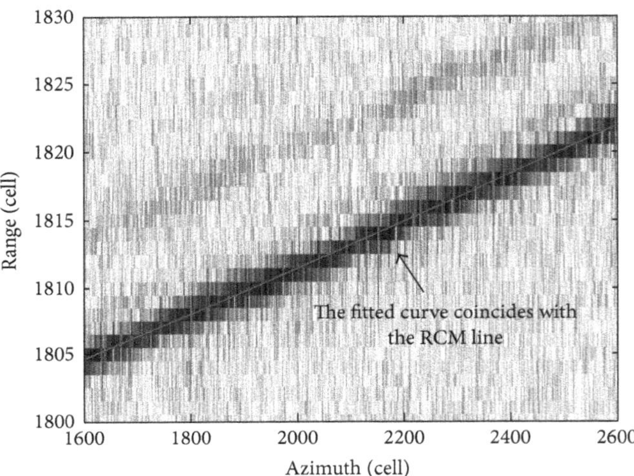

FIGURE 5: Curve fitting result (monostatic SAR). The fitted quadratic curve is coincident with the RCM line (100% for strong points in echo), which can be used to correct the RCM.

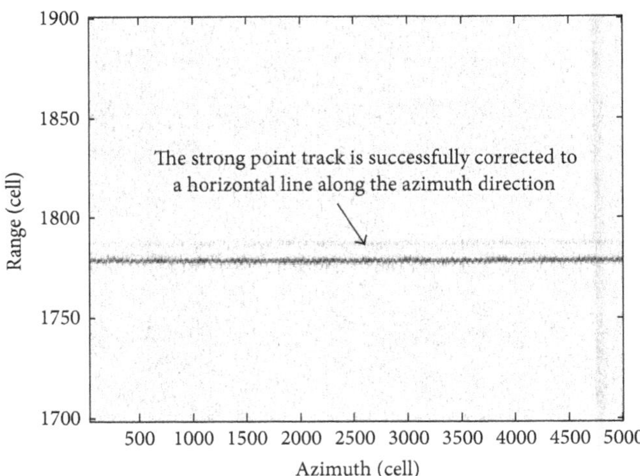

FIGURE 6: CF-RCMC result (monostatic SAR). The horizontal line indicates that the RCM has been successfully corrected.

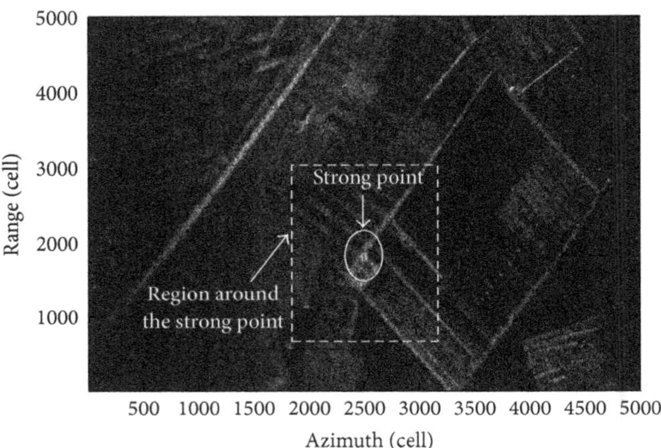

FIGURE 7: Imaging result based on conventional RCMC (monostatic SAR). The whole image is seriously defocused.

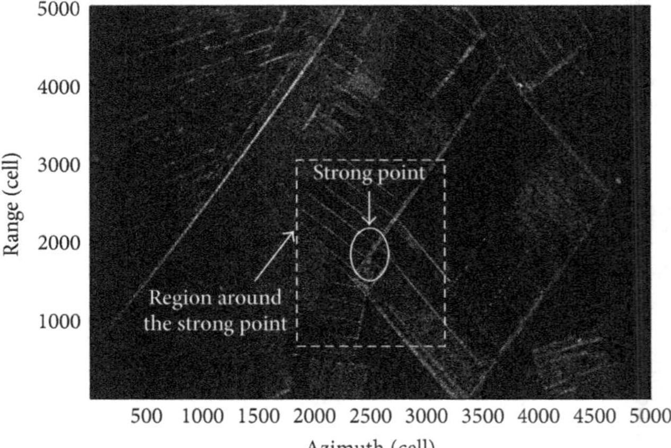

FIGURE 8: Imaging result with CF-RCMC (monostatic SAR). The image is well focused, particularly in the region around the strong point.

which states that the RCM is much more serious than that in the monostatic SAR.

The fitted curve is shown in Figure 10, in which the fitted curve is exactly coincident with the RCM curve of the strong point target.

The RCMC result is illustrated in Figure 11. Without more operations than the monostatic SAR, the track of the strong point target has been corrected to a horizontal straight line along the azimuth direction. To obtain similar RCMC performance, the traditional RCMC method may not be available.

The imaging results based on the conventional RCMC method [19, 20] and the proposed CF-RCMC for bistatic SAR are presented in Figures 12 and 13, respectively. In Figure 12, the defocusing of the image is much more serious than that in the monostatic SAR due to the larger RCM. As seen in Figure 13, it is apparent that the CF-RCMC method

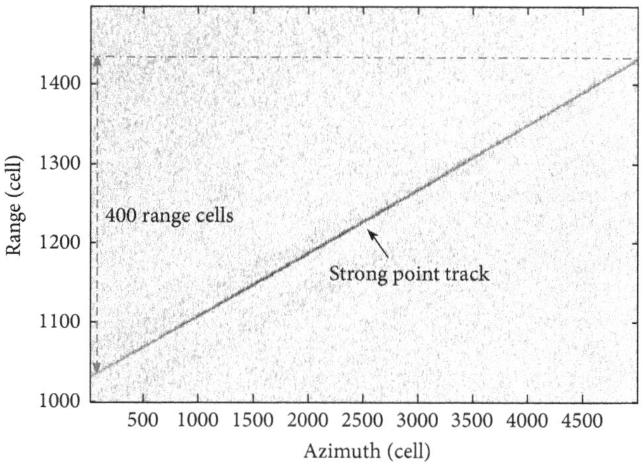

FIGURE 9: Track of a strong point in range-compressed echo (bistatic radar). The amount of RCM is much larger than that in the monostatic SAR.

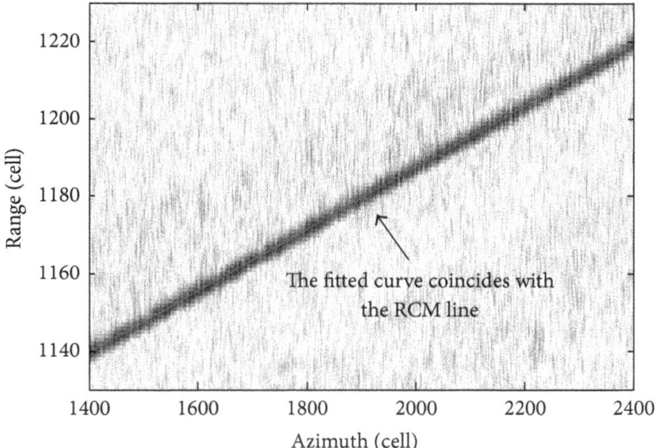

FIGURE 10: Curve fitting result (bistatic radar). The fitted curve is still coincident with the RCM line (100% for strong points in echo).

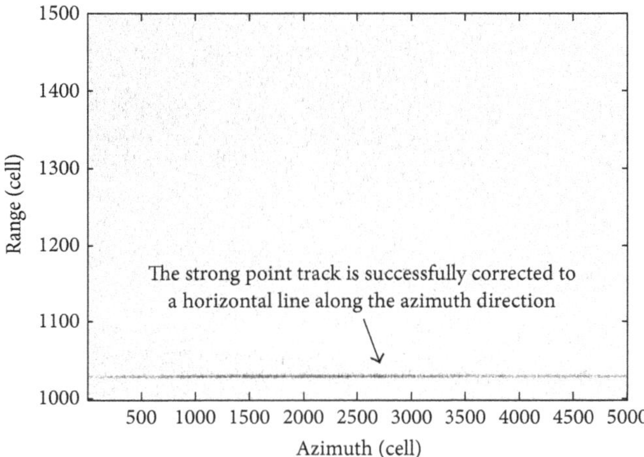

FIGURE 11: CF-RCMC result (bistatic radar). The RCM can be easily and successfully corrected.

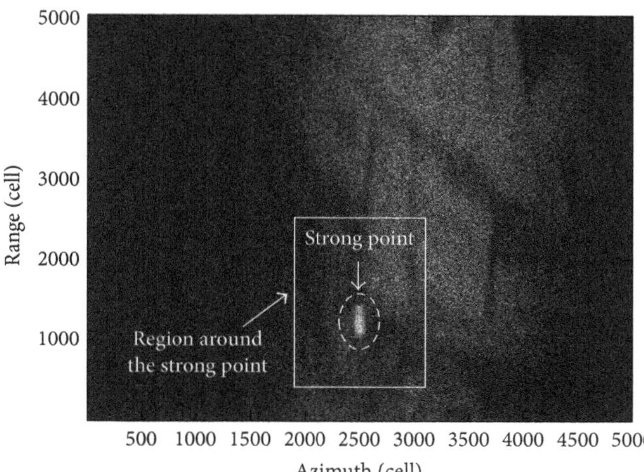

FIGURE 12: Imaging result based on conventional RCMC (bistatic radar). The defocusing of the image is much more serious than that in the monostatic SAR due to the larger RCM.

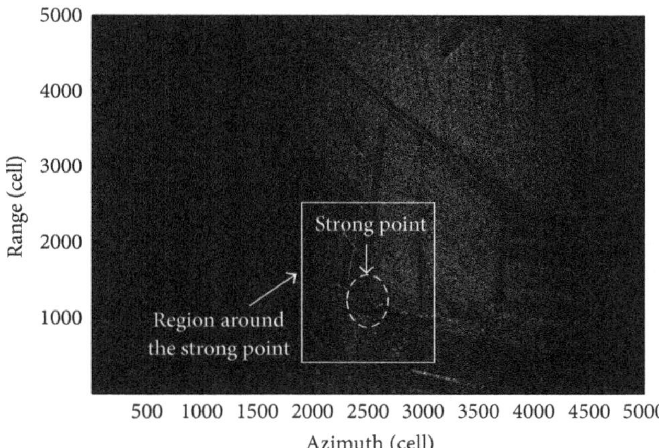

FIGURE 13: Imaging result with CF-RCMC (bistatic radar). By using the RCMC method proposed in this paper, the image is well focused.

greatly improves the focusing quality of the imaging for the bistatic SAR. Particularly, the focusing performance in the area around the selected strong point target is excellent.

4.4. Discussion. In this paper, the proposed CF-RCMC method has been demonstrated to be more effective than the conventional RCMC technique. It is necessary to note that there is an important difference between the existing RCMC and our proposed CF based method. As we know, the traditional RCMC methods generally choose the nearest slant range as references, further derive the migration momentum of the reference points mathematically, and then move the distance curves to the nearest slope distance in time domain or frequency domain [20, 21]. Although these methods are useful in several scenarios, they do not take into account the parametric uncertainty of the system in actual flight cases, which lead to a large error between the migration equation and the actual migration curve. Additionally, another disadvantage is that, under different SAR modes such as bistatic or monostatic, the existing RCMC may not be generalized in applications.

Comparatively, the proposed CF-RCMC in this paper is motivated in a different view. Generally, from the magnitude trajectory of the actual echo strong points, the CF-RCMC aims at adjusting the calibration parameters such that the distance curve can coincide with the trajectory of the strong points of the echo. Then, the migration correction is implemented. The advantages of the proposed method are that, on the one hand, it is robust to the uncertainty of the system parameters and, on the other hand, it can also be adapted to different SAR modes (i.e., monostatic and bistatic).

However, in fact, no matter which RCMC methods are used, the correction results of nonreference points cannot be statistically assessed, and therefore we have to evaluate the effectiveness of RCMC methods from the imaging result after azimuth compression. From our observation in experiment, the regression distance curve coincides with the trajectory of the strengths of the echo completely; that is, the regression accuracy for the strong points of the echo is 100%. However, the regression accuracy of other points (nonstrong points) cannot be estimated due to the fact that the trajectory in echo cannot be observed. In the future, machine learning based methods [27, 28] can be further considered for handling the issues.

5. Conclusions

The correction for range cell migration caused by the motion of the radar relative to the target plays an important role in the RD algorithm in SAR. In this paper, a novel CF-RCMC method based on the polynomial curve fitting under the least squares criterion is proposed. This method is applicable not only to the monostatic SAR but also to the bistatic SAR and even to any kind of SAR configuration. The high flexibility of this method is significantly important for the simplification of CF-RCMC. The experimental results of both the monostatic SAR and the bistatic SAR confirm the effectiveness of the proposed CF-RCMC method.

Conflicts of Interest

The authors declare that there are no conflicts of interest regarding the publication of this paper.

Acknowledgments

This work was supported by Fundamental Research Funds for the Central Universities (no. 106112017CDJQJ168819).

References

[1] C. W. Sherwin, J. P. Ruina, and R. D. Rawcliffe, "Some Early Developments in Synthetic Aperture Radar Systems," *IRE Transactions on Military Electronics*, vol. 6, no. 2, pp. 111–115, 1962.

[2] Y. T. Liu, *Radar Imaging Technology*, Harbin Institute of Technology Press, 1999.

[3] M. Weib, J. H. G. Ender, and C. H. Gierull, "Foreword to the special issue on scientific and technological progress of synthetic aperture radar (SAR)," *IEEE Transactions on Geoscience and Remote Sensing*, vol. 51, no. 8, pp. 4363–4365, 2013.

[4] N. Pierdicca, P. Castracane, and L. Pulvirenti, "Inversion of electromagnetic models for bare soil parameter estimation from multifrequency polarimetric SAR data," *Sensors*, vol. 8, no. 12, pp. 8181–8200, 2008.

[5] X. Huang, B. Huang, and H. Li, "A fast level set method for synthetic aperture radar ocean image segmentation," *Sensors*, vol. 9, no. 2, pp. 814–829, 2009.

[6] H. Li, Y. He, and W. Wang, "Improving ship detection with polarimetric SAR based on convolution between co-polarization channels," *Sensors*, vol. 9, no. 2, pp. 1221–1236, 2009.

[7] X. Zhou, N.-B. Chang, and S. Li, "Applications of SAR interferometry in earth and environmental science research," *Sensors*, vol. 9, no. 3, pp. 1876–1912, 2009.

[8] X. Zhou, G. Wei, S. Wu, and D. Wang, "Three-dimensional ISAR imaging method for high-speed targets in short-range using impulse radar based on SIMO array," *Sensors (Switzerland)*, vol. 16, no. 3, article 364, 2016.

[9] I. Cumming and F. Wong, *Digital Processing of Synthetic Aperture Radar Data: Algorithms and Implementation*, Artech House, 2005.

[10] J. J. M. De Wit, A. Meta, and P. Hoogeboom, "Modified range-doppler processing for FM-CW synthetic aperture radar," *IEEE Geoscience and Remote Sensing Letters*, vol. 3, no. 1, pp. 83–87, 2006.

[11] E. C. Zaugg and D. G. Long, "Generalized frequency-domain SAR processing," *IEEE Transactions on Geoscience and Remote Sensing*, vol. 47, no. 11, pp. 3761–3773, 2009.

[12] H.-S. Shin and J.-T. Lim, "Omega-k algorithm for airborne spatial invariant bistatic spotlight SAR imaging," *IEEE Transactions on Geoscience and Remote Sensing*, vol. 47, no. 1, pp. 238–250, 2009.

[13] S. Jun, M. Long, and Z. Xiaoling, "Streaming BP for non-linear motion compensation SAR imaging based on GPU," *IEEE Journal of Selected Topics in Applied Earth Observations and Remote Sensing*, vol. 6, no. 4, pp. 2035–2050, 2013.

[14] C. Wu, "A digital system to produce imagery from SAR data," in *Proceedings of the Systems Design Driven by Sensors*, pp. 76–968, 1976.

[15] R. Lanari and G. Fornaro, "A short discussion on the exact compensation of the SAR range-dependent range cell migration effect," *IEEE Transactions on Geoscience and Remote Sensing*, vol. 35, no. 6, pp. 1446–1452, 1997.

[16] J. C. Curlander and R. N. McDonough, *Synthetic Aperture Radar: Systems and Signal Processing*, Wiley, Hoboken, NJ, USA, 1991.

[17] R. Bamler, "A comparison of range-Doppler and wavenumber domain SAR focusing algorithms," *IEEE Transactions on Geoscience and Remote Sensing*, vol. 30, no. 4, pp. 706–713, 1992.

[18] C. Wu, K. Y. Liu, and M. Y. Jin, "Modeling and a correlation algorithm for spaceborne SAR signals," *IEEE Transactions on Aerospace and Electronic Systems*, vol. 18, no. 5, pp. 563–575, 1982.

[19] W. Junfeng and L. Xingzhao, "Automatic range-migration correction in sar imaging," in *Proceedings of the 2008 IEEE International Geoscience and Remote Sensing Symposium - Proceedings*, pp. IV1257–IV1260, usa, July 2008.

[20] T. Zeng, W. Yang, Z. Ding, and L. Liu, "Advanced range migration algorithm for ultra-high resolution spaceborne synthetic aperture radar," *IET Radar, Sonar and Navigation*, vol. 7, no. 7, pp. 764–772, 2013.

[21] W. Wang, W. Wu, W. Su, R. Zhan, and J. Zhang, "High squint mode SAR imaging using modified RD algorithm," in *Proceedings of the 2013 IEEE China Summit and International Conference on Signal and Information Processing (ChinaSIP)*, pp. 589–592, July 2013.

[22] S.-X. Zhang, M.-D. Xing, X.-G. Xia, L. Zhang, R. Guo, and Z. Bao, "Focus improvement of high-squint SAR based on azimuth dependence of quadratic range cell migration correction," *IEEE Geoscience and Remote Sensing Letters*, vol. 10, no. 1, pp. 150–154, 2013.

[23] G. Garza and Z. Qiao, "Resolution analysis of bistatic SAR," in *Proceedings of the Radar Sensor Technology XV*, usa, April 2011.

[24] J. Lopez, G. Garza, and Z. Qiao, "Cross-range imaging of SAR data," *Pacific Journal of Applied Mathematics*, vol. 2, 65 pages, 2009.

[25] L. Zhang, H.-L. Li, Z.-J. Qiao, M.-D. Xing, and Z. Bao, "Integrating autofocus techniques with fast factorized back-projection for high-resolution spotlight SAR imaging," *IEEE Geoscience and Remote Sensing Letters*, vol. 10, no. 6, pp. 1394–1398, 2013.

[26] L. Grama and C. Rusu, "Phase approximation by divide-and-conquer piecewise linear fitting of gain," in *Proceedings of the 2009 International Symposium on Signals, Circuits and Systems, ISSCS 2009*, rou, July 2009.

[27] L. Zhang and D. Zhang, "Robust Visual Knowledge Transfer via Extreme Learning Machine-Based Domain Adaptation," *IEEE Transactions on Image Processing*, vol. 25, no. 10, pp. 4959–4973, 2016.

[28] L. Zhang, W. Zuo, and D. Zhang, "LSDT: Latent Sparse Domain Transfer Learning for Visual Adaptation," *IEEE Transactions on Image Processing*, vol. 25, no. 3, pp. 1177–1191, 2016.

18

Reliability Analysis of Network Real-Time Kinematic

Mohammed Ouassou,[1] **Bent Natvig**,[2] **Anna B. O. Jensen**,[3] **and Jørund I. Gåsemyr**[2]

[1]*Norwegian Mapping Authority, Geodetic Institute, 3511 Hønefoss, Norway*
[2]*UiO, Department of Mathematics, Norway*
[3]*KTH Royal Institute of Technology, 10044 Stockholm, Sweden*

Correspondence should be addressed to Mohammed Ouassou; mohammed.ouassou@statkart.no

Academic Editor: Sandro M. Radicella

The multistate reliability theory was applied to the network real-time kinematic (NRTK) data processing chain, where the qualities of the network corrections, baseline residuals, and the associated variance-covariance matrices are considered as the system state vectors. The state vectors have direct influence on the rover receiver position accuracy. The penalized honored stochastic averaged standard deviation (PHSASD) is used to map the NRTK sensitive data, represented by the states vectors to different levels of performance. The study shows that the improvement is possible by identification of critical components in the NRTK system and implementation of some parallelism that makes the system more robust.

1. Introduction

High accuracy positioning with GNSS is carried out using both code and carrier phase data from the GNSS satellites. To obtain position accuracies at the cm or mm level using carrier phase data, an important part of the data processing is to estimate the initial oscillator phase offset, the so-called ambiguity, for each receiver-satellite pair. Resolution of ambiguities requires that the influence of most errors sources in the positioning process is reduced to the cm-level, and high accuracy positioning is therefore often done in a relative mode where the position of a GNSS receiver located in an unknown position (the rover) is determined relative to one or more reference stations located in positions known on beforehand [1].

With relative carrier phase based GNSS positioning the effects of the distance-dependent error sources such as uncertainties in satellite positions and atmospheric effects on the satellite signals induced by the ionosphere and troposphere are reduced. Also the effects of satellite and receiver clock errors in the positioning process are reduced by relative positioning, and all this in combination makes it possible to resolve the ambiguities and thereafter obtain positions for the rover at the cm or mm level.

For high accuracy GNSS positioning in real time, the real time kinematic (RTK) technique has been developed. Traditionally this is based on a reference station transmitting data to the rover where the data is used in estimation of the position of the rover in a relative or differential mode [2].

Using a network of reference stations for RTK, the so-called NRTK technique, provides the opportunity for applying more advanced algorithms for estimation of the distance-dependent errors within the network and thereby possibilities for providing a more robust service. Such operational NRTK services exist in many regions and countries today and have become an indispensable tool in high accuracy navigation and surveying.

A brief description of the NRTK functionality is as follows: the first step is collection of raw observations from the network of reference stations, solving the ambiguities within the reference network, and generating error estimates. Then an interpolation/smoothing scheme is applied to generate the NRTK corrections for the user location. For information on how to avoid a loss of information under interpolation of NRTK data, the interested reader is referred to [3].

The NRTK corrections are then transmitted to rover receivers. Several NRTK techniques exist and the most

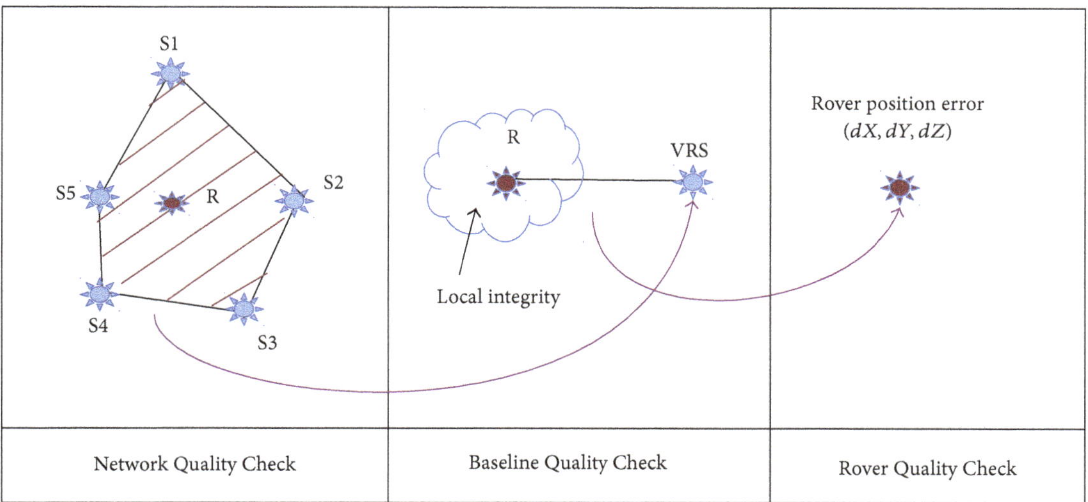

FIGURE 1: NRTK processing chain. The left panel shows NRTK module that produces the network corrections. The curved line indicates the output generated by the network module. Middle panel shows the local and the baseline processing module. Right panel shows the rover position solution module.

commonly used at present are, for instance, the Master Auxiliary Concept (MAC) [4, 5], the Virtual Reference Station (VRS) concept [6], and the FKP techniques [7], as well as the Network Adjustment (NetAdjust) concept developed by Raquet and Lachapelle [8, 9].

Multistate system reliability theory has been a research topic for many years, for instance, extension of the system from two-state to multistate reliability [10, 11] to compute the mean performance level at any given time t and stochastic evaluation and bound computation of multistate coherent systems [12], further, studies on application of reliability analysis to GNSS data processing [2], a comparative GNSS reliability analysis [13], reliability analysis under GNSS weak signals [14], accuracy and reliability of multi-GNSS real-time precise positioning [15], and robust reliability testing in case of signal degradation environment [16, 17].

In order to avoid confusion between the reliability theory definition used in different fields, we refer to the traditional reliability theory to the context of statistical testing based on the theory developed by Baarda [18], while component, system, binary, and multistate reliability terms will be used in case of component and system reliability computation.

Our aim is to provide the user in the field with continuously high quality corrections with the ability to identify the periods for which the reliability of the network RTK performance is reduced in terms of accuracy and availability. Therefore, solution quality indicators describing the reliability of the network RTK are needed to transfer the status of the network to the user in the field. Intensive research has been conducted recently in this field to derive these quality indicators and can be classified into two main classes; (i) spatially correlated (ionosphere, troposphere, and orbital) error indicators; (ii) residuals errors indicators. Most network RTK used quality indicators are residual integrity monitoring (RIM) and irregularity parameters (IP) quality indices [19], residual interpolation uncertainty (RIU) [20], geometry-based quality indicator (GBI) [21], and ionospheric index I95 [20]. An elegant presentation summarizing the network RTK quality indicators can be found in [22].

In order to apply reliability analysis to a NRTK system, the starting point is the decomposition of the block diagrams of NRTK processing chain into simple components and computation of the system reliability. Figure 1 shows three levels of data processing modules, the network, the baseline, and the rover receiver modules, where R denotes rover, S denotes reference station, and (dX, dY, dZ) denote errors in the position X, Y, Z coordinates, respectively. Based on these levels, we can build the reliability block diagrams for the NRTK processing chain and compute the reliability for the entire system.

The rest of the paper is organized as follows:

Section 2: brief introduction to component, system, and binary state reliability theory and deterministic and stochastic reliability.

Section 3: NRTK blocks diagram determination and module reliability computations.

Section 4: multistate reliability theory applied to NRTK processing chain.

Section 5: some test results.

Section 6: procedures used to validate the NRTK system reliability.

Section 7: discussions and conclusion.

Test data used in this investigation is described in Appendix A.

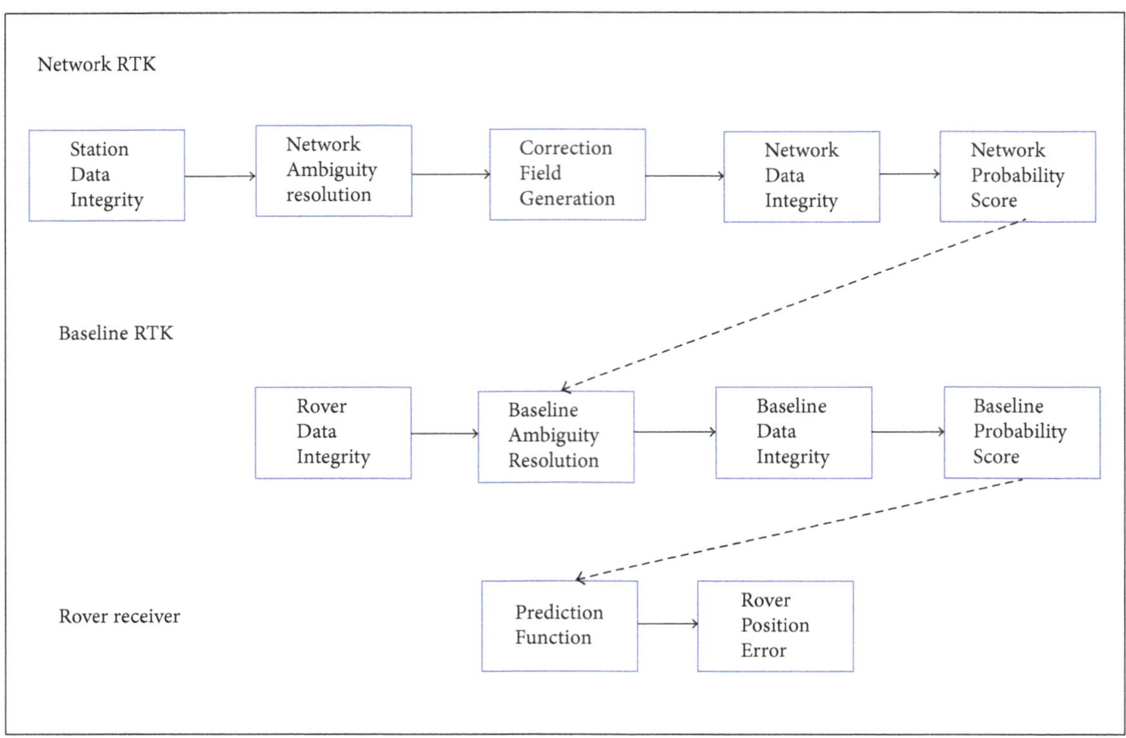

FIGURE 2: Network real-time kinematic processing chain main modules. Dotted lines describe the components dependency from different levels.

2. Reliability Analysis

The aim of this section is to introduce the basic of the component, system, and binary state reliability theory, including deterministic and stochastic reliability, followed by decomposing the block diagrams of the NRTK processing chain into simple components, and compute the entire system reliability. The description of the structural relationship between the components and the system must be defined. Figure 2 illustrates the concept and each main block will be treated separately in the coming sections.

2.1. Structure Functions. In order to construct the entire reliability block diagram of the NRTK data integrity, it is necessary to define the elementary building blocks as Bernoulli indicator function [23, pp. 28-29]. The function will signal if a unit or system is functioning or not.

A random variable X is said to be a Bernoulli random variable if its probability function is given by (1). The indicator and structure functions are given below by following the notation given by Natvig [24]:

$$X_i = \begin{cases} 1, & \text{success with probability } p, \\ 0, & \text{failure with probability } (1-p). \end{cases} \tag{1}$$

2.1.1. Bernoulli Indicator Function. Let x_i denote the indicator function of component, i. Then we have

$$x_i = \begin{cases} 1, & \text{if the } i\text{'th component is functioning,} \\ 0, & \text{otherwise.} \end{cases} \tag{2}$$

2.1.2. System Structure Function. Let the state vector $\mathbf{x} = (x_1, x_2, \ldots, x_n)$ give which components are functioning and which are not.

Let $\phi(\mathbf{x})$ denote the Bernoulli indicator of the state vector, $\mathbf{x}$. Then we have

$$\phi(\mathbf{x}) = \begin{cases} 1, & \text{if the system is functioning} \\ 0, & \text{otherwise} \end{cases} \tag{3}$$

$\phi(\mathbf{x})$ is the structure function of the system.

2.1.3. Series Structure Function. The series structure function $\phi_s(\mathbf{x})$ works *if and only if all components* of the state vector $\mathbf{x}$ are functioning. The series structure function $\phi_s(\mathbf{x})$ reads

$$\phi_s(\mathbf{x}) = \min(x_1, x_2, \ldots, x_n) = \prod_{i=1}^{n} x_i. \tag{4}$$

2.1.4. Parallel Structure Function. The parallel structure function $\phi_p(\mathbf{x})$ works *if and only if at least one* of the components of the state vector $\mathbf{x}$ is functioning. The parallel structure function $\phi_p(\mathbf{x})$ reads

$$\begin{aligned} \phi_p(\mathbf{x}) &= \max(x_1, x_2, \ldots, x_n) = 1 - \prod_{i=1}^{n}(1 - x_i) \\ &= \coprod_{i=1}^{n} x_i, \end{aligned} \tag{5}$$

where $\coprod$ is read "ip" and denotes the parallel coupling operator.

2.1.5. k-out-of-n Structure Function. A system composed of n components which is functioning *if and only if at least k* components are functioning is called a *k-out-of-n structure*

$$\phi_{k,n}(\mathbf{x}) = \begin{cases} 1, & \text{if } \sum_{i=1}^{n} x_i \geq k, \\ 0, & \text{otherwise.} \end{cases} \quad (6)$$

Note that the structure functions given by (4) and (5) are an *n-out-of-n structure* and an *1-out-of-n structure*, respectively. Other structure functions exist, for instance, the bridge Natvig [24, p. 12].

2.2. System Reliability Computation. The structure functions are defined; now is time to compute the system reliability. We move from the deterministic model to the stochastic one by introducing the random variables. Some notations are needed to represent the state vector, structure function, and the reliability. We follow the notations given by Hoyland and Rausand [25, Chaps. 3–5]. We denote the state variables of the n independent component at time t by

$$X_1(t), X_2(t), \ldots, X_n(t). \quad (7)$$

The corresponding state vector and the structure function are denoted, respectively, by

$$\begin{gathered} \mathbf{X}(t) = (X_1(t), X_2(t), \ldots, X_n(t)), \\ \phi(\mathbf{X}(t)). \end{gathered} \quad (8)$$

The probabilities of interest are presented by

$$\begin{aligned} P(\mathbf{X}_i(t) = 1) &= p_i(t) \quad \text{for } i = 1, 2, \ldots, n, \\ P(\phi(\mathbf{X}(t) = 1)) &= p_s(t), \end{aligned} \quad (9)$$

where $p_i(t)$ is the component reliability while $p_s(t)$ is the system reliability. Assuming that the components are independent, then the computation of the reliability of the state vector $\mathbf{X}(t)$ and the system $\phi(\mathbf{X}(t))$ at time t are defined as the expectation operator

$$\begin{aligned} \mathbb{E}(X_i(t)) &= 0 \cdot P(X_i(t) = 0) + 1 \cdot P(X_i(t) = 1) \\ &= p_i(t) \quad \text{for } i = 1, 2, \ldots, n. \end{aligned} \quad (10)$$

Let R denote the system reliability, then we have

$$\begin{aligned} \mathbb{E}(\phi(\mathbf{X}(t))) &= p_s(t) = R(p_1(t), p_2(t), \ldots, p_n(t)) \\ &= R(\mathbf{p}(t)). \end{aligned} \quad (11)$$

To avoid confusion, $p_i = P(X_i(t) = 1)$ is the probability of functioning of the ith component and referred to as the component reliability.

2.2.1. Reliability of Series Structures. The reliability function $R_s(\mathbf{p})$ of the series system of n independent components is given by the expression:

$$\begin{aligned} R_s(\mathbf{p}) &= P(\phi(\mathbf{X}(t) = 1)) \\ &= P\{X_i(t) = 1 \ \forall i = 1, 2, \ldots, n\} = \prod_{i=1}^{n} p_i(t). \end{aligned} \quad (12)$$

If all components have the same $p(t)$, (12) becomes $\{p(t)\}^n$. For $n = 5$ and $p = 0.99$, then the reliability $R_s(\mathbf{p}) = 0.951$.

An important remark is that the reliability of a series structure is at most as reliable as the least reliable component, that is, $R_s(\mathbf{p}) \leq \min_i(p_i(t))$.

2.2.2. Reliability of Parallel Structures. The reliability function $R_p(\mathbf{p})$ of the parallel system of n independent components is given by the expression:

$$\begin{aligned} R_p(\mathbf{p}) &= P(\phi(\mathbf{X}(t) = 1)) \\ &= P\{X_i(t) = 1 \text{ for some } i = 1, 2, \ldots, n\} \\ &= 1 - P\{X_i(t) = 0 \ \forall i = 1, 2, \ldots, n\} \\ &= 1 - \prod_{i=1}^{n}(1 - p_i(t)) \end{aligned} \quad (13)$$

If all components have the same $p(t)$, (13) becomes $1 - \{1 - p(t)\}^n$. For $n = 5$ and $p = 0.99$, then the reliability $R_p(\mathbf{p}) = 1$.

An important remark is that the reliability of a parallel structure is at least as reliable as the most reliable component, that is, $R(\mathbf{p}) \geq \max_i(p_i(t))$.

Details on how to compute the reliability of parallel structure in general are given in Appendix C.

2.2.3. Reliability of k-out-of-n Structures. The reliability function $R_{(k-n)}(\mathbf{p})$ of the k-out-of-n system of n independent components with equal probability $p_i(t) = p$ is given by the expression:

$$\begin{aligned} R_{(k-n)}(\mathbf{p}) &= P(\phi(\mathbf{X}(t) = 1)) = P\left\{\sum_{i=1}^{n} X_i(t) \geq k\right\} \\ &= \sum_{i=k}^{n} \binom{n}{i} \cdot p^i (1 - p)^{n-i}. \end{aligned} \quad (14)$$

3. Reliability in the NRTK Processing Chain

The aim of this section is to determine the structure functions, NRTK module's reliability, and the corresponding block diagrams.

3.1. Considerations around NRTK Data Processing. Some considerations around the NRTK data integrity are introduced below as a background for the design process and to ease the discussions in the following sections. For more information on GNSS data processing, the reader is referred

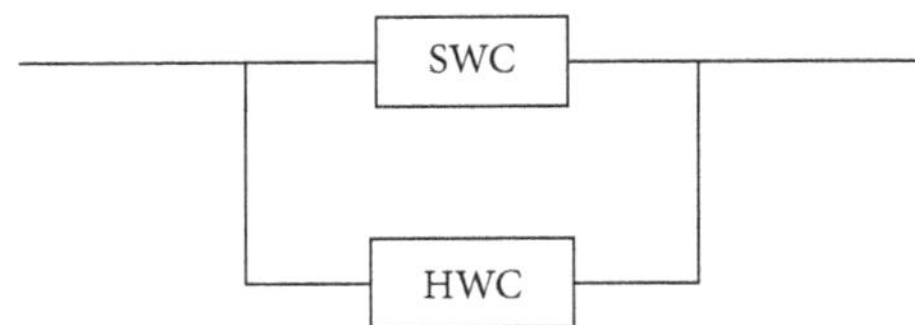

Figure 3: Main block diagram of module $M_{N,1}$.

to [1, 2, 26, 27]. The key to precise positioning is the correct ambiguity resolution and validation. With ambiguities resolved to wrong integer numbers, there will be offsets in the position solution, and with float ambiguities (ambiguities that are not fixed to integer values) the position solution is inaccurate and also very unstable and sensitive to changes in satellite geometry.

Good satellite-receiver geometry, as, for instance, expressed by the so-called DOP factor (dilution of precision), is important to perform successful ambiguity resolution and achieve centimeter level accuracy in real time.

Spatiotemporal models that describe well the variations of the spatially correlated errors in the corrections field are also an important key for reliable NRTK positioning.

Robust estimation algorithms to handle large data sets are also a key factor becoming more important in the future as observations from several GNSS systems to a larger degree will be combined in one processing loop. Today, most NRTK systems operate with data from the American GPS and the Russian GLONASS system. Including data from the European Galileo as well as the Chinese Beidou systems in NRTK operations will soon be the norm for most NRTK services. With satellites from more GNSS systems being available the satellite-receiver geometry on the rover side is improved. This is especially important when the user is operating in constricted environments such as narrow street canyons or forest areas.

3.2. NRTK Corrections Reliability Analysis. The main function of the NRTK is to provide the rover in the field with high quality corrections on an epoch-by-epoch basis.

From Figure 2, five modules $M_{N,i}$ are defined for the network and the reliability of each module will be evaluated. $M_{i,j}$ corresponds to level $i \in \{N, B, R\}$ and module $j \in \{1, 2, \ldots, 5\}$, where N stands for network, B for baseline, and R for rover receiver. For instance, the module $M_{N,1}$ corresponds to the first network module, the Station Data Integrity, as shown in Figure 2, and $M_{R,1}$ denotes the first of the rover modules, i.e., the Prediction Function as shown in Figure 2.

3.2.1. Reference Receivers Data Integrity. Generation of high quality raw observations at reference receivers requires reliable hardware (HWC) and software (SWC) components. Expensive hardware and sophisticated algorithms are keys to achieve this goal. Figure 3 shows the concept and each component will be treated separately in the next sections.

Let software and hardware components be represented by the modules $M_{N,1,X}$ and $M_{N,1,Y}$, respectively.

Software Component of Module $M_{N,1}$ Definition. SWC requires an ensemble of sequential checks on raw observations. This includes

(i) Let $x_{1,1}$ denote the satellite data integrity algorithm. This algorithm will discard the measurements from unhealthy satellite(s) or from satellite(s) for which we do not have the orbital data.

(ii) Let $x_{1,2}$ denote controlled cycle-slip algorithm. This task requires investigation of carrier phase discontinuities by examination of loss of lock (LLI) indicator and signal-to-noise ratio (SNR) flags.

(iii) Let $x_{1,3}$ denote uncontrolled cycle-slips in the observations. The algorithm uses the observation combinations for this purpose. The interested reader is referred to [27, pp. 95–101].

(iv) Let $x_{1,4}$ denote the reference receiver clock offset reset algorithm. Continual corrections are carried out to reduce the effect of the jump. The receiver clock offset (jump with ± 1 ms) must be detected and corrected because they cause jump in carrier phase.

(v) Let $x_{1,5}$ denote outliers detection and repair algorithm.

(vi) Let $x_{1,6}$ denote the low elevation angle. Algorithm prunes satellite(s) based on their low elevation angle.

(vii) Let $x_{1,7}$ denote the minimum observations required to generate the corrections. At least 4 observation types are needed (L_1, L_2, P_1, and P_2). For more information about the observation types provided by satellites, the reader is referred to [28].

(viii) Let $x_{1,8}$ denote the reference receiver clock stability algorithm.

(ix) Let $x_{1,9}$ denote the multipath mitigation algorithm.

(x) Let $x_{1,10}$ denote the reweighting algorithm. All units are parallel coupled.

  (1) Let $x_{[1,10,1]}$ denote a low elevation reweighting algorithm.

  (2) Let $x_{[1,10,2]}$ denote a scintillation reweighting algorithm.

  (3) Let $x_{[1,10,3]}$ denote a signal-to-noise reweighting algorithm.

Figure 5 shows the block diagram of the network reweighting algorithm component $x_{1,10}$.

Block Diagram and Reliability of Software Component. The structure function of the software component is well described by the 5-out-of-10 structure function. This means that the more algorithms check is, the more reliable raw observations become.

In order to produce reliable raw observations of high quality, it is necessary to perform at least five checks from a total of ten. With ten algorithms' check, we can generate high quality raw observations while five algorithms' check

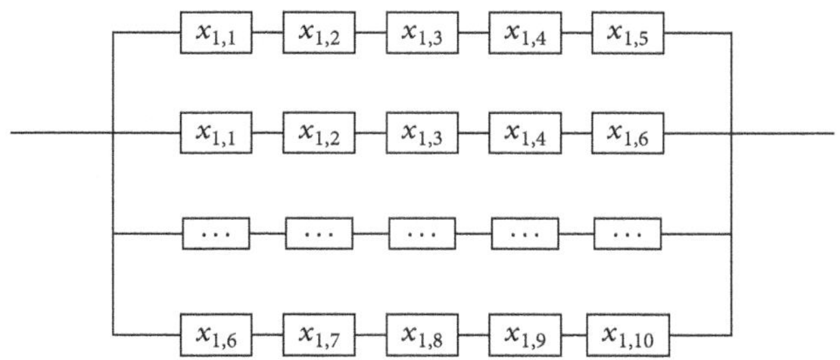

FIGURE 4: Block diagram of software component of network module $M_{N,1}$.

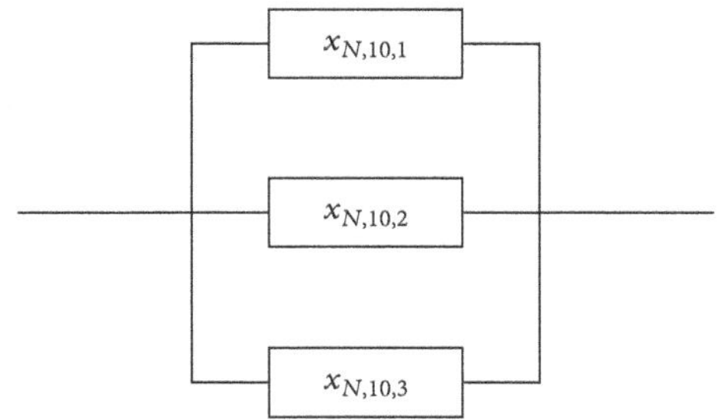

FIGURE 5: Block diagram of the network unit $x_{1,10}$.

will produce acceptable level raw observations. The selection of the algorithms is independent of the order.

The structure function $\Phi(\mathbf{x})$ of Figure 4 is given by the expression

$$\begin{aligned}\Phi(\mathbf{x}) = \max(&\min(x_{1,1}, x_{1,2}, x_{1,3}, x_{1,4}, x_{1,5}),\\ &\min(x_{1,1}, x_{1,2}, x_{1,3}, x_{1,4}, x_{1,6}), \ldots,\\ &\min(x_{1,6}, x_{1,7}, x_{1,8}, x_{1,9}, x_{1,10})).\end{aligned} \quad (15)$$

Assuming that the individual algorithms are independent with equal probability $p(t)$, then the reliability of software component is

$$p_s(t) = P(\mathbf{y}(t) \geq 5) = \sum_{x=5}^{10} \binom{10}{x} p^x (1-p)^{(10-x)}. \quad (16)$$

Due to the fact that $\mathbf{y}(t) \sim \text{binom}(n, p(t))$. The block diagram of the $x_{N,1,10}$ is given by Figure 5.

The corresponding reliability reads

$$\begin{aligned}p_{N,1,10} &= (p_{N,10,1} + p_{N,10,2} + p_{N,10,3})\\ &\quad - (p_{N,10,1}p_{N,10,2} + p_{N,10,1}p_{1,10,3} + p_{N,10,2}p_{N,10,3})\\ &\quad + p_{N,10,1}p_{N,10,2}p_{N,10,3}.\end{aligned} \quad (17)$$

Hardware Component of Module $M_{N,1}$. The qualities of the GNSS receiver, firmware robustness, GNSS antenna, and choke ring quantify the hardware component. Figure 6 shows the block diagram of the hardware component of the network module $M_{N,1}$.

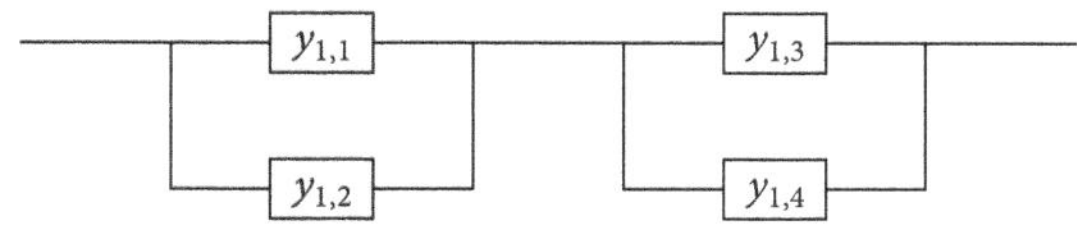

FIGURE 6: Block diagram of hardware component of module $M_{N,1}$. The left block describes the GNSS receiver and the right is for the GNSS antenna.

The elements of the hardware component of the network module $M_{N,1}$ are as follows:

(i) Let $y_{1,1}$ denote the GNSS receiver type.

(ii) Let $y_{1,2}$ denote the rover software known as firmware to decode the GNSS signals.

(iii) Let $y_{1,3}$ denote the GNSS antenna type.

(iv) Let $y_{1,4}$ denote the choke ring that allows better reception of low elevation angle GPS satellites and improved multipath rejection.

(v) Let $y_{1,5}$ denote duplicated system (as discussed in the next section).

The structure function $\Phi(\mathbf{y})$ is given by the expression

$$\begin{aligned}\Phi(\mathbf{y}) &= \min(\max(y_{1,1}, y_{1,2}), \max(y_{1,3}, y_{1,4}))\\ &= (y_{1,1} \coprod y_{1,2})(y_{1,3} \coprod y_{1,4})\\ &= (y_{1,1} + y_{1,2} - y_{1,1}y_{1,2})(y_{1,3} + y_{1,4} - y_{1,3}y_{1,4}).\end{aligned} \quad (18)$$

The reliability of the hardware of the module $M_{N,1}$ reads

$$\begin{aligned}&p_s(t)\\ &\quad = (p_{1,1} + p_{1,2} - p_{1,1}p_{1,2})(p_{1,3} + p_{1,4} - p_{1,3}p_{1,4}).\end{aligned} \quad (19)$$

Hardware Component Improvement. In order to ensure continuous raw data delivery at the station, duplicated hardware components are recommended. This task is accomplished by a parallel coupling of the module $M_{N,1}$, defined by Figure 6. Applying the definition of reliability computation of parallel coupling (Equation (C.2)), the duplicated HWC reliability reads

$$p(t) = p_s(t) + p_s(t) - [p_s(t)]^2 = p_s(t)[2 - p_s(t)], \quad (20)$$

where $p_s(t)$ is given by the Equation (19). The drawback of a duplicated system is the financial issues.

3.2.2. Network Ambiguity Resolution. As mentioned in Section 3.1. The key for precise positioning is the correct ambiguity resolution and validation. The module $M_{N,2}$ is composed of two main parts, the first part is the ambiguity processing algorithms performance, and the second part is the statistical ambiguity quality indicators derived from the processing chain of the ambiguity.

The components of the module $M_{N,2}$ are as follows:

(i) Let $x_{2,1}$ denote the float solution of the ambiguity obtained via least square or Kalman filter.

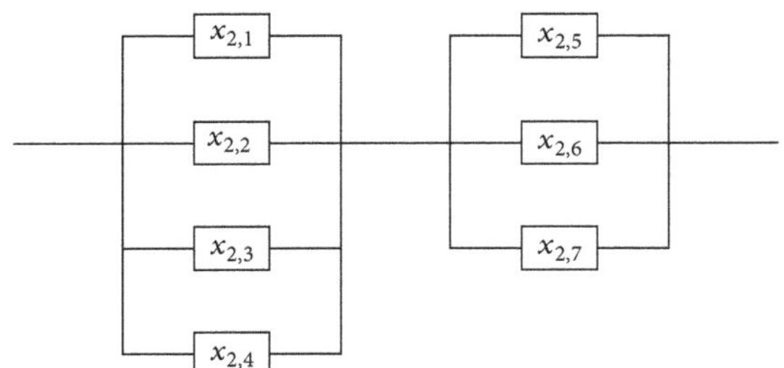

FIGURE 7: Network ambiguity resolution module $M_{N,2}$. The first block of this figure represents the ambiguity processing algorithms and the second block is the ambiguity quality indicators.

(ii) Let $x_{2,2}$ denote the LAMBDA method [29] applied to a float solution to reduce the search space and to obtain a fix solution.

(iii) Let $x_{2,3}$ denote the validation procedures to validate the final solution.

(iv) Let $x_{2,4}$ denote the administration of the ambiguities.

(v) Let $x_{2,5}$ denote the success rate of the ambiguities resolution.

(vi) Let $x_{2,6}$ denote the ambiguity dilution of precision (ADOP) [30]. ADOP measures the precision of the ambiguities and can be viewed as a quality indicator.

(vii) Let $x_{2,7}$ denote time to fix.

The block diagram of the $M_{N,2}$ is given by Figure 7.

Structure Function of $M_{N,2}$. The structure function of the module $M_{N,2}$ reads

$$\begin{aligned}\Phi(\mathbf{x}) &= \min(\max(\Phi_1(x)), \max(\Phi_2(x))) \\ &= \min(\max(x_{2,1}, x_{2,2}, x_{2,3}, x_{2,4}), \\ &\max(x_{2,5}, x_{2,6}, x_{2,7})) = 1 - \left[\prod_{i=1}^{4}(1 - x_{2,i})\right]\left[1\right. \\ &\left. - \prod_{i=5}^{7}(1 - x_{2,i})\right].\end{aligned} \tag{21}$$

Reliability of $M_{N,2}$. Reliability of the module $M_{N,2}$ reads

$$\begin{aligned}p_s(t) &= \left\{1 - \prod_{i=1}^{4}(1 - p_{2,i})\right\}\left\{1 - \prod_{i=5}^{7}(1 - p_{2,i})\right\} \\ &= \min(P_1, P_2) = P_1 P_2\end{aligned} \tag{22}$$

FIGURE 8: Block diagram of network correction quality module $M_{N,3}$.

dropping the index 2 in the expression of $p_s(t)$ and compute P_1 and P_2

$$\begin{aligned}P_1(t) &= \sum_{i=1}^{4} p_i - \prod_{i=1}^{4} p_i \\ &- \{p_1p_2 + p_1p_3 + p_1p_4 + p_2p_3 + p_2p_4 + p_3p_4\} \\ &+ \{p_1p_2p_3 + p_1p_2p_4 + p_1p_3p_4 + p_2p_3p_4\}, \\ P_2(t) &= \sum_{i=5}^{7} p_i + \prod_{i=5}^{7} p_i - \{p_5p_6 + p_5p_7 + p_6p_7\}.\end{aligned} \tag{23}$$

3.2.3. Network Corrections Quality. The quality of the network corrections depends on various parameters, for instance, the estimation algorithms, network status (sparse/dense), the covariance functions, and the smoothing/interpolation algorithms. The components of the module $M_{N,3}$ are as follows:

(i) Let $x_{3,1}$ denote the network reference receivers' separation. Dense network is attractive because the network corrections are better estimated with short distances between reference receivers.

(ii) Let $x_{3,2}$ denote the quality of the estimation algorithm used to estimate the parameter vector $\theta \in \Theta$.

(iii) Let $x_{3,3}$ denote the quality of the covariance function used to model the network correlation errors.

(iv) Let $x_{3,4}$ denote the quality of the interpolation algorithm used to generate the user corrections. Parallel interpolation algorithms will enhance the quality of the user corrections generation and avoid the information loss.

Block Diagram of $M_{N,3}$. The block diagram of the module $M_{N,3}$ is given by Figure 8.

Structure Function of $M_{N,3}$. The structure function of the module $M_{N,3}$ reads

$$\Phi(\mathbf{x}) = \min(x_{3,1}, x_{3,2}, x_{3,3}, x_{3,4}) = \prod_{i=1}^{4} x_{3,i}. \tag{24}$$

Reliability of $M_{N,3}$. Dropping the index 3, the reliability of $M_{N,3}$ reads

$$p_s(t) = \prod_{i=1}^{4} p_{3,i}. \tag{25}$$

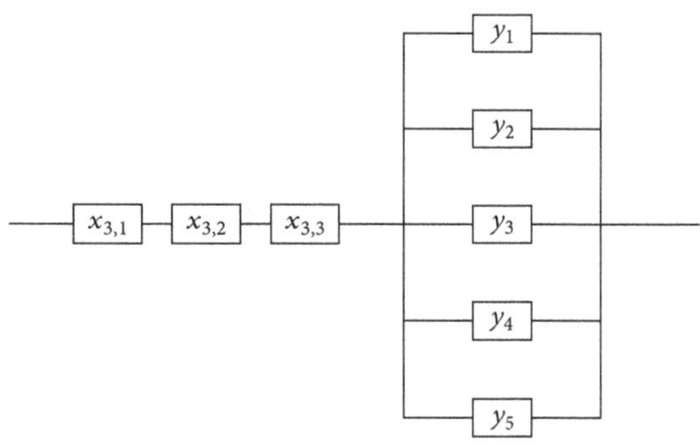

FIGURE 9: Improved block diagram of module $M_{N,3}$. The component $x_{3,4}$ is replaced by a parallel structure function $\phi(y)$.

Amelioration Potential of $M_{N,3}$. Our aim is to provide the user in the field with high quality corrections on an epoch-by-epoch basis. The interpolation/smoothing algorithm plays a central role. We can implement different parallel interpolation/smoothing algorithms that compete about the quality of service parameters. The corrections will be sent from the algorithm with higher score. For more information on this topic, the interested reader is referred to [3].

In this case, replacing the component $x_{3,4}$ with a parallel structure function $\phi(x_{3,4}) = \max(y_1, y_2, \ldots, y_5)$, the computation of the new block diagram is straightforward.

Let q_i denote the functioning probability of unit y_i (Figure 9) for $i = 1, 2, \ldots, 5$. Then the reliability of $x_{3,4}$ reads

$$\begin{aligned} p_{s_{(3,4)}}(t) = &\sum_{i=1}^{5} q_i + \prod_{i=1}^{5} q_i - \{q_1q_2 + q_1p_3 + q_1p_4 + q_1p_5 \\ &+ q_2q_3 + q_2q_4 + q_2q_5 + q_3q_4 + q_3q_5 + q_4q_5\} \\ &+ \{q_1q_2q_3 + q_1q_2q_4 + q_1q_2q_5 + q_2q_3q_4 + q_2q_3q_5 \\ &+ q_3q_4q_5\} - \{q_1q_2q_3q_4 + q_1q_2q_3q_5 + q_2q_3q_4q_5\}. \end{aligned} \quad (26)$$

The improved reliability of module $M_{N,3}$ reads

$$p_s = p_{s_{(3,4)}} \prod_{i=1}^{3} p_{3,i}. \quad (27)$$

3.2.4. Network Data Integrity. The module $M_{N,4}$ responsibility is to carry out the quality control on the corrections field and the corresponding variance-covariance matrices. This includes the following:

(i) Let $x_{4,1}$ denote global test statistics to detect any extremal events that can bias the rover position, corrections field investigation.

(ii) Let $x_{4,2}$ denote the inspection of variance-covariance matrices for Heywood effects algorithm [31].

(iii) Let $x_{4,3}$ denote the application of the imputation algorithm to compute the statistics.

(iv) Let $x_{4,4}$ denote the total variance monitoring algorithm.

(v) Let $x_{4,5}$ denote the generalized variance monitoring algorithm.

Serial coupling is the appropriate choice for the module $M_{N,4}$ and the block diagram is given by Figure 10.

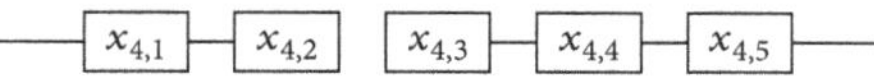

FIGURE 10: Block diagram of module $M_{N,4}$.

Block Diagram of $M_{N,4}$. See Figure 10.

Structure Function of $M_{N,4}$. The structure function of the module $M_{N,4}$ reads

$$\Phi(\mathbf{x}) = \min(x_{4,1}, x_{4,2}, x_{4,3}, x_{4,4}, x_{4,5}) = \prod_{i=1}^{5} x_{4,i}. \quad (28)$$

Reliability of $M_{N,4}$. Reliability of the module $M_{N,4}$ reads

$$p_s(t) = \prod_{i=1}^{5} p_{4,i}. \quad (29)$$

3.2.5. Network Probability Score. The module $M_{N,5}$ computes the network quality indicators in terms of the successfully ambiguities resolution and the quality of the network corrections. This is the first state vector of the system under investigation.

(i) Let $x_{5,1}$ denote quality indicator for the corrections field.

(ii) Let $x_{5,2}$ denote quality indicator for the uncertainty of corrections field.

(iii) Let $x_{5,3}$ denote quality indicator for the ambiguities expressed by ADOP from Section 3.2.2.

(iv) Let $x_{5,4}$ denote the number of common satellites used in the computation.

(v) Let $x_{5,5}$ denote the number of rejected satellites from computation.

Note that $x_{5,1}$ and $x_{5,2}$ are actually the network RTK quality indicators (RIM, IP, RIU, GBI, and ionospheric status indicator), defined in Section 1.

Serial coupling is the appropriate choice for the module $M_{N,5}$ and the block diagram is represented by Figure 11.

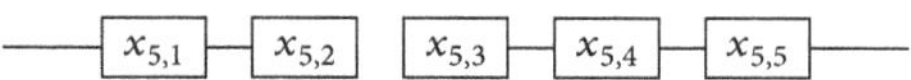

FIGURE 11: Block diagram of module $M_{N,5}$.

Block Diagram of $M_{N,5}$. See Figure 11.

Structure Function of $M_{N,5}$. The structure function of $M_{N,5}$ reads

$$\Phi(\mathbf{x}) = \min(x_{5,1}, x_{5,2}, x_{5,3}, x_{5,4}, x_{5,5}) = \prod_{i=1}^{5} x_{5,i}. \quad (30)$$

Reliability of Module $M_{N,5}$. The reliability of $M_{1,5}$ reads

$$p_s(t) = \prod_{i=1}^{5} p_{5,i}. \quad (31)$$

3.3. NRTK Baseline Reliability Analysis. This is the second level of the NRTK data processing. The corrections generated by the network are involved to generate the computation point (CP), then the unknown rover coordinates are determined relative to the computation point. This method is known as the relative positioning technique.

Similarity between the baseline and the network data processing exists as we see in the coming subsections.

3.3.1. Rover Receiver Data Integrity. The module $M_{B,1}$ is similar to the module $M_{N,1}$ defined in Section 3.2.1. The quality of the raw observations collected by the rover receiver depends strongly on the statistics methods used to check for anomalies. All variables defined in the module $M_{N,1}$ are applicable to the module $M_{B,1}$.

3.3.2. Baseline Ambiguity Resolution. The module $M_{B,2}$ is similar to the module $M_{N,2}$ defined in Section 3.2.2. All variables defined for the module $M_{N,2}$ are applicable to the module $M_{B,2}$.

3.3.3. Baseline Data Integrity. The module $M_{B,3}$ is similar to the module $M_{N,4}$ defined in Section 3.2.4. All variables defined for the module $M_{N,4}$ are applicable to the module $M_{B,3}$.

3.3.4. Baseline Probability Score. The module $M_{B,4}$ is similar to the module $M_{N,5}$ defined in Section 3.2.5. All variables defined for the module $M_{N,5}$ are applicable to the module $M_{B,4}$.

3.4. NRTK Rover Reliability Analysis. The investigation of the rover position error is the final check. The quality is measured in terms of standard deviations of the topocentric coordinates $(\delta e, \delta n, \delta u)$.

3.4.1. Rover Prediction Function. This module $M_{R,1}$ uses information from the double-difference variance-covariance matrix to compute the prediction of the position error. The number of satellites used in the computation is considered as a parameter.

(i) Let $x_{3,1}$ denote inspection of the main diagonal of variance-covariance matrix.

(ii) Let $x_{3,2}$ denote the total variance monitoring algorithm.

(iii) Let $x_{3,3}$ denote the generalized variance monitoring algorithm.

3.4.2. Rover Position Error. This module $M_{R,2}$ computes the standard deviations of the rover position error $(\delta e, \delta n, \delta u)$ and assigns a final score to determine the state of the rover receiver accuracy.

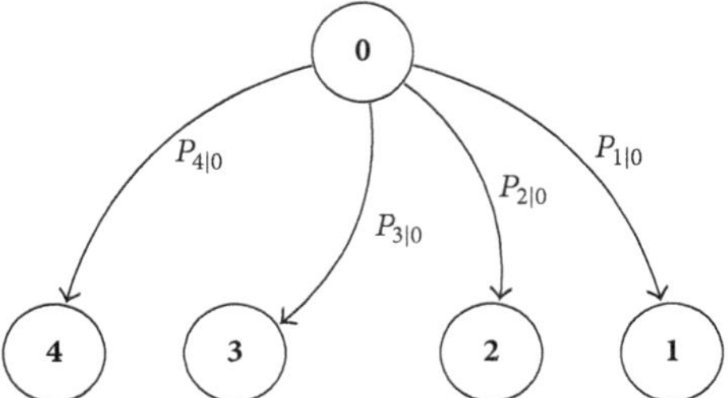

FIGURE 12: Start state of the rover position accuracy.

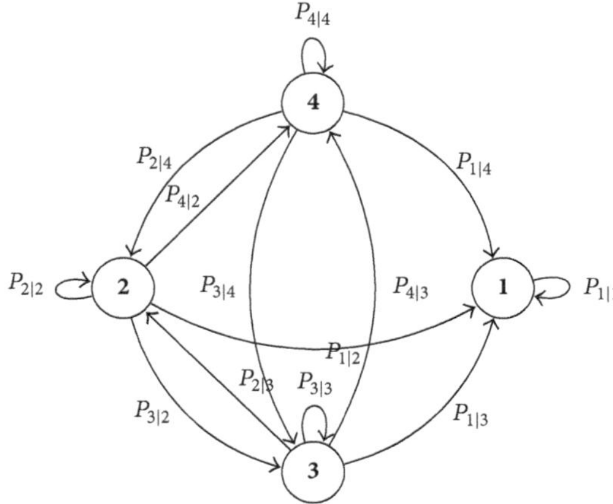

FIGURE 13: State diagram of the rover position accuracy.

4. Multistate Reliability Analysis

Since the rover position accuracy cannot be represented by a binary system with two performance level states as functioning or failed, the multistate system (MSS) approach is chosen to deal with situations where more than two levels of performance are considered. The material used to construct this section is from Natvig [32].

4.1. Definition of NRTK Performance Levels. Based on the values computed from the score modules $M_{N,5}$ (Section 3.2.5) and $M_{B,4}$ (Section 3.3.4), respectively, a single judging number is assigned to determine the performance level of the rover position accuracy.

The states represent level of performance ranging from the perfect functioning level *perfect* down to the complete failure level *catastrophic*. Five states are defined for a NRTK system, namely, *start*, *perfect*, *acceptable*, *rejected*, and *catastrophic* states.

The rover position accuracy is well described by the state diagrams given by Figures 12 and 13, respectively.

(1) State **4**: perfect functioning level. No complications. The user requirements regarding the position accuracy are satisfied.

(2) State **3**: acceptable position accuracy. Minor complications.

(3) State **2**: rejectable position accuracy. Major complications due to the atmosphere, multipath, or algorithms failure. The user requirements are not satisfied.

(4) State **1**: catastrophic state. The NRTK system is down and not delivering the corrections to the user in the field.

(5) State **0**: start state. The process always starts at this state and can reach any other states.

Note that the states **0, 4, 3, 2** are transient states while the state **1** is absorbing state.

Note that the probabilities $P_{i|j}$ must be computed from real data.

4.2. Penalized Honored Stochastic Averaged Variance. Based on state vectors data, our aim is to construct the NRTK state diagram from the network and baseline. The variance-covariance matrix (VCM) is considered as the state vector and the average variance with respect to the number of observed satellites (n_{sat}) is used to compute the quality indicator on an epoch-by-epoch basis. In addition, the number of rejected satellites (n_{rej}) and the geometry factor (DOP) are used to penalize/honor the average variance.

4.2.1. Penalized Average Variance Component. The total average variance computed from the VCM shall be penalized in case of rejection of satellite(s) with bad data and causes the increase of DOP indicator. The penalized function shall look similar to the following:

(1) Penalized least square (PLS) proposed by Green and Silverman [33, p. 5]

$$S(g) = \sum_{i=1}^{n} \{y_i - g(t_i)\}^2 + \alpha \int_a^b \{g''(x)\}^2 \, dx, \tag{32}$$

where y_i is the observations, $g(t_i)$ is the curve we fit to the data, $g''(\cdot)$ is the second derivative of the function $g(\cdot)$, and α is the smoothing parameter and defines the rate of change between the residuals and local variations. Anyway, minimizing $S(g)$ gives the best compromise between smoothness and goodness-of-fit. A large value of α will make the penalty term more in action, while with a small value the first term will be the main contribution.

(2) Information criteria type penalizing the model complexity. Denote by M the model to be investigated and $\dim(M)$ is the length of its parameter vector $\boldsymbol{\theta}$.

Akaike's information criterion (AIC) [34, Chap. 2]

$$\text{AIC}(M) = 2\,\text{log-likelihood}_{\max}(M) - 2\dim(M). \tag{33}$$

The Bayesian information criterion (BIC) of Schwarz (1978) takes the form of a penalized log-likelihood function where the penalty is equal to the logarithm of the sample size times the number of estimated parameters in the model [34, Chap. 3]

$$\begin{aligned}\text{BIC}(M) = {} & 2\,\text{log-likelihood}_{\max}(M) \\ & - (\log n)\dim(M).\end{aligned} \tag{34}$$

4.2.2. Honored Average Variance Component. Detection and rejection of satellite(s) with bad data are a good thing. The check algorithms shall be honored as long as the DOP values remain in the acceptable region.

The value of the horizontal dilution of precision (HDOP) is expected to be less or equal 2.0, that is, HDOP ≤ 2.0.

Forming a new stochastic variable $T_j = \text{HDOP}_j - \mu_{\text{HDOP}}$, then we can monitor the values of T_j over time. Note that μ_{HDOP} corresponds to the mean value of the HDOP in time span Δt.

4.2.3. Balanced Average Variance. Our aim is to put together the pieces defined in Sections 4.2.1 and 4.2.2, respectively, and find a way to balance between the satellite(s) rejection n_{rej} and the HDOP value.

The exponential reweighting algorithm type is an option. The algorithm places more importance to more recent data by discounting older data in an exponential manner. For epoch j, let $k_j = ((n_{\text{obs},j} - n_{\text{rej},j}) - n_{\text{const}})/n_{\text{sys}}$, where $n_{\text{const}} \geq 8$ is the user defined parameter and corresponds to the minimum number of satellites required to compute a reliable solution and preserve a good HDOP value, and $n_{\text{sys}} = 31$ is the total satellites in GPS constellation.

A suitable stabilization factor $\eta \in [0, 1]$ is chosen such that the penalized honored average variance (σ_{phav}) reads

$$\begin{aligned}\sigma_{\text{phasd}}(\eta, k, T) = {} & \underbrace{\left\{\frac{1}{n_{\text{sat}}}\sum_{i=1}^{n_{\text{sat}}} c_{i,i}\right\}^{1/2}}_{\text{first-term}} \\ & + \underbrace{\frac{1}{n_j}\sum_{j=1}^{n_j}\{\eta k_j + (1-\eta) T_j\}}_{\text{second-term}},\end{aligned} \tag{35}$$

where $c_{i,i}$ are diagonal elements of the covariance matrix of the baseline, $n_{\text{obs},j}$ is the number of satellites with valid data used in the computation, $n_{\text{rej},j}$ is the number of rejected satellites by the algorithms, and n_j is the window size which is user defined.

4.2.4. Penalized Honored Average Variance Validation. The parameter vector of (35) is $\boldsymbol{\theta} = (\eta, k, T)$. Our aim is to study the variation of the second term of (35) and try to get some valid answers.

Full details of the penalized honored average standard deviation algorithm are given in Appendix B.

4.3. NRTK Residuals Contribution. NRTK residuals generated by the network and baseline data processing are considered as state vectors and will be to construct the state diagram. The procedure is defined as follows:

(i) choose the time window $\Delta t = 10$ seconds;

(ii) compute the standard deviation of the residuals σ_{res}. Figure 18 shows the concept;

(iii) choose a suitable strategy to map the computed values of σ_{res}.

5. NRTK Reliability Results

The aim of this section is to present the results from the analysis. The level of performance ranging from the perfect functioning level *perfect* down to the complete failure level *catastrophic* shall be determined from the data.

5.1. Horizontal Dilution of Precision (HDOP). The geometry of the visible satellites is considered as an important factor in achieving high quality results especially for point positioning and kinematic surveying. Anyway, the geometry changes with time due to the relative motion of the user and satellites. A measure of the instantaneous geometry is the dilution of precision (DOP) factor.

The DOP values are computed from the variance-covariance matrix in the ECEF coordinate system and converted to the topocentric local coordinate system with its axes along the local north, east, and up (i.e., vertical) by rotational matrix R by applying the law of covariance propagation.

The DOP value can be defined in various ways; PDOP value in the local system is identical to the value in the global system. In addition to the PDOP, two further DOP definitions are used; HDOP, the dilution of precision in the horizontal position, and VDOP, denoting the corresponding value for the vertical component. The interested reader is referred to (Hoffmann-Wellenhof et al., 2008) [1, pp. 262–270]

$$\begin{aligned} \text{GDOP} &= \sqrt{\sigma_e^2 + \sigma_n^2 + \sigma_u^2 + \sigma_t^2}, \\ \text{PDOP} &= \sqrt{\sigma_e^2 + \sigma_n^2 + \sigma_u^2}, \\ \text{HDOP} &= \sqrt{\sigma_e^2 + \sigma_n^2}, \\ \text{VDOP} &= \sqrt{\sigma_u^2}. \end{aligned} \tag{36}$$

Acceptable horizontal DOP value is HDOP $\leq$ 2.0. Figure 14 shows the computed HDOP for the analyzed data set. In addition, Figure 15 shows the viewed satellites. Clearly, the number varies between 8 and 10 satellites.

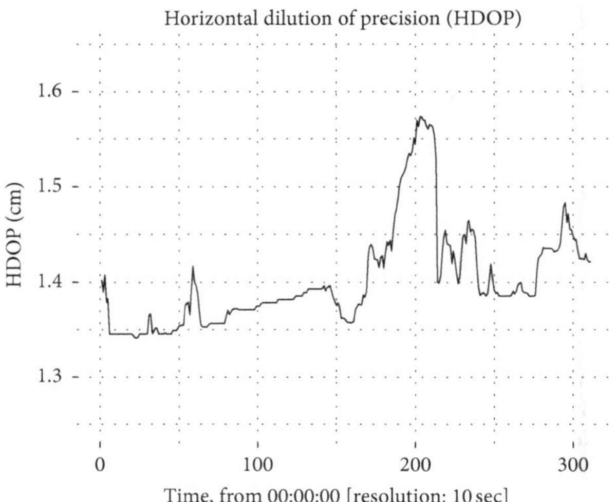

FIGURE 14: Computed HDOP values. Baseline of ~41 km, year: 2014, DOY: 85.

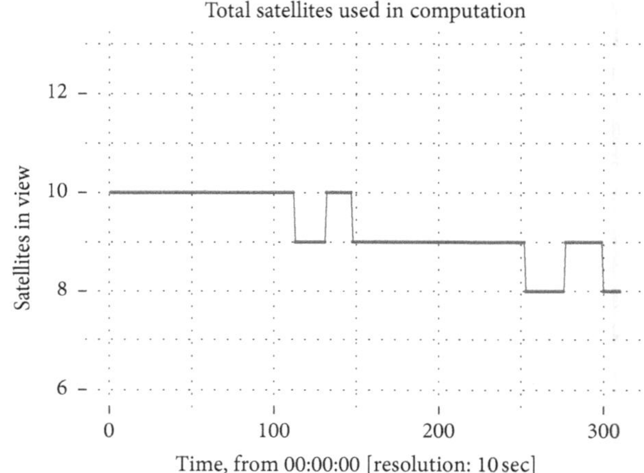

FIGURE 15: Number of satellites used in the computation. Baseline of ~41 km, year: 2014, DOY: 85. Plotted as red dots connected by blue lines.

5.2. Rover Level of Performance Prediction. We will predict the rover level of performance ranging from the perfect functioning level *perfect* down to the complete failure level *catastrophic*. Three classification lines are chosen in order to separate the computed average standard deviations using (35) into four decisions regions based on the values of $\sigma_{\text{pred}} \in \{0.2, .03, .4\}$. Figure 16 shows the concept.

The computation of σ_{pred} using (35) proceeds as follows:

(i) Based on the sliding window size Δt, form a data matrix from the baseline residuals. Compute the variance-covariance matrix VCM_{res} and get and sort in ascending order the diagonal elements D.

(ii) Compute the averaged standard deviation, the first term of (35)

$$\sigma_{\text{avg}} = \left\{ \frac{1}{n_{\text{sat}}} \sum_{i=1}^{n_{\text{sat}}} c_{i,i} \right\}^{1/2}. \tag{37}$$

(iii) Generate a random number $n_{\text{rej}} \in \{0, 1, 2\}$, and compute K_j

$$K_j = \frac{\left(n_{\text{obs},j} - n_{\text{rej},j} - n_{\text{const}}\right)}{n_{\text{tot}}}. \tag{38}$$

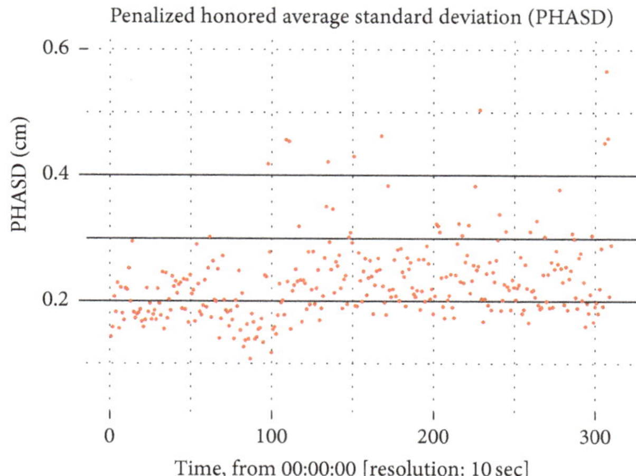

FIGURE 16: Computed penalized honored average standard deviation. Horizontal lines are used for classification. Baseline of ~41 km, year: 2014, DOY: 85.

(iv) Compute the second term of (35)

$$\sigma_{\text{pen}} = \frac{1}{n_j}\sum_{j=1}^{n_j}\left\{\eta k_j + (1-\eta)\,T_j\right\}, \tag{39}$$

where T_j is the HDOP computed from the solution variance-covariance matrix. The computed HDOP values are shown in Figure 14.

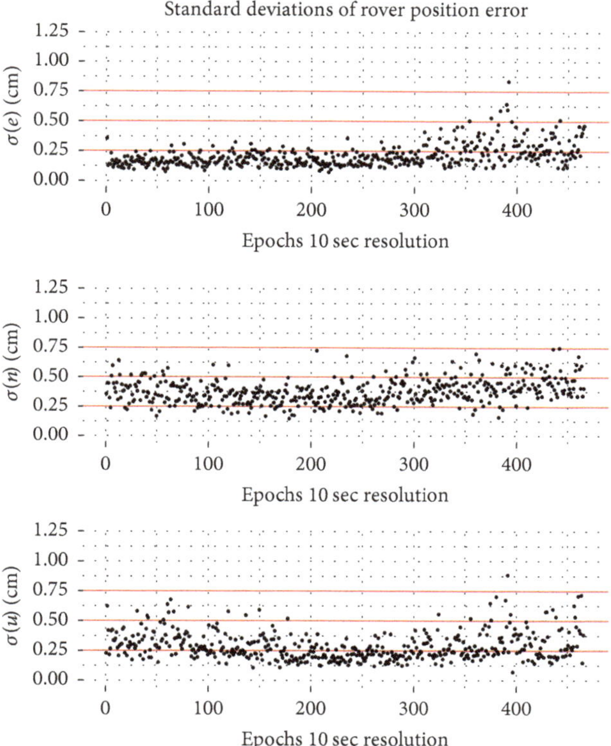

FIGURE 17: Standard deviations of the rover position errors $(\sigma_e, \sigma_n, \sigma_u)$ in the topocentric coordinate system. Horizontal lines are used for classification. Baseline of ~1 km, year: 2013, DOY: 152.

6. NRTK Reliability Validation

The aim of this section is to introduce the procedures used to validate the NRTK system reliability. The level of performance ranging from the perfect functioning level *perfect* down to the complete failure level *catastrophic* shall be determined from data.

6.1. NRTK State Diagram Definition. As we mentioned in the introduction, the key of the NRTK method is the measurement of the distance-dependent errors. The variations in the ionospheric and tropospheric fields are assumed constant in a period of time $\Delta t < 10$ s.

Since we have access to the rover position error in the topocentric coordinates system $(\delta_e, \delta_n, \delta_u)$, the state diagram is computed as follows:

(i) Choose the time window $\Delta t = 10$ seconds.

(ii) For each component, compute the standard deviation of the rover position error $\sigma_p = (\sigma_e, \sigma_n, \sigma_u)$. Figure 17 shows the concept.

(iii) Based on the user requirements and the computed values of σ_p, map this value to the performance defined levels, namely, the states $S = \{0, 1, 2, 3, 4\}$.

(iv) Compute transitions probabilities $P_{i|j}$ from data. This task is accomplished by counting the frequencies and computing the associated probability, that is, $p_s = N_s/N_T$, where N_s is the number of times we are visiting the state s and N_T is the total number of events.

6.2. States Transition Probabilities. We have computed the transition states probability based on Figure 18, where the threshold T_h values are defined, respectively, $T_h \in \{0.25, .5, .75\}$ cm. Note that $\sum_{j=1}^{4} P_{i,j} = 1$, for $i = 1, \ldots, 4$

Current State

$$\text{Next state}\quad \begin{matrix} & 4 & 3 & 2 & 1 \\ 4 & \\ 3 & \\ 2 & \\ 1 & \end{matrix}\begin{bmatrix} 0.596 & 0.378 & 0.024 & 0.001 \\ 0.234 & 0.668 & 0.098 & 0.001 \\ 0.077 & 0.69 & 0.231 & 0.001 \\ 0 & 0 & 0 & 1 \end{bmatrix} = \mathbf{P}. \tag{40}$$

The limiting distribution is obtained by matrix multiplication of the transition matrix:

$$\lim_{n\to\infty} \mathbf{P}_{ij}^{n} = \pi_j, \quad j \geq 0, \tag{41}$$

$$\mathbf{P}^{(50)} = \begin{pmatrix} 0.3312 & 0.5443 & 0.0798 \\ 0.3321 & 0.5459 & 0.0800 \\ 0.3315 & 0.5449 & 0.0799 \end{pmatrix}. \tag{42}$$

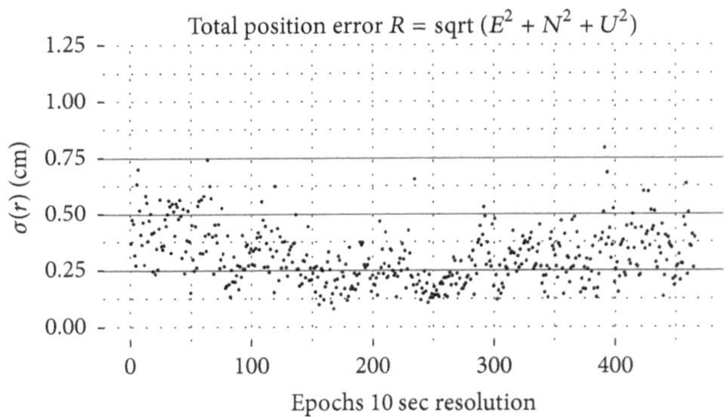

FIGURE 18: Standard deviation of the rover position error. Horizontal lines are threshold values. Baseline of ~1 km, year: 2013, DOY: 152.

We see that each row of $\mathbf{P}^{(50)}$ has almost identical entries; this confirms that (41) converges to some values as $n \longrightarrow \infty$. It seems that the existence of a limiting probability that the process p will be in state j after a large number of transitions n, and the value is independent of the initial state.

For this data set, the probability of being in state 4 is 33.12%, state 3 is 54.59%, and state 2 is 8.0%.

6.3. State Diagram. Information from the network (Section 3.2) and the baseline (Section 3.3) is combined in such way that a single judging number is mapped to the performance levels with states $S = \{0, 1, 2, 3, 4\}$ and the transitions probabilities $P_{i|j}$ are obtained. This task is accomplished by counting the frequencies and computing the associated probability, that is, $p_s = N_s/N_T$, where N_s is the number of times we are visiting the state s and N_T is the total number of events.

In order to define a single judging number, various schemes are considered, for instance, an averaging scheme or to assign different weights to each component.

7. Discussions and Conclusion

An improvement of the rover position accuracy can be achieved by applying procedures for multistate reliability analysis at the system and user level in NRTK. More concretely, the network corrections, baseline residuals, and the associated variance-covariance matrices are considered as the system states and have a direct influence on the rover position accuracy.

The use of the multistate reliability analysis will help us to get some concrete answers to the following problems

(i) can we trust the corrections provided by NRTK to the user?

(ii) at which level?

(iii) what are the amelioration potentials?

The weaknesses and the strengths of the system have to be identified and the amelioration potential can be achieved by modifying the serial critical components coupling into paralleled one with a cost effectiveness.

The methods tested make it possible to identify the NRTK critical component with bad data so this can be eliminated or downweighted in the positioning process leading to an improvement in the rover position from epoch to epoch.

It is expected that the suggested approach will reduce the number of wrong or inaccurate rover positions encountered by NRTK users in field, which subsequently will lead to a more efficient work flow for NRTK users.

8. Discussions

The rover position accuracy is well described by the state diagram. Based on the values computed by the score modules from the NRTK and the baseline, a single judging number is assigned that determines the performance rate of the rover position accuracy $\rho = (e^2 + n^2 + u^2)^{1/2}$, measured in terms of standard deviations σ_ρ.

The computation of probability of the rover position accuracy in time span $\Delta t = 10$ seconds is carried out as follows:

(1) Network RTK corrections score: based on algorithms efficiency defined in Section 3.2, a probability $p_1 \in [0, 1]$ is assigned to the quality of the corrections.

(2) Baseline score: based on algorithms efficiency defined in Section 3.3, a probability $p_3 \in [0, 1]$ is assigned to the quality of the baseline residuals.

(3) Rover raw observations score: based on algorithms efficiency used to edit the rover raw observations, a probability $p_2 \in [0, 1]$ is assigned to the quality of the rover raw observations.

The validation process is carried out by computing the standard deviations of the rover position error $(\sigma_n, \sigma_e, \sigma_u)$ of topocentric coordinates. A single number, σ_ρ, is assigned and the corresponding state is obtained. Equation (43) shows the mapping used in this investigation

$$\begin{aligned} F : [0, 1] \times [0, 1] \times [0, 1] &\longmapsto [0, 1] \\ F(p_1, p_2, p_3) &\longmapsto p \in [0, 1] \end{aligned} \quad (43)$$

On future work, a monitor station will be used to revalidate our approach for quality control and to carry out classification with

(i) empirical mapping function between the observation and the position domains;

(ii) classification boundaries determination in the observation and position domains.

Appendix

A. Test Data

Data used in this investigation is from the Norwegian RTK network known as CPOS operated by the Norwegian Mapping Authority (NMA). The test area is from the Rogaland region in the south west of Norway. Reference receivers are

Table 1: Subnetwork reference receivers' characteristics.

Site	4-chars ID	Receiver type	Antenna type
Tonstad	TNSC	TRIMBLE NETR9	TRM55971.00
Sirevag	SIRC	TRIMBLE NETR9	TPSCR3_GGD
Stavanger	STAS	TRIMBLE NETR9	TRM55971.00
Akrahamn	AKRC	TRIMBLE NETR9	TPSCR3_GGD
Lysefjorden	LYSC	TRIMBLE NETR9	TRM55971.00
Prestaasen	PREC	TRIMBLE NETR9	TPSCR3_GGD

Table 2: Distances in subnetwork [Km].

Sites	TNSC	SIRC	STAS	AKRC	LYSC	PREC
TNSC	X	56.32	75.00	109.60	44.61	95.23
SIRC	-	X	58.38	91.26	68.50	112.96
STAS	-	-	X	35.83	45.72	64.41
AKRC	-	-	-	X	73.51	65.60
LYSC	-	-	-	-	X	51.45
PREC	-	-	-	-	-	X

Table 3: Reference receiver coordinates, Euref89 *XYZ*.

Sites	*X*	*Y*	*Z*
TNSC	3302221.359	388315.600	5424777.872
SIRC	3323397.670	336993.537	5415277.838
STAS	3275753.912	321110.865	5445041.883
AKRC	3254758.852	295601.453	5458918.670
LYSC	3269684.205	366420.447	5446037.395
PREC	3227088.927	353649.666	5471909.728

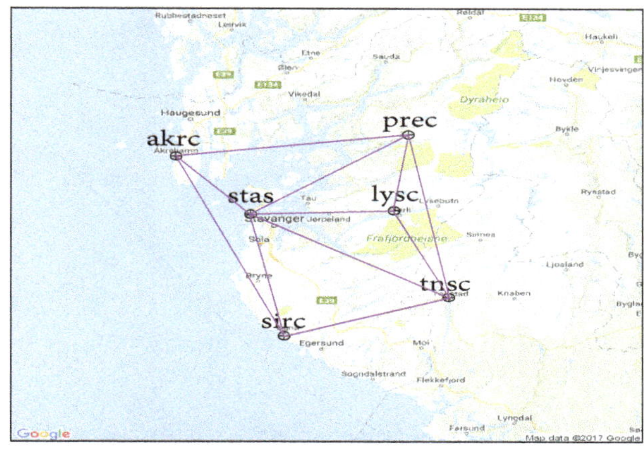

Figure 19: Test area used in this investigation, from Rogaland region. Composed of 6 reference receivers.

equipped with *Trimble NetR9* receivers, tracking GPS, and GLONASS satellite signals. Baselines vary between 35 and 112 km and the height difference between the sites is about 225 m. Tables 1, 2, and 3 give a full description of subnetwork while Figure 19 shows the location of reference receivers. The data used for testing was collected on day of year (doy) 152 in 2013 and doy 85 in 2014, respectively.

B. Penalized Honored Average Standard Deviation

$$\sigma_{\text{phasd}}(\eta, k, T) = \underbrace{\left\{\frac{1}{n_{\text{sat}}}\sum_{i=1}^{n_{\text{sat}}} c_{i,i}\right\}^{1/2}}_{\text{first-term}} + \underbrace{\frac{1}{n_j}\sum_{j=1}^{n_j}\left\{\eta k_j + (1-\eta)T_j\right\}}_{\text{second-term}}. \tag{B.1}$$

Algorithm Recipes. The recipes of the algorithm given by (B.1) read as follows:

(1) Compute the effective local coverage (ELC) for epoch j which is defined by the expression: $K_j = (n_{\text{obs},j} - n_{\text{rej},j} - n_{\text{const}})/n_{\text{sys}}$ where

(i) $n_{\text{obs},j}$ is total number of satellites used in the estimation process.

(ii) $n_{\text{rej},j}$ is total number of satellites rejected by the algorithms.

(iii) n_{const} is user defined parameter. Default is set to 8.

(iv) n_{sys} is total number of satellites in the GNSS constellation. For the GPS, the value is set to 31.

(2) Position domain quality indicator T_j which is defined as the HDOP

$$T_j = \left\{\sigma_n^2 + \sigma_e^2\right\}_j^{1/2} - \left\{\frac{1}{n_j}\sum_{i=1}^{n_j}\left(\sigma_{n,i}^2 + \sigma_{e,i}^2\right)\right\}^{1/2} = \text{HDOP}_j - \mu_{\text{HDOP}}. \tag{B.2}$$

(3) Stabilization factor η combines and balances between T_j and K_j. A reasonable combination is to binomial/exponential trial:

$$\frac{1}{n_j}\sum_{j_1}^{n_j}\left(\eta k_j + (1-\eta)T_j\right), \tag{B.3}$$

where n_j is the window size and is user defined. Default is set to 10.

(4) Operation level: the second term in (B.1) shall operate on the same level as the first term and shall have the same unit. This is accomplished by adjustment of parameters η, K_j, and T_j.

(5) Stabilization factor η can be implemented by the Danish method.

C. Reliability Computation Technique

In this section we present how the computation of the reliability of parallel structure is carried out step by step.

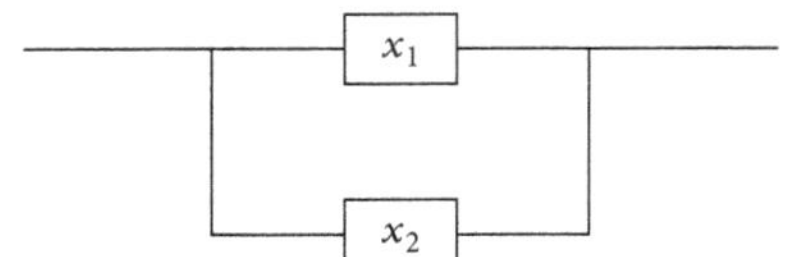

FIGURE 20: Block diagram of two parallel components.

C.1. Block Diagram. See Figure 20.

C.2. Structure Function. The structure function $\Phi(\mathbf{x})$ of the Figure 20 reads

$$\Phi(\mathbf{x}) = \max(x_1, x_2) = x_1 + x_2 - x_1 x_2. \quad \text{(C.1)}$$

The corresponding reliability reads

$$R(\Phi(x)) = p_1 + p_2 - p_1 p_2. \quad \text{(C.2)}$$

In case of more than two components, the computation of the reliability function $R(\Phi(x))$ is straightforward. The first step is to divide the whole system components into two main components and applying the formula given by (C.2). The next step is to substitute each individual reliability function with the corresponding terms. The last step is simple calculations.

Conflicts of Interest

The authors declare that there are no conflicts of interest regarding the publication of this paper.

Acknowledgments

The international GNSS Service (IGS) is acknowledged for providing geodetic infrastructure and geodetic products used in this work. The authors would also like to thank Jon G. Gjevestad for proofreading of this paper.

References

[1] B. Hofmann-Wellenhof, H. Lichtenegger, and E. Wasle, *GNSS—Global Navigation Satellite Systems: GPS, GLONASS, Galileo, and More*, Springer, Vienna, Austria, 2008.

[2] A. Leick, *GPS Satellite Surveying*, John Wiley & Sons, 2015.

[3] M. Ouassou, A. B. O. Jensen, J. G. O. Gjevestad, and O. Kristiansen, "Next generation network real-time kinematic interpolation segment to improve the user accuracy," *International Journal of Navigation and Observation*, vol. 2015, Article ID 346498, 15 pages, 2015.

[4] H. J. Euler, C. R. Keenan, B. E. Zebhauser, and G. Wübbena, "Study of a simplified approach in utilizing information from permanent reference station arrays," in *Proceedings of the National Technical Meeting of the Satellite Division of the Institute of Navigation (ION GPS '01)*, vol. 104, pp. 371–391, 2001.

[5] F. Takac and O. Zelzer, "The relationship between network RTK solutions MAC, VRS, PRS, FKP and i-MAX," in *Proceedings of the 21st International Technical Meeting of the Satellite Division of the Institute of Navigation (ION GNSS '08)*, pp. 348–355, September 2008.

[6] H. Landau, U. Vollath, and X. Chen, "Virtual reference station systems," *Journal of Global Positioning Systems*, vol. 1, no. 2, pp. 137–143, 2002.

[7] G. Wuebbena, A. Bagge, G. Seeber, V. Boeder, and P. Hankemeier, "Reducing distance dependent errors for real-time precise DGPS applications by establishing reference station networks," in *Proceedings of the 9th International Technical Meeting of the Satellite Division of the Institute of Navigation (ION GPS '96)*, vol. 9, pp. 1845–1852, September 1996.

[8] J. Raquet and G. Lachapelle, "Development and testing of a kinematic carrier-phase ambiguity resolution method using a reference receiver network," *Navigation*, vol. 46, no. 4, pp. 283–295, 1999.

[9] J. F. Raquet, "Development of a method for kinematic GPS carrier-phase ambiguity resolution using multiple reference receivers," UCGE Reports Number 20116, 1998.

[10] J. Xue and K. Yang, "Dynamic reliability analysis of coherent multistate systems," *IEEE Transactions on Reliability*, vol. 44, no. 4, pp. 683–688, 1995.

[11] Y. Gu and J. Li, "Multi-state system reliability: a new and systematic review," *Procedia Engineering*, vol. 29, pp. 531–536, 2012.

[12] F. Ohi, *Stochastic Evaluation of Multi-State Coherent Systems*, Omohi College, Nagoya Institute of Technology, Nagoya, Japan, 2012.

[13] M. Hossam-E-Haider, A. Tabassum, R. H. Shihab, and C. M. Hasan, "Comparative analysis of GNSS reliability: GPS, GALILEO and combined GPS-GALILEO," in *Proceedings of the International Conference on Electrical Information and Communication Technology (EICT '13)*, pp. 1–6, IEEE, February 2014.

[14] S. Satyanarayana, D. Borio, and G. Lachapelle, "C/N0 estimation: design criteria and reliability analysis under global navigation satellite system (GNSS) weak signal scenarios," *IET Radar, Sonar & Navigation*, vol. 6, no. 2, pp. 81–89, 2012.

[15] X. Li, M. Ge, X. Dai et al., "Accuracy and reliability of multi-GNSS real-time precise positioning: GPS, GLONASS, BeiDou, and Galileo," *Journal of Geodesy*, vol. 89, no. 6, pp. 607–635, 2015.

[16] A. Angrisano, C. Gioia, S. Gaglione, and G. Del Core, "GNSS reliability testing in signal-degraded scenario," *International Journal of Navigation and Observation*, vol. 2013, Article ID 870365, 12 pages, 2013.

[17] H. Kuusniemi, A. Wieser, G. Lachapelle, and J. Takala, "User-level reliability monitoring in urban personal satellite-navigation," *IEEE Transactions on Aerospace and Electronic Systems*, vol. 43, no. 4, pp. 1305–1318, 2007.

[18] W. Baarda, *A Testing Procedure for Use in Geodetic Networks*, vol. 2, no. 5, Netherlands Geodetic Commission, 1968.

[19] X. Chen, H. Landau, and U. Vollath, "New tools for network RTK integrity monitoring," in *Proceedings of the 16th International Technical Meeting of the Satellite Division of the Institute of Navigation (ION GPS/GNSS '03)*, pp. 1355–1360, Oregon Convention Center, Portland, Oreg, USA, 2003.

[20] L. Wanninger, "Ionospheric disturbance indices for RTK and network RTK positioning," in *Proceedings of the 17th International Technical Meeting of the Satellite Division of the Institute of Navigation (ION GNSS '04)*, pp. 2849–2854, Long Beach Convention Center, Long Beach, Calif, USA, September 2004.

[21] P. Alves, I. Geisler, N. Brown, J. Wirth, and H.-J. Euler, "Introduction of a geometry-based network RTK quality indicator," in *Proceedings of the 18th International Technical Meeting of the*

Satellite Division of the Institute of Navigation (ION GNSS '05), pp. 2552–2563, September 2005.

[22] D. Prochniewicz, R. Szpunar, and J. Walo, "A new study of describing the reliability of GNSS Network RTK positioning with the use of quality indicators," *Measurement Science and Technology*, vol. 28, no. 1, Article ID 015012, 2017.

[23] S. M. Ross, *Introduction to Probability Models*, Academic Press, 8th edition, 2003.

[24] B. Natvig, *Reliability Analysis with Technological Applications*, Department of Mathematics, University of Oslo, 3rd edition, 1998.

[25] A. Hoyland and M. Rausand, *System Reliability Theory: Models, Statistical Methods, and Applications*, vol. 396 of *Wiley Series in Probability and Statistics—Applied Probability and Statistics Section*, Wiley, 2004.

[26] E. Kaplan and C. Hegarty, *Understanding GPS: Principles and Applications*, Artech House Mobile Communications Series, Artech House, 2nd edition, 2005.

[27] G. Xu, *GPS: Theory, Algorithms and Applications*, Springer, Berlin, Germany, 2007.

[28] W. Gurtner and L. Estey, RINEX: The Receiver Independent Exchange Format Version 2.11, 2007.

[29] P. J. Teunissen, P. J. De Jonge, and C. C. J. M. Tiberius, "Performance of the LAMBDA method for fast GPS ambiguity resolution," *Navigation*, vol. 44, no. 3, pp. 373–383, 1997.

[30] D. Odijk and P. J. G. Teunissen, "ADOP in closed form for a hierarchy of multi-frequency single-baseline GNSS models," *Journal of Geodesy*, vol. 82, no. 8, pp. 473–492, 2008.

[31] H. B. Heywood, "On finite sequences of real numbers," *Proceedings of the Royal Society of London Series A*, vol. 134, no. 824, pp. 486–501, 1931.

[32] B. Natvig, *Multistate Systems Reliability Theory with Applications*, Wiley Series in Probability and Statistics, Wiley, 2011.

[33] P. J. Green and B. W. Silverman, *Nonparametric Regression and Generalized Linear Models: A Roughness Penalty Approach*, Chapman & Hall/CRC Monographs on Statistics & Applied Probability, Taylor & Francis, 1993.

[34] G. Claeskens and N. L. Hjort, *Model Selection and Model Averaging*, Cambridge Series in Statistical and Probabilistic Mathematics, Cambridge University Press, Cambridge, UK, 2008.

19

Compensatory Analysis and Optimization for MADM for Heterogeneous Wireless Network Selection

Jian Zhou[1,2] **and Can-yan Zhu**[1]

[1] *Institute of Intelligent Structure and System, Soochow University, Soochow 215006, China*
[2] *Department of Information Engineering, Suzhou Global Institute of Software Technology, Soochow 215163, China*

Correspondence should be addressed to Can-yan Zhu; qiwuzhu@suda.edu.cn

Academic Editor: Rajesh Khanna

In the next-generation heterogeneous wireless networks, a mobile terminal with a multi-interface may have network access from different service providers using various technologies. In spite of this heterogeneity, seamless intersystem mobility is a mandatory requirement. One of the major challenges for seamless mobility is the creation of a network selection scheme, which is for users that select an optimal network with best comprehensive performance between different types of networks. However, the optimal network may be not the most reasonable one due to compensation of MADM (Multiple Attribute Decision Making), and the network is called pseudo-optimal network. This paper conducts a performance evaluation of a number of widely used MADM-based methods for network selection that aim to keep the mobile users always best connected anywhere and anytime, where subjective weight and objective weight are all considered. The performance analysis shows that the selection scheme based on MEW (weighted multiplicative method) and combination weight can better avoid accessing pseudo-optimal network for balancing network load and reducing ping-pong effect in comparison with three other MADM solutions.

1. Introduction

With the emerging and development of all kinds of wireless access technology, including 2G, 3G, WLAN, WiMax (World Interoperability for Microwave Access), and MBWA (Mobile Broadband Wireless Access), wireless networks overlap and complement each other, forming a hybrid wireless network called heterogeneous wireless networks [1]. To support seamless mobility while a mobile station roams within a heterogeneous wireless network, VHO (Vertical Handoff) necessity estimation and decision to select a best target network are two important aspects of the overall mobility framework. The handoff necessity estimation is important in order to keep the unnecessary handoffs and their failures at a low level. On the other hand, to maximize the end-users' satisfaction level, the decision to select the best network among other available candidates plays an important role as well. The network selection process consists of three major subservices: (1) network monitoring monitors the current network conditions (network availability, signal strength, current call connection, etc.) and provides the data gathered together with information related to the user preferences, current running applications on the user's mobile device, and their QoS requirements; (2) handover decision handles the network selection process (which ranks the candidate networks and selects the best target) and is initiated either by an automatic trigger for handover for an existing call connection or by a request for a new connection on the mobile device; and (3) handover execution: once a new target network is selected, the connection is set up on the target candidate network (and the old connection torn down).

Network selection algorithm has become a more complex problem and combines multiple systems' attributes to choose the target network that offers the highest overall performance. This approach is considered optimal as compared to the other traditional approaches that rely on a single system's attributes like RSS (Received Signal Strength) or available bandwidth to make handoff decisions. As all of these parameters present different ranges and units of measurements, they need to be normalized in order to make them comparable.

Utility functions are used for normalization to map all the parameters into dimensionless units within the range [0, 1] [2–7]. This normalized information is then used in the decision making process in order to compute a ranked list of the best available network choices. MADM including SAW, MEW, GRA, and TOPSIS is widely used as score function methods for network selection [8–15]. User or network operator preferences for the main trade-off criteria can be represented by the use of different weights in weighted score functions. The candidate network with the highest score is selected as the target network if that differs from the current network connection (or it is for a new connection); it prompts handover execution (or new network connection setup).

However, the optimal network with best comprehensive performance may not be the most reasonable one due to compensation of MADM; we call the network pseudo-optimal network. For example, if network with best comprehensive performance and heavy load is selected, which may further aggravate network congestion, end-user cannot enjoy good network quality. Moreover, it is argued that an appropriate MADM should not make end-user access pseudo-optimal network in [10]. Hence, performance of accessing pseudo-optimal network is firstly analyzed for SAW (Simple Additive Weighting Method), TOPSIS (Technique for Order Preference by Similarity to Ideal Solution), GRA (Grey Relational Analysis), and MEW, then network selection based on MEW and combinational weight is proposed. It can be seen from simulation that the proposed algorithm can make end-user better avoid assessing pseudo-optimal network and has better performance in network load balance and reducing ping-ping effect.

2. Compensatory Analysis for MADM

2.1. Common MADM. MADM algorithms can be divided into compensatory and noncompensatory ones. Noncompensatory algorithms are used to find acceptable alternatives which satisfy the minimum cutoff. On the contrary, compensatory algorithms combine multiple attributes to find the best alternative. Most MADM algorithms that have been studied for the network selection problem are compensatory algorithms.

SAW is widely used by most studies of the network selection problem using cost or utility functions, generally given by

$$U_{\text{SAW}} = \sum_{j=1}^{n} w_j u_{ij}, \tag{1}$$

where w_j represents the weight of attribute c_j and u_{ij} represents the adjusted value of attribute c_j of the network r_i.

MEW is to calculate the coefficient by multiplicative operation, given by

$$U_{\text{MEW}} = \prod_{j=1}^{n} u_{ij}^{\,w_j}. \tag{2}$$

Other two MADM algorithms used for network selection are TOPSIS and GRA, which both consider the distance from the evaluated network to one or multiple reference networks. Coefficient of TOPSIS can be calculated as

$$U_{\text{TOPSIS}} = \frac{D^{\alpha}}{D^{\alpha} + D^{\beta}}, \tag{3}$$

where D^{α} and D^{β} represent the Euclidean distances from the current network to the worst and best reference networks, respectively, given by

$$D^{\alpha} = \sqrt{\sum_{j=1}^{n} w_j^2 \left(u_{ij} - v_j^{\alpha}\right)^2}, \tag{4}$$

$$D^{\beta} = \sqrt{\sum_{j=1}^{n} w_j^2 \left(u_{ij} - v_j^{\beta}\right)^2}, \tag{5}$$

where v_j^{α} and v_j^{β} represent the values of attribute c_j of the worst and best reference networks, respectively.

Different from TOPSIS, GRA uses only the best reference network to calculate the coefficient, given by

$$U_{\text{GRA}} = \frac{1}{\sum_{j=1}^{n} w_j \left|u_{ij} - v_j^{\beta}\right| + 1}. \tag{6}$$

TABLE 1: Network selection based on SAW.

Utility	w_j	Network A	Network B
$U(C)$	1/3	0.5	0.8
$U(E)$	1/3	0.5	0.8
$U(L)$	1/3	0.5	0.05
Aggregate utility		0.5	0.55

2.2. Compensatory Analysis. According to the principle of network selection, the optimal network is the network with best comprehensive performance, but it may be not the most reasonable one due to compensation of MADM. For example, the result is the fact that end-user chooses best network from networks A and B by SAW as shown in Table 1, where decision attributes are cost (C), power consumption (E), and load (L). Assume that only subjective weight is considered.

Table 1 shows that the comprehensive performance of network B is superior to network A, and end-user will choose network B as access network. However, load utility of network B in Table 1 is 0.05 which is close to 0; namely, network B is not suitable for new access due to its heavy load, and otherwise it may lead to congestion for network B and not balance load between networks A and B. So network A should be the reasonable choice for the above situation. But end-user chooses network B due to performance compensation between attributes, which means the excellent performance of cost and power consumption compensates the bad performance of load in network B, and we call the network like network B pseudo-optimal network.

The results that end-user chooses best network from networks A and B by TOPSIS and GRA are shown in Tables 2 and 3, respectively. It can be seen from Tables 2 and 3 that

TABLE 2: Network selection based on TOPSIS.

Utility	w_j	Network A	Network B
$U(C)$	1/3	0.5	0.8
$U(E)$	1/3	0.5	0.8
$U(L)$	1/3	0.5	0.05
Aggregate utility		0.5	0.5333

TABLE 3: Network selection based on GRA.

Utility	w_j	Network A	Network B
$U(C)$	1/3	0.5	0.8
$U(E)$	1/3	0.5	0.8
$U(L)$	1/3	0.5	0.05
Aggregate utility		0.6667	0.6897

TABLE 4: Network selection based on MEW ($w_l = 1/3$).

Utility	w_j	Network A	Network B
$U(C)$	1/3	0.5	0.8
$U(E)$	1/3	0.5	0.8
$U(L)$	1/3	0.5	0.05
Aggregate utility		0.5	0.3175

TABLE 5: Network selection based on MEW ($w_l = 0.05$).

Utility	w_j	Network A	Network B
$U(C)$	0.5	0.5	0.8
$U(E)$	0.45	0.5	0.8
$U(L)$	0.05	0.5	0.05
Aggregate utility		0.5	0.6964

TOPSIS and GRA also choose network B with heavy load like SAW, and they have the same limitation.

The result that end-user chooses best network from networks A and B by MEW is shown in Table 4. It can be seen from Table 4, unlike SAW, TOPSIS, and GRA, that end-user chooses network A with light load as access network. However, if we readjust attribute weight in Table 4 and reestimate the comprehensive performance of networks A and B by MEW, as shown in Table 5, it can be seen from Table 5 that the comprehensive performance of network B is superior to network A again; end-user will choose network B as access network. Obviously, MEW cannot ensure that end-user can always avoid accessing the pseudo-optimal network.

The reason that end-user chooses optimal network from network A to network B when attribute weight of load is from one-third to 0.05 is shown in Figure 1. In Figure 1, the expression of two curves is 0.5^w and 0.05^w, respectively, difference of utility value between two curves is larger when w is larger, and difference is close to 0 when w is smaller; hence, the comprehensive performance of network B will be reduced well due to multiplicative features of MEM and network A is selected as optimal network when difference of load utility between networks A and B is larger because weight of load is one-third. However, network B is selected as optimal network again when difference of load utility between networks A and B is smaller because weight of load is 0.05, and network A with light load is not selected. Through the above analysis, to make end-user avoid accessing the network with attribute with poor performance, weight of the attribute should be adjusted in real-time to prevent its value from being too small, and MEW should be used as decision making method to rank alternative network.

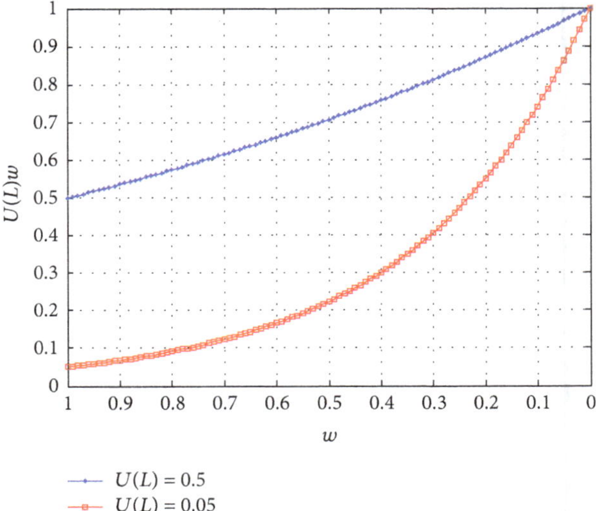

FIGURE 1: Performance comparison of load with different weight.

3. Network Selection Based on MEW and Combination Weight

As described in Section 2, attribute weight should be adjusted in real-time to make end-user avoid accessing the pseudo-optimal network, while objective weight method can calculate attribute weight according to attribute value of alternative network and recalculate attribute weight when attribute data changes. However, subjective weight must also be considered to reflect experience and subjective importance for attribute of decision makers. Hence, combination weight which integrates the subjective weight and objective weight will be considered. The steps for calculating combination weight are as follows:

Step 1. Objective weight is determined by entropy weighting method, calculated as follows:

(1) Construct normalized matrix $\mathbf{R}$, given by

$$\mathbf{R} = \begin{bmatrix} u_{11} & \cdots & u_{1j} & \cdots & u_{1n} \\ \vdots & & \vdots & \cdots & \vdots \\ u_{i1} & \cdots & u_{ij} & \cdots & u_{in} \\ \vdots & & \vdots & \cdots & \vdots \\ u_{m1} & \cdots & u_{mj} & \cdots & u_{mn} \end{bmatrix}. \quad (7)$$

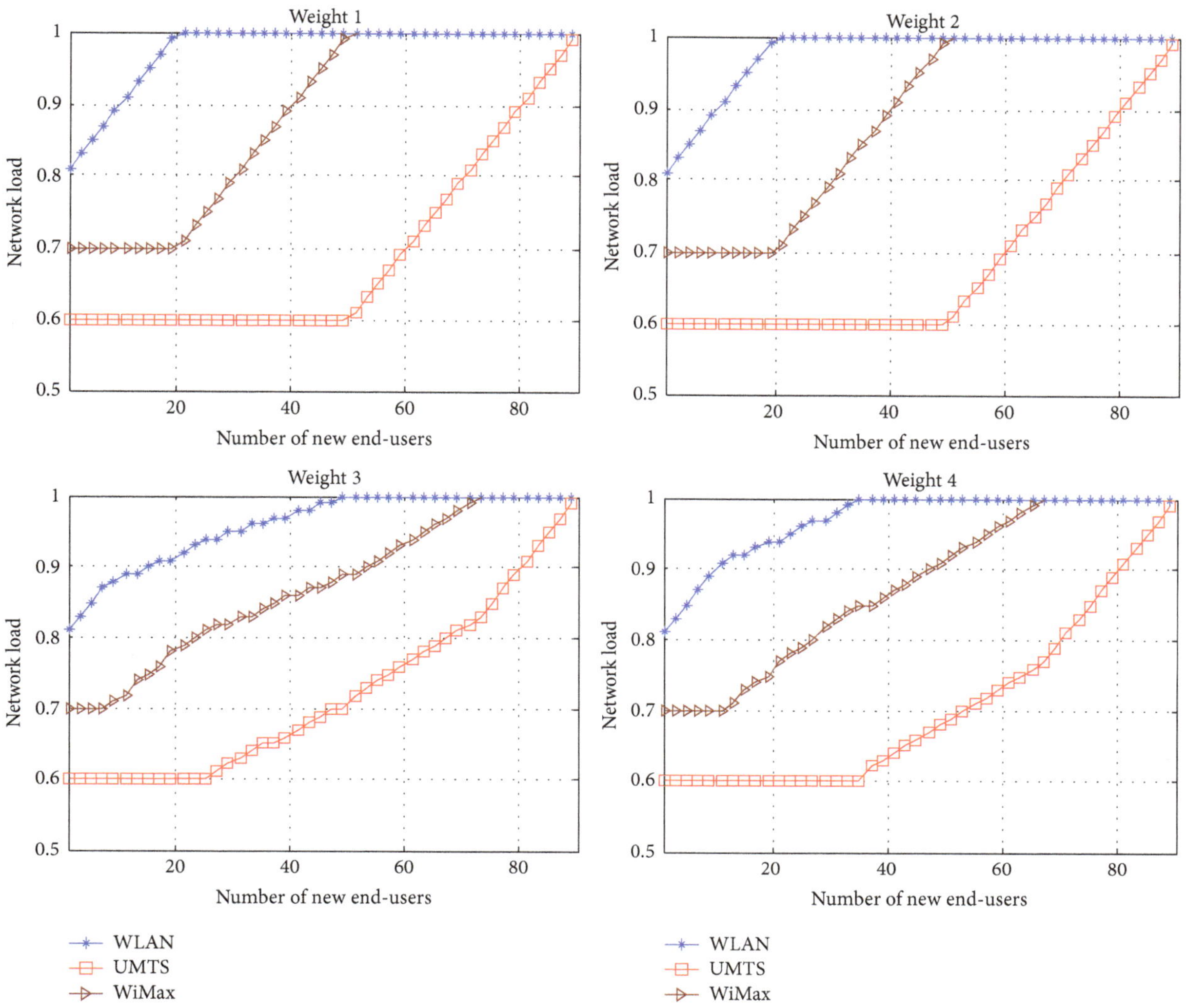

FIGURE 2: Change of network load by SAW.

(2) Calculate information entropy of attribute c_j, given by

$$E_j = \frac{-1}{\ln m} \sum_{i=1}^{m} u_{ij} \ln u_{ij}. \tag{8}$$

(3) Calculate objective weight of attribute c_j, given by

$$w_{oj} = \frac{1 - E_j}{\sum_{k=1}^{n} (1 - E_k)}. \tag{9}$$

Step 2. Subjective weight is determined by experience and assigned directly in this paper, denoted by w_{sj}.

Step 3. For considering subjective weight and objective weight, combination weight can be expressed as

$$w_{cj} = \alpha w_{oj} + \beta w_{sj}, \tag{10}$$

where α and β meet $\alpha + \beta = 1$ and $\alpha, \beta \geq 0$.

Considering that weighted attribute values determined by subjective weight and objective weight should be consistent, optimal mathematical model can be constructed to solve α and β. Deviation degree of subjective and objective evaluation value of alternative networks is given by

$$d_i = \sum_{j=1}^{n} \left(\alpha u_{ij} w_{oj} - \beta u_{ij} w_{sj} \right)^2 . \tag{11}$$

The smaller the value of d_i is, the more consistent the subjective and objective evaluations tend to be. Hence, optimal mathematical model can be constructed as follows:

$$\begin{aligned} \min \quad & \sum_{i=1}^{m} d_i = \sum_{i=1}^{m} \sum_{j=1}^{n} \left(\alpha u_{ij} w_{sj} - \beta u_{ij} w_{oj} \right)^2 \\ \text{s.t.} \quad & \alpha + \beta = 1 \\ & \alpha, \beta \geq 0 \end{aligned} \tag{12}$$

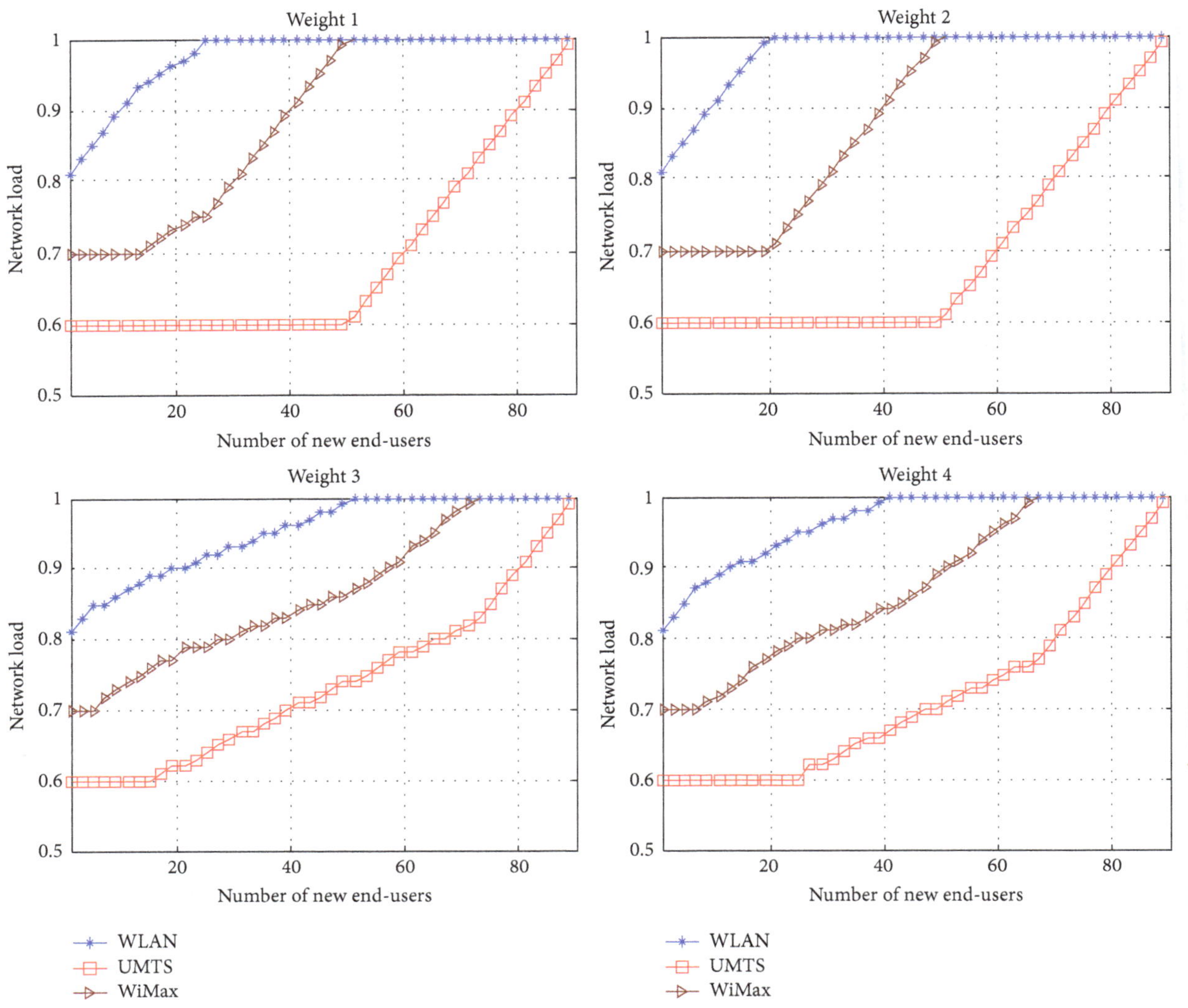

FIGURE 3: Change of network load by TOPSIS.

and its solution can be obtained as

$$\alpha = \frac{\sum_{i=1}^{m}\sum_{j=1}^{n} u_{ij}^2 w_{oj}\left(w_{sj}+w_{oj}\right)}{\sum_{i=1}^{m}\sum_{j=1}^{n} u_{ij}^2 \left(w_{sj}+w_{oj}\right)^2}$$

$$\beta = \frac{\sum_{i=1}^{m}\sum_{j=1}^{n} u_{ij}^2 w_{sj}\left(w_{sj}+w_{oj}\right)}{\sum_{i=1}^{m}\sum_{j=1}^{n} u_{ij}^2 \left(w_{sj}+w_{oj}\right)^2}. \tag{13}$$

Substitute (13) into (10); combination weight can be determined.

Step 4. Substitute w_{cj} into (2); rank for alternative networks can be obtained by MEW.

4. Simulation and Analysis

In this section, two groups of simulations are used to validate performance of the proposed algorithm; one simulation is for performance evaluation of network load balance and the other is for performance evaluation of reducing ping-pong effect.

TABLE 6: Measurement value of attribute and parameter setting for utility function.

Attribute	WLAN	UMTS	WiMax	x_α	x_β
C (cent/Mb)	10	50	30	0	100
E (w)	2	6	3	0	10
L (%)	80	60	70	0	100
D (kbps)	220	400	350	200	800

Notes: x_α and x_β are lower limit and upper limit of linear utility function.

In simulation environment, three networks WLAN, UMTS, and WiMax are selected as alternative networks, and cost, power consumption, load, and data rate (B) are decision attribute used for wireless network selection. Linear utility function is adopted as utility function for all attributes. Measurement value of attribute and parameter setting for utility function is shown in Table 6, and measurement

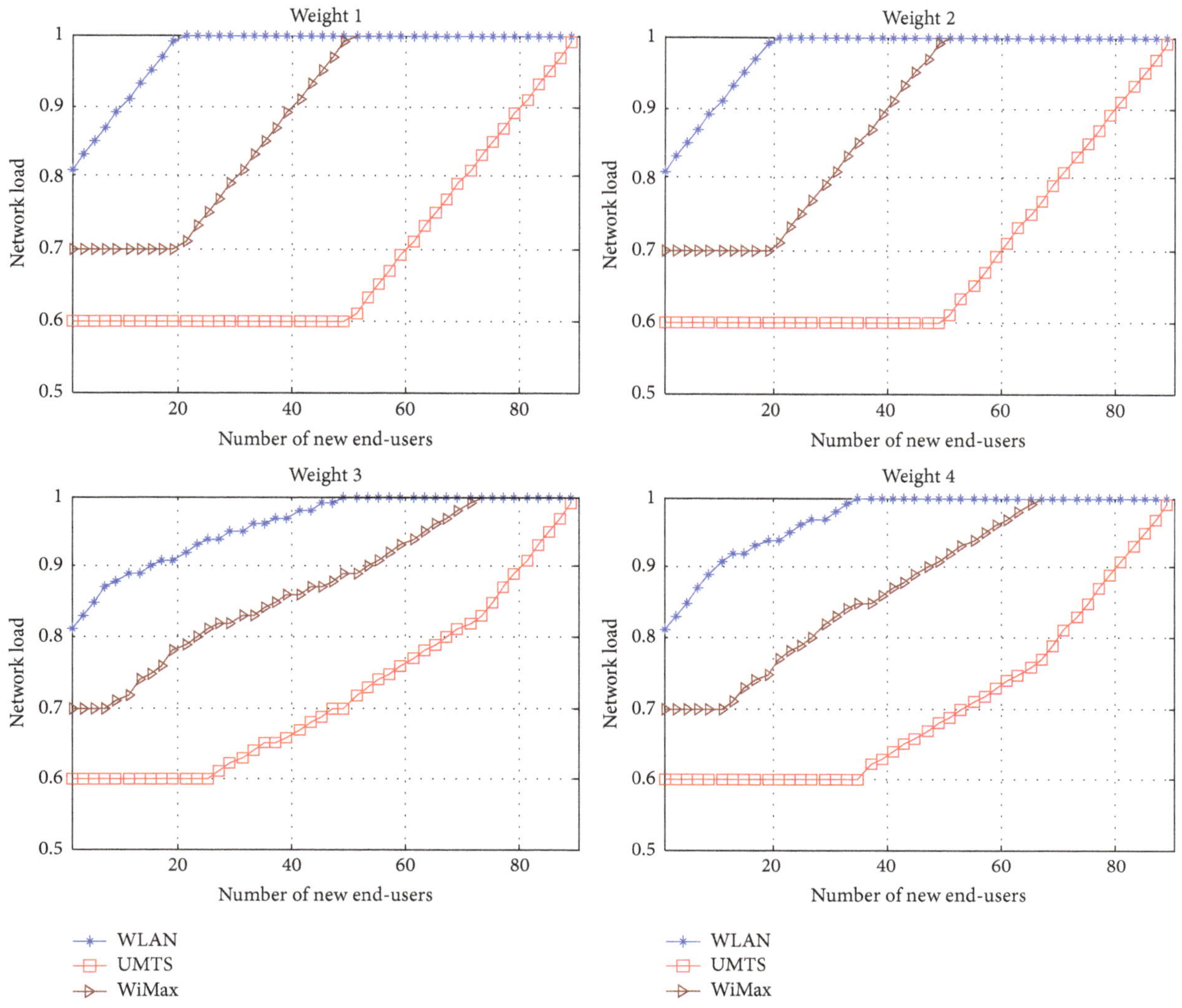

FIGURE 4: Change of network load by GRA.

TABLE 7: Utility used for load balance simulation.

Utility	WALN	UMTS	WiMax
$U(C)$	0.9	0.5	0.7
$U(E)$	0.8	0.4	0.7
$U(L)$	0.2	0.4	0.3

values of load and data rate will change after beginning of simulation.

4.1. Simulation for Performance of Network Load Balance. According to Table 6, price, power consumption, and load are selected as decision attributes; utility and weight of decision attributes are shown in Tables 7 and 8, respectively.

Assume that there are new requests coming constantly in overlapping area of WLAN, UMTS, and WiMax, and end-user accesses the optimal network based on network selection algorithm. Then, Figures 2–5 show change of network load of alternative network when end-user selects optimal network by SAW, TOPSIS, GRA, and MEW, respectively, where attribute weight is in case of weight 1, weight 2, weight 3, and weight 4, respectively. In Figures 2–5, there are a number of new end-users on horizontal coordinates and network load on vertical coordinates.

TABLE 8: Weight for network attributes.

Weight	(C, E, L)
Weight 1	Only subjective weight (1/3, 1/3, 1/3)
Weight 2	Only subjective weight (0.5, 0.45, 0.05)
Weight 3	Combination of subjective weight (1/3, 1/3, 1/3) and objective weight computed by (9)
Weight 4	Combination of subjective weight (0.5, 0.45, 0.05) and objective weight computed by (9)

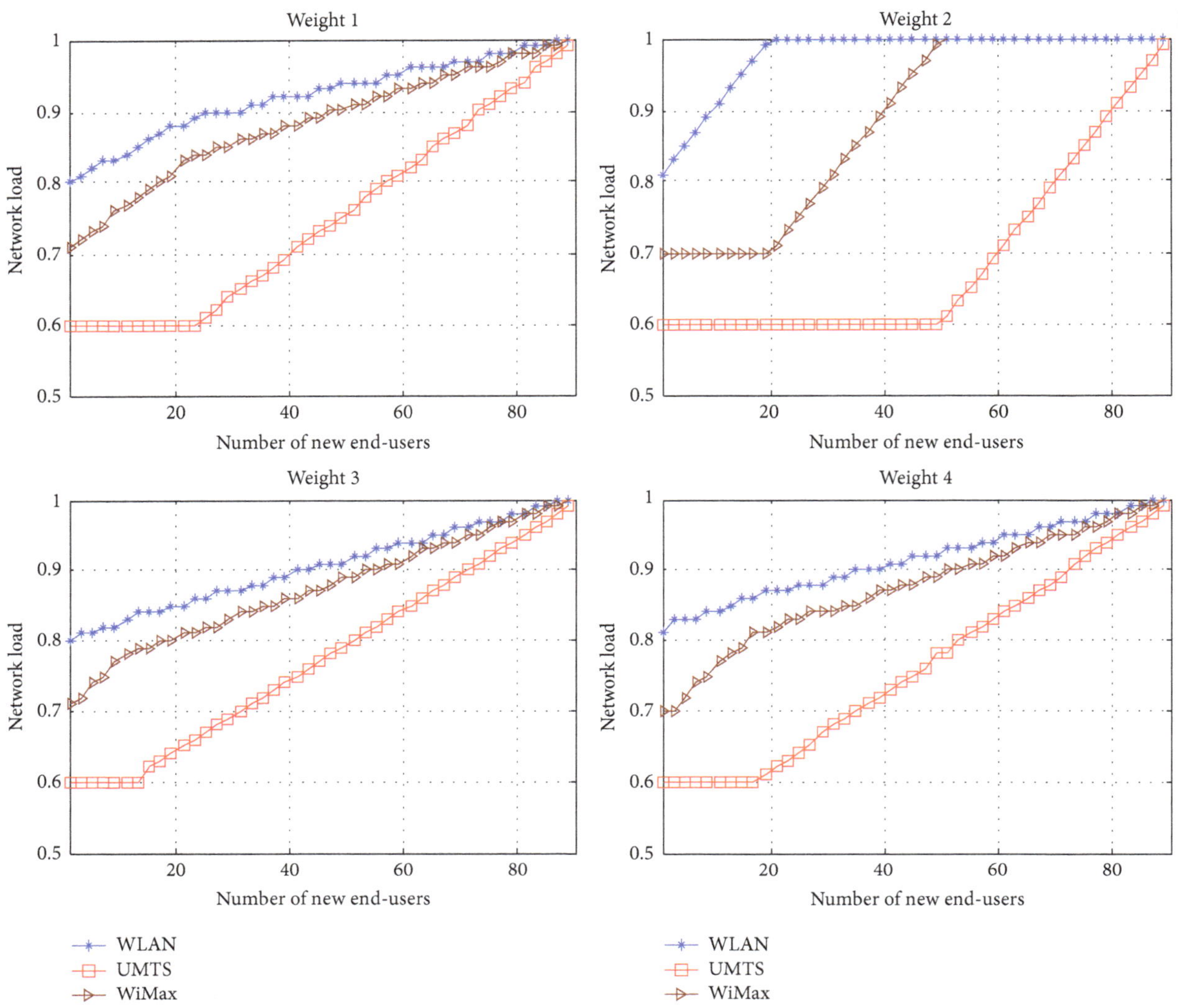

FIGURE 5: Change of network load by MEW.

Figure 2 shows change of network load when optimal network is selected by SAW. The order of optimal network selected by new end-users is much the same for four weight cases. It is failed to balance network load well, and it is easy to cause congestion for network with heavy load so it cannot provide good quality for end-user.

Changes of network load when optimal network is selected by TOPSIS and GRA, respectively, are shown in Figures 3 and 4, where the performance is like SAW.

Figure 5 shows change of network load when optimal network is selected by MEW. It can be seen from Figure 5 that network load can be balanced better by MEW when attribute weight is in case of weight 1, weight 3, and weight 4. Hence, no matter what value of subject weight is, end-user can avoid better accessing of the pseudo-optimal network and balance network load between three alternative networks by the scheme based on MEW and combination weight.

4.2. Simulation for Performance of Ping-Pong Effect. According to Table 6, price, power consumption, and data rate are selected as decision attributes; utility of decision attributes is shown in Table 9 and weight of decision attributes is as in Table 8.

TABLE 9: Utility used for ping-pong effect simulation.

Utility	WALN	UMTS	WiMax
$U(C)$	0.9	0.5	0.7
$U(E)$	0.8	0.4	0.7
$U(B)$	0.0625	0.125	0.25

Assume that data rate will vary by around 50 kbps after start of simulation. Figures 6–9 show performance of ping-pong effect of alternative network when end-user selects optimal network by SAW, TOPSIS, GRA, and MEW, respectively, where attribute weight is in case of weight 1, weight 2, weight 3, and weight 4, respectively. In Figures 6–9, there are a number of data rate fluctuations on horizontal coordinates and a number of network handoffs on vertical coordinates.

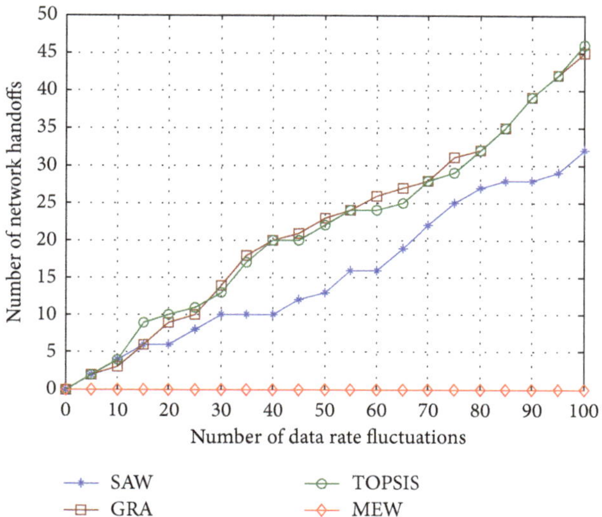

FIGURE 6: Network handoffs in case of weight 1.

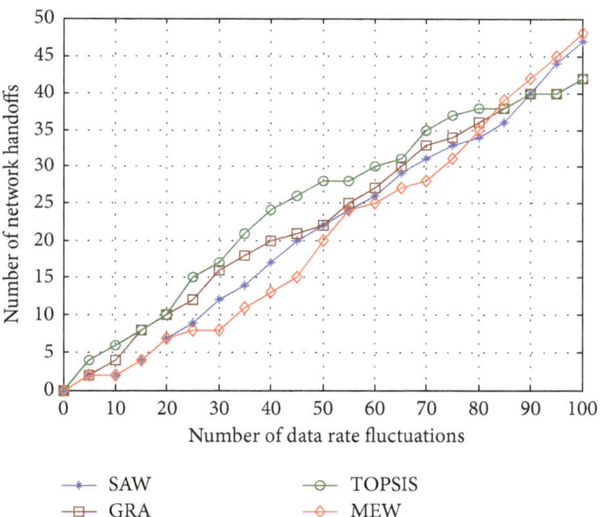

FIGURE 7: Network handoffs in case of weight 2.

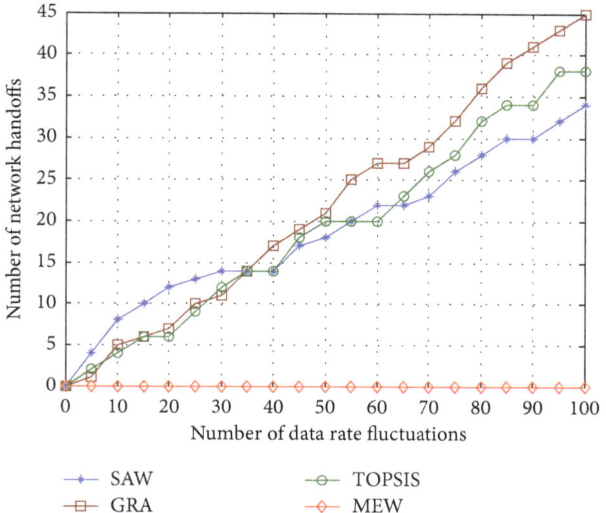

FIGURE 8: Network handoffs in case of weight 3.

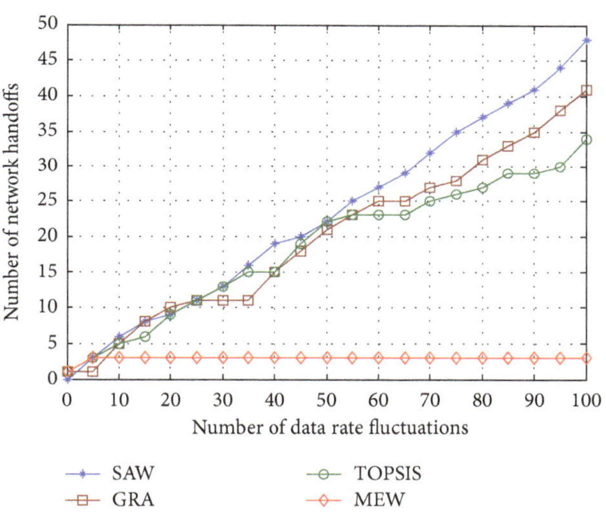

FIGURE 9: Network handoffs in case of weight 4.

It can be seen from Figures 6–9 that network handoff is more frequent when optimal network is selected by SAW, TOPSIS, and GRA, because data rate of WLAN may be lower than the minimum requirements of end-user for data rate fluctuation, and end-user must select other eligible networks. However, there are fewer network handoffs when optimal network selected by MEW is in case of weight 1, weight 3, and weight 4. Hence, no matter what value of subject weight is, end-user can avoid better accessing of the pseudo-optimal network and reduce network handoffs between alternative networks by the scheme based on MEW and combination weight, which is a better solution for ping-pong effect caused by data rate fluctuations.

4.3. Analysis and Comparison of Algorithm Complexity. In this section, network selection algorithm based on SAW, TOPSIS, GRA, and MEW is implemented on DSP device, respectively, and computational overhead of network selection process is computed. In Table 10, mean computational overhead of one network selection process is showed when three networks are alternative and three decision attributes are chosen.

TABLE 10: Comparison of computational overhead.

	SAW	TOPSIS	GRA	MEW
Weight 3	1.471 us	1.562 us	1.521 us	1.575 us
Weight 4	1.470 us	1.560 us	1.522 us	1.578 us

It can be seen from Table 10 that computational overhead of MEW is about 7%, 0.8%, and 3% more than SAW, TOPSIS, and GRA, respectively, but the performance of balancing network load and reducing ping-pong effect of MEW is vastly better than the others.

5. Conclusions

In this paper, network selection algorithm based on MEW and combination weight is proposed for the problem that pseudo-optimal network may be chosen because of compensation of attribute performance when optimal network is selected by SAW, TOPSIS, and GRA. Simulation shows that the proposed algorithm not only makes end-user avoid accessing pseudo-optimal network, but also balances network load to prevent network congestion and reduce ping-pong effect, which improves system performance.

Competing Interests

The authors declare that they have no competing interests.

Acknowledgments

This work is financially supported by NSFC no. 61071214.

References

[1] M. Corici, J. Fiedler, T. Magedanz, and D. Vingarzan, "Access network discovery and selection in the future wireless communication," *Mobile Networks and Applications*, vol. 16, no. 3, pp. 337–349, 2011.

[2] Q.-T. Nguyen-Vuong, G. D. Yacine, and N. Agoulmine, "On utility models for access network selection in wireless heterogeneous networks," in *Proceedings of the IEEE Network Operations and Management Symposium (NOMS '08)*, pp. 44–151, Salvador, Brazil, April 2008.

[3] I. Chamodrakas and D. Martakos, "A utility-based fuzzy TOPSIS method for energy efficient network selection in heterogeneous wireless networks," *Applied Soft Computing Journal*, vol. 12, no. 7, pp. 1929–1938, 2012.

[4] L. Wang and G.-S. G. S. Kuo, "Mathematical modeling for network selection in heterogeneous wireless networks—a tutorial," *IEEE Communications Surveys and Tutorials*, vol. 15, no. 1, pp. 271–292, 2013.

[5] R. Trestian, O. Ormond, and G.-M. Muntean, "Enhanced power-friendly access network selection strategy for multimedia delivery over heterogeneous wireless networks," *IEEE Transactions on Broadcasting*, vol. 60, no. 1, pp. 85–101, 2014.

[6] P. Kosmides, A. Rouskas, and M. Anagnostou, "Utility-based RAT selection optimization in heterogeneous wireless networks," *Pervasive and Mobile Computing*, vol. 12, no. 6, pp. 92–111, 2014.

[7] R. Chai, H. Zhang, X. Dong, Q. Chen, and T. Svensson, "Optimal joint utility based load balancing algorithm for heterogeneous wireless networks," *Wireless Networks*, vol. 20, no. 6, pp. 1557–1571, 2014.

[8] J. Liu and X.-N. Li, "Handover algorithm for WLAN/cellular networks with analytic hierarchy process," *Journal on Communications*, vol. 34, no. 2, pp. 65–72, 2013.

[9] S. Zhang and Q. Zhu, "Heterogeneous wireless network selection algorithm based on group decision," *Journal of China Universities of Posts and Telecommunications*, vol. 21, no. 3, pp. 1–9, 2014.

[10] Q.-T. Nguyen-Vuong, N. Agoulmine, E. H. Cherkaoui, and L. Toni, "Multicriteria optimization of access selection to improve the quality of experience in heterogeneous wireless access networks," *IEEE Transactions on Vehicular Technology*, vol. 62, no. 4, pp. 1785–1800, 2013.

[11] R. Trestian, O. Ormond, and G.-M. Muntean, "Performance evaluation of MADM-based methods for network selection in a multimedia wireless environment," *Wireless Networks*, vol. 21, no. 5, pp. 1745–1763, 2014.

[12] L. Xu and Y. Li, "A network selection scheme based on topsis in heterogeneous network environment," *Journal of Harbin Institute of Technology*, vol. 21, no. 1, pp. 43–48, 2014.

[13] M. Drissi and M. Oumsis, "Performance evaluation of multi-criteria vertical handover for heterogeneous wireless networks," in *Proceedings of the 1st International Conference on Intelligent Systems and Computer Vision (ISCV '15)*, pp. 1–5, IEEE, Fez, Morocco, March 2015.

[14] R. Verma and N. P. Singh, "GRA based network selection in heterogeneous wireless networks," *Wireless Personal Communications*, vol. 72, no. 2, pp. 1437–1452, 2013.

[15] N. P. Singh and B. Singh, "Vertical handoff decision in 4G wireless networks using multi attribute decision making approach," *Wireless Networks*, vol. 20, no. 5, pp. 1203–1211, 2014.

20

The 2D Spectral Intrinsic Decomposition Method Applied to Image Analysis

Samba Sidibe, Oumar Niang, Abdoulaye Thioune, Abdoul-Dalibou Abdou, and Ndeye Fatou Ngom

Laboratoire de Traitement de l'Information et Systèmes Intelligents, Ecole Polytechnique de Thiès, BP A10, Thiès, Senegal

Correspondence should be addressed to Oumar Niang; oniang@ucad.sn

Academic Editor: M. Jamal Deen

We propose a new method for autoadaptive image decomposition and recomposition based on the two-dimensional version of the Spectral Intrinsic Decomposition (SID). We introduce a faster diffusivity function for the computation of the mean envelope operator which provides the components of the SID algorithm for any signal. The 2D version of SID algorithm is implemented and applied to some very known images test. We extracted relevant components and obtained promising results in images analysis applications.

1. Introduction

The need of components extraction and reconstruction in signal and image processing in time frequency analysis is very strong for many fields of application. Notorious methods that have been proposed include Fourier technics, wavelet decomposition, and Empirical Mode Decomposition. While Fourier transform is localized in frequency, wavelets are localized in both time and frequency; EMD is autoadaptive. EMD decomposes a signal in AM-FM components called Intrinsic Mode Functions (IMF) and a residue. This nonlinear and nonstationary decomposition works on 1D signals [1] and 2D signals such as images [2, 3]. The EMD algorithm is based on a procedure called sifting process which iteratively uses the upper and lower envelopes to extract IMFs. To create a mathematical model to compute envelopes directly, an envelope operator has been proposed in [4] and from this operator a new decomposition method called Spectral Intrinsic Decomposition, SID, was proposed in [5].

The SID method allows decomposing any signal into a superposition of Spectral Proper Mode Functions (SPMFs) [5]. This method has been presented in a 1D version and depends on an operator interpolating the characteristics points of the signal to be decomposed. In this paper, the two-dimensional version of the Spectral Intrinsic Decomposition for images analysis is introduced. An algorithm for a faster spectral decomposition is proposed and illustrated with some images. We first recall the SID principle in one dimension and propose a faster method to determine the signal characteristics points in Section 2. Than an algorithm of the two-dimensional SID is presented in Section 3. Applications on grayscale images are depicted in Section 3.1.

2. The Spectral Intrinsic Decomposition Method

The Spectral Intrinsic Decomposition Method decomposes any signal into a combination of eigenvectors of a Partial Differential Equation (PDE) interpolation operator as presented in [5, 6].

2.1. The PDEs System Interpolator. For a given signal s_0, the upper (s^+) and lower (s^-) envelope are the asymptotic solution of the following PDEs system:

$$\frac{\partial s^{\pm}(x,t)}{\partial t} = -g^{\pm}(x,t)\left(\alpha \frac{\partial^2 s^{\pm}(x,t)}{\partial^2 x} + (1-\alpha)\frac{\partial^4 s^{\pm}(x,t)}{\partial^4 x}\right), \quad (1)$$

$$s^{\pm}(x,0) = s_0.$$

α is the tension parameter which ranges from 0 to 1. g^+ and g^- are diffusivity functions for upper and lower envelope which are equal to zero at characteristic points of s_0 and range from zero to one. $g^{\pm}$ based on Maximum Curvature Points (MCP) can be computed as follows:

$$g^{\pm}(x) = \frac{1}{9}\left[\left|\operatorname{sgn}\left(\frac{\partial^3 s_0(x)}{\partial^3 x}\right)\right| \pm \operatorname{sgn}\left(\frac{\partial^2 s_0(x)}{\partial^2 x}\right) + 1\right]^2, \quad (2)$$

where sgn denotes the sign function.

Equation (1) is resolved numerically in its discrete implicit unconditionally stable scheme as follows:

$$S^{k+1} = S^k + \Delta t A S^{k+1}, \quad S^0 = S_0, \quad (3)$$

where Δt is the time step, $S^{k+1} = s(x, k\Delta t)$ is signal value at step $k+1$, and A is a matrix formed with finite difference approximation coefficients of second- and fourth-order differential operators (resp., D_2 and D_4), as $A = G(\alpha D_2 - (1-\alpha)D_4)$, with G being the diagonal matrix constructed with discrete version of stopping function values $g(x)$, exactly, $G(i,i) = g(i)$.

So the explicit form leads to the following numerical resolution:

$$S^{k+1} = (I - \Delta t A)^{-1} S^k, \quad S^0 = s_0, \ k \geq 0 \quad (4)$$

with I being the identity matrix. Finally (1) can be decomposed into a linear system from implicit numerical scheme (4) by

$$S^{(k+1)} = L^{-1} S^k, \quad S^0 = s_0, \ k \geq 0. \quad (5)$$

L is given by $L = I - \Delta t A$.

The operator matrix, L, has real-valued eigenvalues that are always greater than or equal to 1. Then, eigenvalues, λ_n, of L^{-1} are always smaller than or equal to 1 ($0 < \lambda_n \leq 1$); see [4].

2.2. On the Asymptotic Solution. Iterative scheme (5) can be rewritten in terms of initial solution s_0 as

$$S^k = \left(L^{-1}\right)^k s_0, \quad k \geq 1. \quad (6)$$

After convergence (see [7]), the asymptotic solution, S^{∞}, is given by

$$S^{\infty} = \left(L^{-1}\right)^{\infty} s_0. \quad (7)$$

Let V be a matrix of L^{-1}'s sequence of eigenvectors (V_n) and $\mathscr{D}$ a diagonal matrix having L^{-1}'s sequence of eigenvalues (λ_n), at the diagonal. So we have the following decomposition $L^{-1} = V\mathscr{D}V^{-1}$. It is easy to see that

$$\left(L^{-1}\right)^k = \left(V\mathscr{D}V^{-1}\right)^k = VD^kV^{-1}. \quad (8)$$

So, the asymptotic solution in (7) is given by

$$S^{\infty} = \left(V\mathscr{D}^{\infty}V^{-1}\right) S_0. \quad (9)$$

The asymptotic eigenvalue matrix $\mathscr{D}^{\infty}$ is a diagonal matrix with eigenvalues $\lambda_n^{\infty} = 1$ only at loci where matrix G is zeroed, and $\lambda_n^{\infty} = 0$, where $g[n] > 0$. Finally, the asymptotic solution of the PDE interpolator system is a linear combination of fixed vector point of upper and lower envelope operators.

2.3. A Faster Stopping Function for Discrete Signal. For image processing we will consider region boundaries as characteristic points. The characteristic points of the upper envelope will be the local maximums and the limits of the regions where the value of the gray level of the pixel is equal to or greater than the gray level of all the pixels in their neighborhood represented, for example, by a rectangular window.

We define the diffusion function g_M^- for lower envelope to be equal to 1 everywhere except in characteristic points of the lower envelope where it will be equal to 0.

Similarly the characteristic points of the lower envelope are local minimums and region boundaries where the pixel value is equal to or less than the gray level of all pixels in their neighbors. We define the diffusion function g_M^+ for upper envelope to be equal to 1 everywhere except in characteristics points of the upper envelope where it will be equal to 0.

The diffusivity function called stopping function $g_M^{\pm}$ is calculated by using morphological dilation and erosion operations [8].

Let $b = [-1, 0, 1]$ be a structured element; the grayscale dilation of s by b at x is given by

$$[s \oplus b](x) = \max_{a \in b} \{s(x+a)\}. \quad (10)$$

The grayscale erosion of s by b at x is given by

$$[s \ominus b](x) = \min_{a \in b} \{s(x+a)\}. \quad (11)$$

$s \oplus b$ is equal to s at local maxima and when s is locally constant; $s \ominus b$ is equal to s at local minima and when s is locally constant.

g_M^+ is zeroed at points which are invariant to morphological dilation and variant to morphological erosion; g_M^+ is equal to 1 for any other point.

Similarly, g_M^- is zeroed at points which are invariant to morphological erosion and variant to morphological dilation; g_M^- is equal to 1 for any other point.

Let

$$A = \{x \mid [s \oplus b](x) = s(x)\},$$
$$B = \{x \mid [s \ominus b](x) = s(x)\}. \quad (12)$$

We have

$$g_M^+(x) = \begin{cases} 0, & \text{if } x \in A \setminus B, \\ 1, & \text{otherwise,} \end{cases}$$
$$g_M^-(x) = \begin{cases} 0, & \text{if } x \in B \setminus A, \\ 1, & \text{otherwise} \end{cases} \quad (13)$$

TABLE 1

Image	Width	Height	Time for $g^{\pm}$	Time for $g_M^{\pm}$
Figure 4(a)	300	300	0,4684466	0.320695
Figure 3(a)	400	266	0,826974833	0,2877035
Figure 2(a)	400	272	0,807968	0,284108

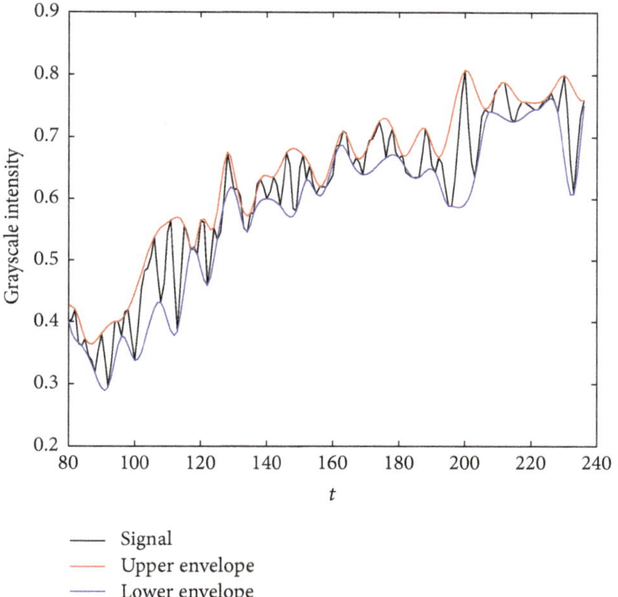

FIGURE 1: The effectiveness of envelope computation using g_M instead of g.

g_M functions are faster than g to compute and give satisfying results for the computation of envelope operators for real images (Table 1).

Figure 1 shows an example of calculating the envelope using g_M, g_M^+ for upper envelope and g_M^- for lower envelope.

2.4. The Spectral Intrinsic Decomposition. In the following, E denotes either the upper or lower envelope operator. The upper and lower envelope of the signal are calculated with the eigenvectors associated with eigenvalue $\lambda = 1$. Hence, S^{∞} in formula (9) is a linear combination of all 1-eigenvectors of the envelope operator E weighted by the signal amplitude. Instead of focusing only on the envelope calculus, we now consider all the set of eigenvalues of the envelope operator E.

The Spectral Intrinsic Decomposition procedure is defined as the calculus of all the SPMFs for a given signal. $E = L^{-1}$. The same procedure can be performed with the lower envelope. The eigendecomposition of E gives $[V_E, L_E] = \text{eig}(E)$, where $V_E = [V_1, \ldots, V_{\text{size}(s0)}]$ and $L_E = [L_1, \ldots, L_{\text{size}(s0)}]$ (with the possibility of zeros to complete the size of the vector) are, respectively, the set of eigenvectors and the set of eigenvalues of E. The coefficient reconstruction of s_0 is given by $C = L_E V_E^{-1} s_0^{\perp}$, with $s_0^{\perp}$ being the transpose of s_0.

```
(1) M ← transform image to data
(2) NL ← number of lines of M
(3) NC ← number of columns of M
(4) M_R ← NL by NC zeroed matrix
(5) for i ← 1, NL do
(6)     s_i ← line i of M
(7)     compute g_i^+ from s_i
(8)     compute E_i
(9)     decompose E_i, [V_{E_i}, L_{E_i}] ← eig(E)
(10)    compute C_i, C_i = L_{E_i} V_{E_i}^{-1} s_i'
(11)    for j ← 1, NC do
(12)        M_R(i, j) ← V_i(j) * C_i(j)
(13)    end for
(14) end for
(15) recompose image from M_R
```

ALGORITHM 1: Pseudocode for decomposition and recomposition.

Hence s_0 is computed by the formula $s0 = VC$ (Algorithm 2). The Spectral Intrinsic Decomposition of s_0 is given as follows:

$$s_0 = \sum_{k=1}^{N} V_k C_k, \quad N = \text{size}(s_0). \tag{14}$$

This decomposition is intrinsic and depends only on the position of the characteristic points of s_0 that define the diffusivity function in the interpolation operator. We notice that the SID with lower envelope works like the SID with the upper envelope and has the same reconstruction ability.

2.5. Advantage and Disadvantage of SID. The SID is adaptive and depends on the position of the characteristics points of the signal. It is autoadaptive and works for nonlinear and nonstationary signal. SID can decompose an IMF and can be used to separate mixing mode [9] in EMD.

However, the main disadvantage of the proposed SID algorithm is the computation time when the size of signal is a large. This is due to matrix inversion in the algorithm. Thus a faster algorithm is proposed in the next section.

3. The 2D Spectral Intrinsic Decomposition Algorithm

In this section we present, in Algorithm 1, the 2D SID procedure and implement it for images decomposition-recomposition.

3.1. The Algorithm of Image Decomposition and Recomposition by SID. Let us consider an image as represented by a matrix M *(Line (1))*. Each row i of the matrix M *(Line (5))* can be seen as a one-dimensional signal s_i. Thus, we can apply the spectral decomposition of s_i *(Lines (6) to (10))* as described in [4].

Each line can be recomposed to build a matrix M_R *(Lines (11) to (13))* and finally reconstruct the image from M_R *(Line (15))*. Image decomposition and recomposition are described in Algorithm 1. We can also make decomposition

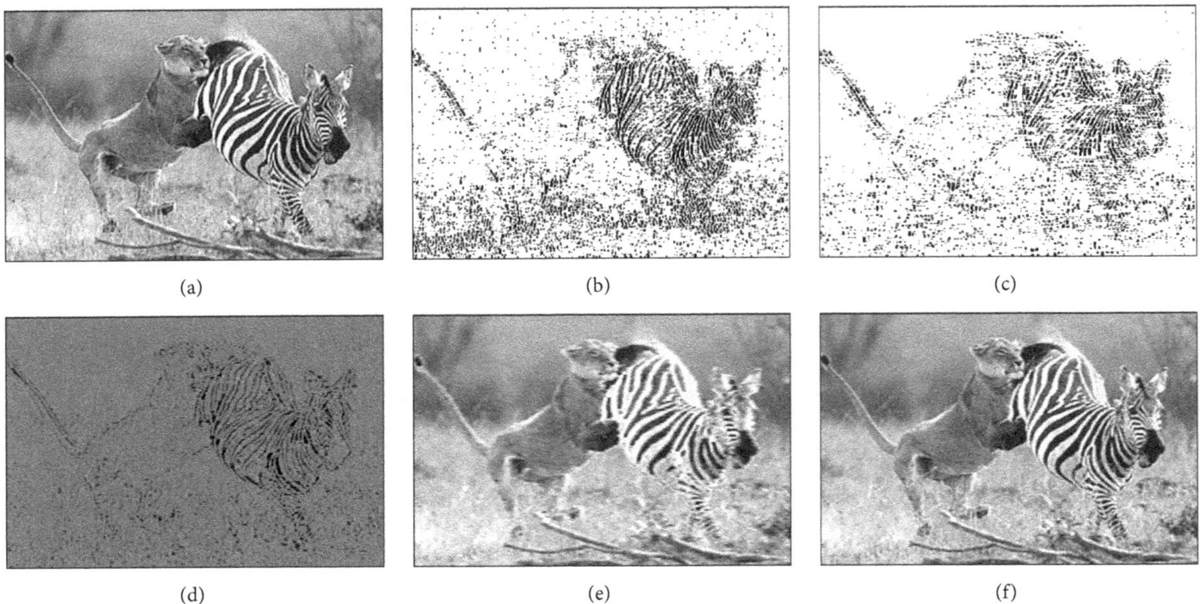

Figure 2: Decomposition (a), component extraction and representation of high frequency component ($\lambda \in [0, 0.1]$) (b), medium frequency component ($\lambda \in [0.5, 0.7]$) (c), high and medium frequency component ($\lambda < 1$) (d), low frequency component ($\lambda = 1$) (e), and recomposition of all components (f).

```
[λ_min, λ_max] ← eigenvalues interval to extract
M ← transforms image to data
NL ← number of lines of M
NC ← number of columns of M
M_R ← NL by NC zeroed matrix
for i ← 1, NL do
    S_i ← line i of M
    compute g_i^+ from s_i
    compute E_i
    decompose E_i, [V_{E_i}, L_{E_i}] ← eig(E)
    L_{R_i} ← L_{E_i}
    for j ← 1, NC do
        if L_{R_i}(j) ≤ λ_min or L_{R_i}(j) ≥ λ_max then
            L_{R_i}(j) ← 0
        end if
    end for
    compute C_{R_i}, C_{R_i} ← L_{R_i} V_{E_i}^{-1} S_i'
    for j ← 1, NC do
        M_R(i, j) ← M_R(i, j) + V_{E_i}(j) * C_{R_i}(j)
    end for
end for
```

Algorithm 2: Pseudocode to extract SPMFs between two eigenvalues.

along columns and catch more properties of the image to be analyzed.

The elementary components of a spectral decomposition are the modulation of eigenvectors by their coefficients as can be seen in (14). It is clear that these components depend on eigenvalues of the envelope operator.

The range of smallest eigenvalues catches higher frequencies contents of the reconstructed signal with the smallest modulated amplitude.

So we can associate components which have similar frequency and amplitude by summing the components that have same eigenvalues; this method works well for signal in one dimension. For images, it is possible numerically to have missing eigenvalues in many lines. Hence, to avoid this drawback, elementary components which have the same eigenvalues will be associated with the same belonging to a specific range of eigenvalues.

3.2. Application to Image Components Extraction. In the following, 2D SID is applied to very known images test to demonstrate the ability of this pectoral intrinsic decomposition for images analysis, particularly in components extraction. In Figures 4(a), 3(a), and 2(a), we have a representation of high frequency components on Figures 4(b), 3(b), and 2(b); we note that they are more sensitive to small variations of the intensity of the image than low frequencies components in Figures 4(c), 3(c), and 2(c).

4. Conclusion

In this paper, we have presented a new method for autoadaptive image representation called two-dimensional version of the Spectral Intrinsic Decomposition. A new faster diffusivity function in the computation of the mean envelope operator is also provided. In future works, we will investigate how to use the Spectral Proper Mode Functions (SPMFs) to do signals classification or to treat other aspects of image processing, like edge detection, segmentation, and so on.

(a) (b) (c)

(d) (e) (f)

FIGURE 3: Decomposition (a), component extraction and representation of high frequency component ($\lambda \in [0, 0.1]$) (b), medium frequency component ($\lambda \in [0.5, 0.7]$) (c), high and medium frequency component ($\lambda < 1$) (d), low frequency component ($\lambda = 1$) (e), and recomposition of all components (f).

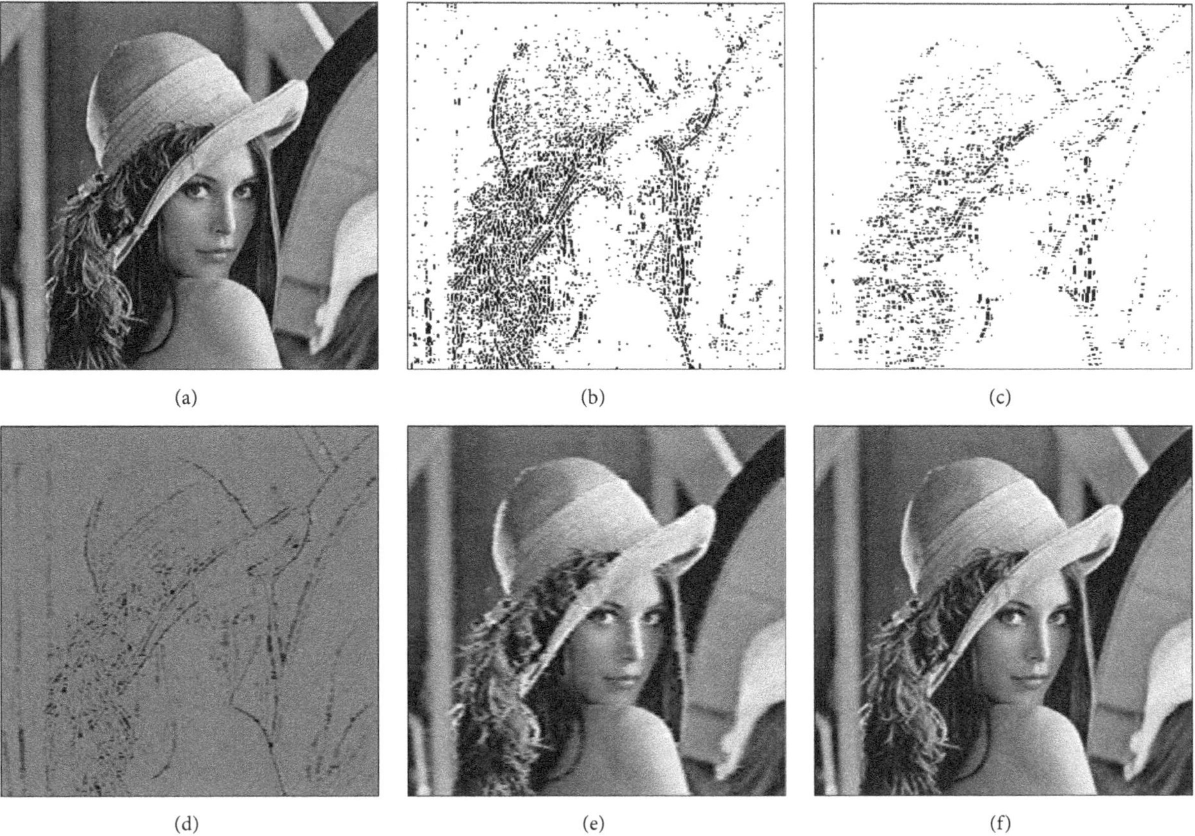

(a) (b) (c)

(d) (e) (f)

FIGURE 4: Decomposition (a), component extraction and representation of high frequency component ($\lambda \in [0, 0.1]$) (b), medium frequency component ($\lambda \in [0.5, 0.7]$) (c), high and medium frequency component ($\lambda < 1$) (d), low frequency component ($\lambda = 1$) (e), and recomposition of all components (f).

Conflicts of Interest

The authors declare that they have no conflicts of interest.

References

[1] N. E. Huang, Z. Shen, S. R. Long et al., "The empirical mode decomposition and the Hilbert spectrum for nonlinear and non-stationary time series analysis," *Proceedings A*, vol. 454, no. 1971, pp. 903–995, 1998.

[2] J. C. Nunes, Y. Bouaoune, E. Delechelle, O. Niang, and P. Bunel, "Image analysis by bidimensional empirical mode decomposition," *Image and Vision Computing*, vol. 21, no. 12, pp. 1019–1026, 2003.

[3] F. B. Arfia, A. Sabri, M. B. Messaoud, and M. Abid, "The Modified Bidimensional Empirical Mode Decomposition for Color Image Decomposition," in *Proceedings of the World Congress on Engineering*, vol. 2, 2011.

[4] O. Niang, É. Deléchelle, and J. Lemoine, "A spectral approach for sifting process in empirical mode decomposition," *Signal Processing, IEEE Transactions on*, vol. 58, no. 11, pp. 5612–5623, 2010.

[5] O. Niang, A. Thioune, É. Deléchelle, and J. Lemoine, "Spectral Intrinsic Decomposition Method for Adaptive Signal Representation," *ISRN Signal Processing*, vol. 2012, pp. 1–10, 2012.

[6] O. Niang, *Décomposition Modale Empirique: Contribution à la Modélisation Mathématique et Application en Traitement du Signal et de lImage [Ph.D. thesis]*, Université Paris XII Val de Marne, 2007.

[7] O. Niang, A. Thioune, É. Deléchelle, M. T. Niane, and J. Lemoine, "About a Partial Differential Equation-Based Interpolator for Signal Envelope Computing: Existence Results and Applications," *ISRN Signal Processing*, vol. 2013, pp. 1–18, 2013.

[8] L. Vincent, "Morphological grayscale reconstruction in image analysis: applications and efficient algorithms," *IEEE Transactions on Image Processing*, vol. 2, no. 2, pp. 176–201, 1993.

[9] Y. Gao, G. Ge, Z. Sheng, and E. Sang, "Analysis and solution to the mode mixing phenomenon in EMD," in *Proceedings of the 1st International Congress on Image and Signal Processing, CISP 2008*, pp. 223–227, May 2008.

Permissions

 Every chapter published in this book has been scrutinized by our experts. Their significance has been extensively debated. The topics covered herein carry significant findings which will fuel the growth of the discipline. They may even be implemented as practical applications or may be referred to as a beginning point for another development.

The contributors of this book come from diverse backgrounds, making this book a truly international effort. This book will bring forth new frontiers with its revolutionizing research information and detailed analysis of the nascent developments around the world.

We would like to thank all the contributing authors for lending their expertise to make the book truly unique. They have played a crucial role in the development of this book. Without their invaluable contributions this book wouldn't have been possible. They have made vital efforts to compile up to date information on the varied aspects of this subject to make this book a valuable addition to the collection of many professionals and students.

This book was conceptualized with the vision of imparting up-to-date information and advanced data in this field. To ensure the same, a matchless editorial board was set up. Every individual on the board went through rigorous rounds of assessment to prove their worth. After which they invested a large part of their time researching and compiling the most relevant data for our readers.

The editorial board has been involved in producing this book since its inception. They have spent rigorous hours researching and exploring the diverse topics which have resulted in the successful publishing of this book. They have passed on their knowledge of decades through this book. To expedite this challenging task, the publisher supported the team at every step. A small team of assistant editors was also appointed to further simplify the editing procedure and attain best results for the readers.

Apart from the editorial board, the designing team has also invested a significant amount of their time in understanding the subject and creating the most relevant covers. They scrutinized every image to scout for the most suitable representation of the subject and create an appropriate cover for the book.

The publishing team has been an ardent support to the editorial, designing and production team. Their endless efforts to recruit the best for this project, has resulted in the accomplishment of this book. They are a veteran in the field of academics and their pool of knowledge is as vast as their experience in printing. Their expertise and guidance has proved useful at every step. Their uncompromising quality standards have made this book an exceptional effort. Their encouragement from time to time has been an inspiration for everyone.

The publisher and the editorial board hope that this book will prove to be a valuable piece of knowledge for researchers, students, practitioners and scholars across the globe.

List of Contributors

Xiaotian Bi, Ang Ren, Simeng Li, Mingming Han and Qingquan Li
Shandong Provincial Key Laboratory of UHV Transmission Technology & Equipment School of Electrical Engineering, Shandong University, Jinan 250061, China

Mingjiu Wang and Shu Fan
Collaborative Innovation Center for Network Security Enforcement and Public Security Informatization, Criminal Investigation Police University of China, Shenyang 110854, China
Audio-Visual and Image Technology Department, Criminal Investigation Police University of China, Shenyang 110854, China

Jian Qi, Qun Sun, Chong Wang and Linlin Chen
School of Mechanical & Automotive Engineering, Liaocheng University, Liaocheng 252059, China

Xiaoliang Wu
Department of Medical Equipment, Liaocheng People's Hospital, Liaocheng 252000, China

Jing-bo Zhuang, Zhen-miao Deng, Yi-shan Ye, Yi-xiong Zhang and Yan-yong Chen
School of Information Science and Engineering, Xiamen University, Xiamen 361005, China

Xinhe Zhang and Yuehua Zhang
School of Electronic and Information Engineering, University of Science and Technology Liaoning, Anshan 114051, China

Chang Liu
National Key Laboratory of Science and Technology on Communications, University of Electronic Science and Technology of China, Chengdu 611731, China

Hanzhong Jia
State Grid Liaoning Information and Communication Company, Shenyang 110006, China

Yunjian Zhang, Zhenmiao Deng, Jianghong Shi, Linmei Ye and Maozhong Fu
School of Information Science and Engineering, Xiamen University, Fujian 361005, China

Chen Zhao
Faculty of Science, University of Auckland, Auckland 1052, New Zealand

Wenkang Gong, Qi Liu, Wenhao Du, Weichen Xu and Gang Wang
School of Information Science and Engineering, Northeastern University, Shenyang, Liaoning 110003, China

Yong Bai, Mengxing Huang and Zhuhua Hu
College of Information Science & Technology, Hainan University, Haikou 570228, China
State Key Laboratory of Marine Resource Utilization in South China Sea, Haikou 570228, China

Mingshan Xie
College of Information Science & Technology, Hainan University, Haikou 570228, China
State Key Laboratory of Marine Resource Utilization in South China Sea, Haikou 570228, China
College of Network, Haikou College of Economics, Haikou 571127, China

Fang Liu and Meng Liu
School of Information Science and Engineering, Shenyang Ligong University, Shenyang 110159, China

Shoushui Wei, Yutao Long and Chengyu Liu
School of Control Science and Engineering, Shandong University, Jinan 250061, Chin

Yatao Zhang
School of Control Science and Engineering, Shandong University, Jinan 250061, China
School of Mechanical, Electrical & Information Engineering, Shandong University, Weihai 264209, China

Zhenyu Hu, Qiuye Wang, Congcong Ming, Lai Wang, Yuanqing Hu and Jian Zou
School of Information and Mathematics, Yangtze University, Hubei 434020, China

Qian Gao and Chong Shen
State Key Laboratory of Marine Resources Utilization in South China Sea, Hainan University, Haikou, Hainan 570228, China
College of Information Science and Technology, Hainan University, Haikou, Hainan 570228, China

Kun Zhang
State Key Laboratory of Marine Resources Utilization in South China Sea, Hainan University, Haikou, Hainan 570228, China
College of Information Science and Technology, Hainan University, Haikou, Hainan 570228, China
College of Ocean Information Engineering, Hainan Tropical Ocean University, Sanya, Hainan 572022, China

Qunying Zhang, Shengbo Ye, Zhiwu Xu, Jie Chen and Guangyou Fang
Key Laboratory of Electromagnetic Radiation and Sensing Technology, Chinese Academy of Sciences, Beijing 100190, China

Zhenghuan Xia
Key Laboratory of Electromagnetic Radiation and Sensing Technology, Chinese Academy of Sciences, Beijing 100190, China
University of Chinese Academy of Sciences, Beijing 100049, China

Hejun Yin
Chinese Academy of Sciences, Beijing 100864, China

Moinuddin Bhuiyan
Department of Physics, University of Windsor, 401 Sunset Avenue, Windsor, ON, Canada N9B 3P4

Roman Gr. Maev
Department of Physics, University ofWindsor, 401 Sunset Avenue, Windsor, ON, Canada N9B 3P4
Tessonics Corp., 2019 Hazel Street, Birmingham, MI 48009, USA
Institute for Diagnostic Imaging Research, University of Windsor, 401 Sunset Avenue, Windsor, ON, Canada N9B 3P4

Eugene V. Malyarenko
Tessonics Corp., 2019 Hazel Street, Birmingham, MI 48009, USA

Mircea A. Pantea and Dante Capaldi
Institute for Diagnostic Imaging Research, University ofWindsor, 401 Sunset Avenue, Windsor, ON, Canada N9B 3P4

Alfred E. Baylor
Detroit Medical Center, 4201 St. Antoine Street, Detroit, MI 48201, USA

Xueyuan Hao and Xiaohong Yan
Nanjing University of Posts and Telecommunications, Nanjing 210003, China

Hanbing Wang, Hui Li and Bin Li
Department of Electronics and Information, Northwestern Polytechnical University, Xi'an, Shaanxi 710129, China

Renxuan Hao
School of Electronic Engineering, University of Electronic Science and Technology of China, Chengdu, China

Tan Guo
College of Communication Engineering, Chongqing University, Chongqing, China

Mohammed Ouassou
Norwegian Mapping Authority, Geodetic Institute, 3511 Hønefoss, Norway

Bent Natvig and Jørund I. Gåsemyr
UiO, Department of Mathematics, Norway

Anna B. O. Jensen
KTH Royal Institute of Technology, 10044 Stockholm, Sweden

Can-yan Zhu
Institute of Intelligent Structure and System, Soochow University, Soochow 215006, China

Jian Zhou
Institute of Intelligent Structure and System, Soochow University, Soochow 215006, China
Department of Information Engineering, Suzhou Global Institute of Software Technology, Soochow 215163, China

Samba Sidibe, Oumar Niang, Abdoulaye Thioune, Abdoul-Dalibou Abdou and Ndeye Fatou Ngom
Laboratoire de Traitement de l'Information et Syst`emes Intelligents, Ecole Polytechnique deThiés, BP A10, Thiés, Senegal

Index

www.ingramcontent.com/pod-product-compliance
Lightning Source LLC
Chambersburg PA
CBHW050457200326
41458CB00014B/5213
* 9 7 8 1 6 3 2 3 8 8 6 0 5 *